Dynamic Reconfigurable Network-on-Chip Design:
Innovations for Computational Processing and Communication

Jih-Sheng Shen
National Chung Cheng University, Taiwan

Pao-Ann Hsiung
National Chung Cheng University, Taiwan

INFORMATION SCIENCE REFERENCE

Hershey · New York

Director of Editorial Content:	Kristin Klinger
Director of Book Publications:	Julia Mosemann
Acquisitions Editor:	Lindsay Johnston
Development Editor:	Joel Gamon
Publishing Assistant:	Myla Harty
Typesetter:	Deanna Jo Zombro
Production Editor:	Jamie Snavely
Cover Design:	Lisa Tosheff
Printed at:	Yurchak Printing Inc.

Published in the United States of America by
Information Science Reference (an imprint of IGI Global)
701 E. Chocolate Avenue
Hershey PA 17033
Tel: 717-533-8845
Fax: 717-533-8661
E-mail: cust@igi-global.com
Web site: http://www.igi-global.com/reference

Library of Congress Cataloging-in-Publication Data

Dynamic reconfigurable network-on-chip design : innovations for computational
processing and communication / Jih-Sheng Shen and Pao-Ann Hsiung, editors.
    p. cm.
  Includes bibliographical references and index.
  Summary: "This book is on the topic of reconfigurable network-on-chip, which is a culmination of growing trends in the two hot research areas, namely reconfigurable computing and network-on-chip"--Provided by publisher.
  ISBN 978-1-61520-807-4 (hardcover) -- ISBN 978-1-61520-808-1 (ebook) 1.
Networks on a chip. I. Shen, Jih-Sheng, 1980- II. Hsiung, Pao-Ann.
  TK5105.546.D96 2010
  621.3815'31--dc22
                                    2009052397

British Cataloguing in Publication Data
A Cataloguing in Publication record for this book is available from the British Library.

All work contributed to this book is new, previously-unpublished material. The views expressed in this book are those of the authors, but not necessarily of the publisher.

Table of Contents

Section 1
Introduction to Reconfigurable Network-on-Chip

Chapter 1

    Leandro Möller, Pontifícia Universidade Católica do Rio Grande do Sul (FACIN-PUCRS), Brasil
    Ismael Grehs, Pontifícia Universidade Católica do Rio Grande do Sul (FACIN-PUCRS), Brasil
    Ewerson Carvalho, Pontifícia Universidade Católica do Rio Grande do Sul (FACIN-PUCRS), Brasil
    Rafael Soares, Pontifícia Universidade Católica do Rio Grande do Sul (FACIN-PUCRS), Brasil
    Ney Calazans, Pontifícia Universidade Católica do Rio Grande do Sul (FACIN-PUCRS), Brasil
    Fernando Moraes, Pontifícia Universidade Católica do Rio Grande do Sul (FACIN-PUCRS), Brasil

Chapter 2

    Mário P. Véstias, INESC-ID/ISEL/IPL, Portugal
    Horácio C. Neto, INESC-ID/IST/UTL, Portugal

Chapter 3

    Rachid Dafali, European University of Brittany, France
    Jean-Philippe Diguet, CNRS, Lab-STICC (UMR 3192), France

Chapter 4

    Wei-Wen Lin, National Chung Cheng University, Taiwan, R.O.C.
    Jih-Sheng Shen, National Chung Cheng University, Taiwan, R.O.C.
    Pao-Ann Hsiung, National Chung Cheng University, Taiwan, R.O.C.

Section 5
State-of-the-Art Reconfigurable NoC Designs

Detailed Table of Contents

Section 1
Introduction to Reconfigurable Network-on-Chip

Chapter 1

Leandro Möller, Pontifícia Universidade Católica do Rio Grande do Sul (FACIN-PUCRS), Brasil
Ismael Grehs, Pontifícia Universidade Católica do Rio Grande do Sul (FACIN-PUCRS), Brasil
Ewerson Carvalho, Pontifícia Universidade Católica do Rio Grande do Sul (FACIN-PUCRS), Brasil
Rafael Soares, Pontifícia Universidade Católica do Rio Grande do Sul (FACIN-PUCRS), Brasil
Ney Calazans, Pontifícia Universidade Católica do Rio Grande do Sul (FACIN-PUCRS), Brasil
Fernando Moraes, Pontifícia Universidade Católica do Rio Grande do Sul (FACIN-PUCRS), Brasil

The chapter describes how a reconfigurable NoC architecture is designed and two proof-of-concept examples illustrate the proposed architecture. The design consists of both hardware and software parts.

Chapter 2

Mário P. Véstias, INESC-ID/ISEL/IPL, Portugal
Horácio C. Neto, INESC-ID/IST/UTL, Portugal

The chapter describes how reconfigurable routers may be designed for reconfigurable NoCs. Static and runtime adaptive routers are compared and it is shown that runtime adaptive routers are essential for today's complex system-on-chip designs.

Chapter 3

Rachid Dafali, European University of Brittany, France
Jean-Philippe Diguet, CNRS, Lab-STICC (UMR 3192), France

The chapter describes the relation between reconfigurable NoC and the OSI network layers. Further, dynamic reconfiguration administration, network infrastructure reconfiguration and network protocol reconfiguration are all discussed. A self-adaptive network interface architecture is also proposed.

Wei-Wen Lin, National Chung Cheng University, Taiwan
Jih-Sheng Shen, National Chung Cheng University, Taiwan
Pao-Ann Hsiung, National Chung Cheng University, Taiwan

The authors propose three communication architectures for interconnecting an NoC with a microprocessor bus-based conventional architecture. It is shown that the shared memory approach is a good tradeoff between performance and memory space consumption.

Section 2
Design Methods for Reconfigurable NoC Design

Vincenzo Rana, Politecnico di Milano, Italy
Marco D. Santambrogio, Politecnico di Milano, Italy
Alessandro Meroni, Politecnico di Milano, Italy

The chapter discusses a tile-based approach to reconfigurable NoC architecture design. Existing architectures are leveraged for the development of reconfigurable NoC architecture.

Imran Rafiq Quadri, CNRS, France
Majdi Elhaji, CNRS, France
Samy Meftali, CNRS, France
Jean-Luc Dekeyser, CNRS, France

The chapter describes the adaptation of MARTE, an OMG standard for real-time and embedded systems, to complex NoC-connected partially dynamically reconfigurable systems.

Vincenzo Rana, Politecnico di Milano, Italy
Marco Domenico Santambrogio, Politecnico di Milano, Italy
Simone Corbetta, Politecnico di Milano, Italy

The chapter describes the design and implementation issues for reconfigurable NoC such as the placement of bus macros for connecting static and reconfigurable parts of the NoC. A layered approach is also presented for solving switching, routing, and communication protocol design issues.

Section 3
High-Level Programming of Reconfigurable NoC-Based SoCs

Chapter 8

The chapter describes Gannet, a programming model and framework which makes software and hardware integration easier. The inherent parallelism and dynamic reconfigurability can be fully utilized by a user programming with Gannet.

Section 4
Simulation Framework for Fast Reconfigurable NoC Emulation

Chapter 9

The chapter describes an emulation method for exploring the different communication design alternatives corresponding to adapting routers, network interfaces, and processing cores.

Section 5
State-of-the-Art Reconfigurable NoC Designs

Chapter 10

The chapter describes how hardware reconfigurability can be exploited in terms of switching, routing, and packet size. A new architecture for reconfigurable NoC is proposed and evaluated against fixed NoCs.

Chapter 11

Reliability Aware Performance and Power Optimization in DVFS-Based
Aditya Yanamandra, The Pennsylvania State University, USA
Soumya Eachempati, The Pennsylvania State University, USA
Vijaykrishnan Narayanan, The Pennsylvania State University, USA
Mary Jane Irwin, The Pennsylvania State University, USA

The authors propose a dynamically reconfigurable data protection scheme in NoC, while minimizing power and performance overheads. Reconfigurability further enhances this protection scheme by providing both flexibility and cost reduction.

Chapter 12

Björn Osterloh, Technical University of Braunschweig, Germany
Harald Michalik, Technical University of Braunschweig, Germany
Björn Fiethe, Technical University of Braunschweig, Germany

The authors propose a specialized NoC architecture that can withstand the disruption effects of radiation induced particles in space. High reliability is a critical issue in this domain and reconfigurable NoC can be leveraged to achieve this.

Chapter 13

Mehdi Modarressi, Sharif University of Technology, Iran
Hamid Sarbazi-Azad, Sharif University of Technology, Iran

The authors propose an NoC with reconfigurable topology such that it can be configured dynamically to fit the traffic pattern requirements of a set of applications. Experiments demonstrate the performance improvement and power reduction brought about by such an adaptation scheme between topology and traffic pattern.

Foreword

Computers, now mostly systems on chip, are moving away from the desktop to embedded systems, currently more than 98%. At a €60 billion global market growing 14% per year, over 4 billion embedded processors were sold in 2008 and over 40 billion are predicted by 2020. Rapidly growing market demands are massively accelerating the complexity of systems on chip having to run many computationally very intensive tasks in multicore systems often including accelerators. As the main subject of this book, leading researchers present cutting-edge solutions meeting a variety of unique challenges for multicore embedded software like real time, security, safety, reliability, swap, energy efficiency, area consumption and heterogeneity.

Many intellectual property blocks have to be integrated onto a single chip. However, the most important performance bottleneck has shifted from computation to on-chip communication, which requires ultra-low latencies in order to cope with immense data bandwidth needs. This rapidly growing challenge has to be mastered by powerful and flexible Networks on Chip (NoC). Here the book gives an valuable update on recent developments in theory and practice, highlighting important emerging issues like communication protocols, interconnect architectures, interfacing and security requirements.

This book describes, how the efficiency of both homogeneous and heterogeneous solutions can be substantially improved by runtime changes, where the most flexible and efficient alternative is the NoC implemented on Field Programmable Gate Arrays (FPGAs): a reconfigurable NoC (rNoC), also massively reducing the NRE cost compared to an ASIC-based NoC. The book focuses on how communication between hardware and software can be made efficient for rNoCs, and covers reconfiguration issues as well as different communication architectures for rNoC, like FIFO-based and shared-memory architectures etc. Because of the trend to substantial re-use, the availability of reconfigurability, along with processors, on-chip bus, digital and mixed-signal connectivity IP at several levels of abstractions shortens the prototyping time and even supports pre-silicon software development and verification.

This book is the first reference book fully covering newest developments even going beyond mere reconfigurability, where devices support also dynamic partial reconfiguration such, that IP blocks can be configured onto FPGA also at run time of other blocks, without interfering with modules already running. Such dynamically reconfigurable or runtime reconfigurable NoCs (drNoCs or rrNoCs) are the next generation of NoCs which reconfigure themselves in terms of switching, routing and packet size with the changing communication requirements of the system at run time, thus utilizing the maximum available channel bandwidth. They represent a new set of benefits in terms of area overhead, performance, power consumption, fault tolerance and quality of service compared to the previous generation where the architecture is decided at design time. The book introduces all this in detail.

More error-prone and defective components due to transistor scaling, as well as dynamic voltage and frequency scaling techniques result in voltage variations causing variable error rates across the chip, in some applications also by the influences of radiation induced particles. Several chapters of the book investigate dynamically reconfigurable error protection schemes in a NoC to achieve reliability. The book is a comprehensive treasure trove of newest methodologies in MPSoC development.

Reiner Hartenstein
http://hartenstein.de

Preface

This is a first book on the topic of reconfigurable network-on-chip, which is a culmination of growing trends in the two hot research areas, namely reconfigurable computing and network-on-chip. While reconfigurable computing has brought immense flexibility in on-chip processing, network-on-chip has brought similar flexibility in on-chip communication. The integration of these two areas of research will reap the benefits of both and is a promising future design paradigm for multiprocessor systems-on-chip.

Design issues related to reconfigurable network-on-chip are numerous, some of which include the following. How reconfigurable computing techniques are to be integrated into the network-on-chip design flow? How must one reconfigure the processing elements attached to a network-on-chip? How must one design a reconfigurable router? How must a set of given tasks be assigned to the processing elements? How must one schedule the arrival of tasks or the configuration of processing elements in a network-on-chip? How can one leverage reconfiguration techniques to avoid crosstalk interferences in a network-on-chip? How can one design low-power network-on-chip using reconfiguration techniques? How can one design a programming model for the processing elements attached to a network-on-chip? How can an operating system be designed to take advantage of the computing and communication flexibilities brought about by run-time reconfiguration and network-on-chip? The above list is just a partial one and from this one can guess the importance of a book like this. This is the single reference, where most of the above issues will be discussed and elaborated on.

Since most of the above design issues are still in the research and development stage. This book presents the state-of-the-art techniques that can handle the integration of reconfiguration and network-on-chip. Besides a description of the techniques, the book also describes specific application domains where the techniques have actually been applied with success. Thus, the book can be seen as a collection of techniques and applications for reconfigurable network-on-chip.

The book can be used by students, researchers, and engineers. Students interested in reconfigurable system design, network-on-chip, system-on-chip, multiprocessor design, router design, parallel computing should be interested in this book because it can provide reference materials to study all the mentioned topics. This book is recommended for researchers interested in solving the design issues related to network-on-chip and reconfigurable systems. Last but not the least, engineers will find a wealth of techniques and applications in this book related to network-on-chip and reconfigurable system design.

The book is basically divided into five sections as follows:

SECTION 1: INTRODUCTION TO RECONFIGURABLE NETWORK-ON-CHIP

This section introduces reconfigurable network-on-chip from the design perspective, including a summary of all the issues and possible solutions. It also discusses the design of reconfigurable routers and network interfaces. This sectiont has four chapters.

- **The chapter titled:** *"A NoC-Based Infrastructure to Enable Dynamic Self Reconfigurable Systems"* by Möller, Grehs, Carvalho, Soares, Calazans, and Moraes from Brazil describes how a reconfigurable NoC architecture is designed and two proof-of-concept examples illustrate the proposed architecture. The design consists of both hardware and software parts.
- **The chapter titled:** *"Dynamically Reconfigurable Networks-on-Chip using Runtime Adaptive Routers"* by Véstias and Neto from Portugal describes how reconfigurable routers may be designed for reconfigurable NoCs. Static and runtime adaptive routers are compared and it is shown that runtime adaptive routers are essential for today's complex system-on-chip designs.
- **The chapter titled:** *"Keys for Administration of Reconfigurable NoC: Self-Adaptive Network Interface Case Study"* by Dafali and Diguet from France describes the relation between reconfigurable NoC and the OSI network layers. Further, dynamic reconfiguration administration, network infrastructure reconfiguration and network protocol reconfiguration are all discussed. A self-adaptive network interface architecture is also proposed.
- **The chapter titled:** *"An Efficient Hardware/Software Communication Mechanism for Reconfigurable NoC"* by Lin, Shen, and Hsiung from Taiwan proposes three communication architectures for interconnecting an NoC with a microprocessor bus-based conventional architecture. It is shown that the shared memory approach is a good tradeoff between performance and memory space consumption.

SECTION 2: DESIGN METHODS FOR RECONFIGURABLE NOC DESIGN

This section is mainly about modeling and design methods for reconfigurable NoC. This section consists of three chapters.

- **The chapter titled:** *"Design Methodologies and Mapping Algorithms for Reconfigurable NoC-Based Systems"* by Rana, Santambrogio, and Meroni from Italy discusses a tile-based approach to reconfigurable NoC architecture design. Existing architectures are leveraged for the development of reconfigurable NoC architecture.
- **The chapter titled:** *"From MARTE to Reconfigurable NoCs: A Model Driven Design Methodology"* by Quadri, Elhaji, Meftali, and Dekeyser from France describes the adaptation of MARTE, an OMG standard for real-time and embedded systems, to complex NoC-connected partially dynamically reconfigurable systems.
- **The chapter titled:** *"Dynamic Reconfigurable NoCs: Characteristics and Performance Issues"* by Rana, Santambrogio, and Corbetta from Italy describes the design and implementation issues for reconfigurable NoC such as the placement of bus macros for connecting static and reconfigurable parts of the NoC. A layered approach is also presented for solving switching, routing, and communication protocol design issues.

SECTION 3: HIGH-LEVEL PROGRAMMING OF RECONFIGURABLE NOC-BASED SOCS

This part discusses how NoC-based system-on-chip is to be programmed. This section contains only one chapter.

- **The chapter titled:** *"High-Level Programming of Dynamically Reconfigurable NoC-Based Heterogeneous Multicore SoCs"* by Vanderbauwhede from UK describes Gannet, a programming model and framework which makes software and hardware integration easier. The inherent parallelism and dynamic reconfigurability can be fully utilized by a user programming with Gannet.

SECTION 4: SIMULATION FRAMEWORK FOR FAST RECONFIGURABLE NOC EMULATION

This section is mainly about how NoC can be emulated. This section consists of only one chapter.

- **The chapter titled:** *"Dynamic Reconfigurable NoC (DRNoC) Architecture: Application to Fast NoC Emulation"* by Krasteva, de la Torre, and Riesgo from Spain describes an emulation method for exploring the different communication design alternatives corresponding to adapting routers, network interfaces, and processing cores.

SECTION 5: STATE-OF-THE-ART RECONFIGURABLE NOC DESIGNS

This section consists of four examples on different reconfigurable NoCs. Some of them focus on power reduction, some on crosstalk reduction, and yet another on how topology is adapted.

- **The chapter titled:** *"Dynamically Reconfigurable NoC for Future Heterogeneous Multi-Core Architectures"* by Ahmad, Ahmadinia, and Arslan from UK describes how hardware reconfigurability can be exploited in terms of switching, routing, and packet size. A new architecture for reconfigurable NoC is proposed and evaluated against fixed NoCs.
- **The chapter titled:** *"Reliability Aware Performance and Power Optimization in DVFS-Based On-Chip Networks"* by Yanamandra, Eachempati, Narayanan, and Irwin from USA proposes a dynamically reconfigurable data protection scheme in NoC, while minimizing power and performance overheads. This is an increasingly important topic due to the variations in chip fabrication brought about by deep-submicron technologies. NoC being a data communication mechanism must support such protection schemes. Reconfigurability further enhances this protection scheme by providing both flexibility and cost reduction.
- **The chapter titled:** *"SpaceWire Inspired Network-on-Chip Approach for Fault Tolerant System-on-Chip Designs"* by Osterloh, Michalik, and Fiethe from Germany proposes a specialized NoC architecture that can withstand the disruption effects of radiation induced particles in space. This is a very interesting application domain in which dynamically reconfigurable NoC technology brings

all the advantages that were unforeseen before in this area. High reliability is a critical issue in this domain and reconfigurable NoC can be leveraged to achieve this.

- **The chapter titled:** *"A High-Performance and Low-Power On-Chip Network with Reconfigurable Topology"* by Modarressi and Sarbazi-Azad from Iran proposes an NoC with reconfigurable topology such that it can be configured dynamically to fit the traffic pattern requirements of a set of applications. Experiments demonstrate the performance improvement and power reduction brought about by such an adaptation scheme between topology and traffic pattern.

This book is a collective effort by a group of expert representatives in the area of reconfigurable network-on-chip design. We believe that the technologies, issues, and solutions presented in this book will be of immense help to researchers and engineers in various fields of application such as real-time embedded systems, system-on-chip, multimedia, and networking.

The integration of reconfigurable computing techniques and network-on-chip communication infrastructure is inevitable. This book provides a basic reference for such an integration. Though there are still lots of research issues to be solved in this integration, we are already at a stage where we need more convergence on how our technologies have affected this well-known communication structure.

Any queries related to this book can be addressed to one of the authors.

Jih-Sheng Shen and Pao-Ann Hsiung
National Chung Cheng University
Taiwan

November 2009

Acknowledgment

The editors of this book would like to thank all the people who have been directly or indirectly involved in the development of this book. First of all, we would like to thank all the authors who submitted chapters. Though the contributions were all of high quality, due to page constraints, we could not include all the chapters. Special thanks goes to all the authors whose work were finally included into the book. Next we will like to thank the editorial board members and all the reviewers who worked hard to read all the chapter submissions, provide construction suggestions, and help with the selection process. Each chapter was reviewed by at least three editorial board members or other reviewers as assigned by the board members.

We will like to give a big thanks to Professor Reinald Hartenstein, Germany who graciously accepted our invitation to write a foreword for this book. He took the pains in not only going through the chapters, but also wrote different versions of the foreword. His remarks will indeed bright light to the people wondering what this book is all about.

The authors would like to thank all the members of the Embedded Systems Laboratory, National Chung Cheng University, Taiwan, who were always supportive of this work and helped with the related logistics.

Last but not the least, Pao-Ann Hsiung would like to thank his wife Nancy Chang, daughters Alice, Phoebe, and Tiffany and son Roger for all the support that they provided during the course of editing this book. Jih-Sheng Shen would like to thank his advisor, Dr. Pao-Ann Hsiung, for giving him the opportunity to co-edit the important publication. He would also like to acknowledge the support from his father Hua-Shan Shen and his mother Yueh-Chuan Pan. The importance of their support during the period cannot be neglected. The authors would also like to thank the support from IGI Global and the assistant book development editor Joel A. Gamon.

As a final note, the authors would also like to thank the National Science Council, Taiwan for the research project grants that made possible the work included within this book.

Jih-Sheng Shen and Pao-Ann Hsiung
National Chung Cheng University
Taiwan

November 2009

Section 1
Introduction to Reconfigurable Network-on-Chip

Chapter 1
A NoC–Based Infrastructure to Enable Dynamic Self Reconfigurable Systems

Leandro Möller
Pontifícia Universidade Católica do Rio Grande do Sul (FACIN-PUCRS), Brasil

Ismael Grehs
Pontifícia Universidade Católica do Rio Grande do Sul (FACIN-PUCRS), Brasil

Ewerson Carvalho
Pontifícia Universidade Católica do Rio Grande do Sul (FACIN-PUCRS), Brasil

Rafael Soares
Pontifícia Universidade Católica do Rio Grande do Sul (FACIN-PUCRS), Brasil

Ney Calazans
Pontifícia Universidade Católica do Rio Grande do Sul (FACIN-PUCRS), Brasil

Fernando Moraes
Pontifícia Universidade Católica do Rio Grande do Sul (FACIN-PUCRS), Brasil

ABSTRACT

Platform-based designed SoC includes one or more processors, RTOS, intellectual property blocks, memories and an interconnection infrastructure. An associated advantage of processor is flexibility at the software level. Hardware is not flexible. Thus, dedicated IP blocks must be inserted at design time. An alternative is to provide the platform with reconfigurable hardware blocks with sufficient capacity to implement any envisaged dedicated IP block. Dynamic self-reconfigurable systems (DSRSs) introduce flexibility to hardware. In DSRSs, IP blocks are loaded according to application demand, an approach that potentially reduces area, power consumption and total system cost. Platform-based design associated to dynamic reconfiguration techniques provide both hardware and software flexibility. The contributions of this work are: (i) DSRS architecture proposal; (ii) straightforward DSRS design flow for this architecture;

DOI: 10.4018/978-1-61520-807-4.ch001

(iii) NoC specifically designed to support dynamic hardware reconfiguration; (iv) two proof-of-concept case studies. Results point that among the best implementation choices for DSRS are those that employ NoCs as communication infrastructure, adopt the use of software configuration controllers, make use of unidirectional LUT-based IP interfaces and dispose of an internal port for reconfiguration.

INTRODUCTION

Platform-based design (Keutzer at al., 2000) is a method to implement complex SoCs, avoiding chip design from scratch. Several IPs other than processors compose SoCs. Examples are communication interfaces, memory controllers and hardware accelerators. These IPs as well as processor may be implemented directly in silicon or using reconfigurable hardware technology. Using the second option, it becomes possible to: (i) improve system performance, by migrating critical tasks to hardware; (ii) build products in smaller devices, thus reducing costs; (iii) extend product life cycle; (iv) update hardware after system manufacturing.

In order to accomplish (i) and (ii), reconfigurable hardware must allow partial and dynamic reconfiguration. Systems using these characteristics are called Dynamically Reconfigurable Systems (DRSs). The main drawback of DRSs is their reconfiguration time. To minimize this drawback, DRSs may be built with the capacity to manage their own reconfiguration process. This can be achieved through the availability of internal reconfiguration ports. Such systems are named Dynamic Self-Reconfigurable Systems (DSRSs) (Van den Branden et al., 2005). DSRSs are the target architecture of this work.

One natural implementation choice for DSRSs are dedicated ASICs, with embedded reconfigurable areas (Mrabet et al., 2006). As the goal of this paper is to propose an infrastructure for DSRS, fine-grain reconfigurable FPGAs are used here as a device platform for proof-of-concept purposes. Current FPGAs are clearly limited in terms of useful silicon area, since most of the silicon area

is used for programming purposes. In addition, DSRSs may waste a significant amount of this useful silicon to implement the necessary infrastructure. Despite these drawbacks, FPGAs are certainly adequate to prototype the infrastructure proposed herein, serving to demonstrate its benefits and limitations.

An important issue in current SoC design is the implementation of its communication infrastructure. Present SoCs require using scalable communication infrastructures, with shorter wires to minimize power consumption (Dally & Towles, 2001). NoCs are an alternative to busses, with several advantages, as stated in (Benini & De Micheli, 2002). However, few works (Marescaux et al., 2004) have suggested mixing reconfigurable IPs and NoCs.

This paper has four goals. First, to propose an infrastructure for DSRSs, identifying which are its required components. The second goal is to present a straightforward design flow supporting DSRSs. The third goal is to describe a NoC actively supporting the process of partial and dynamic IP reconfiguration. The last goal is to depict proof-of-concept case studies, comparing area overhead and reconfiguration time.

The rest of this work presents a conceptual architecture for self reconfigurable systems and some implementation alternatives. The definition of the proposed DSRS architecture and a NoC architecture called ARTEMIS are presented. A practical design flow to build DSRSs and two DSRS case studies are discussed throughout the next Sections. Finally, some conclusions and directions for future work are presented.

RELATED WORK IN DRS

This Section provides a historical landscape of the state of the art in reconfigurable devices, processors and systems using the timeline of Figure 1. A previous review on reconfigurable architectures can be found in (Hartenstein, 2001), while reconfigurable processors are reviewed in (Barat et al., 2002).

In 1985, Xilinx launched the first FPGA, the XC2064 device. It was a small device, with a capacity of around 1K equivalent gates. This device and several of its successors implemented glue logic or prototyped small circuits. Since then, devices have grown in terms of gate count and embedded resources. For example, Virtex-4 families devices may implement circuits with around 16 Mgates, excluding up to two hard core processors, up to 10Mbits of RAM memory, more than 100 multipliers, clock managers and SERDES (serializer/deserializer) devices. These complex reconfigurable devices are employed in final products, and not only for rapid prototyping. Most industrial reconfigurable devices are fine grain architectures, being 1-bit LUTs the element

defining the system function. Academic architectures, as KressArray (Hartenstein & Kress, 1995), RAW (Waingold et al. 1997) and DReAM (Becker et al., 2000), propose coarse grain reconfigurable devices, using ALUs or simple processors as the basic building block.

Reconfigurable devices must be partially and dynamically reconfigurable to enable DRS implementation. Among such devices, the following are partially and dynamically reconfigurable: Algotronix CAL1024, Plessey ERA60100, Concurrent Logic CLi6000, National CLAy-10 and CLAy-31, Atmel AT6000 and AT40K (Atmel, 2004), Xilinx XC6200 family and all Virtex and Spartan families (Xilinx, 2004). From these, only Atmel and all Xilinx Virtex and Spartan families are currently commercialized.

Most works in reconfigurable processors and reconfigurable systems use Virtex devices to prototype the proposed architectures. This is justified, since these are the only available devices with partial and dynamic reconfiguration features with a sufficient gate count to implement complex SoCs. In the 90's, research in reconfigurable architectures focused in reconfigurable processors. The main

Figure 1. Reconfigurable technology evolution: systems, processors and devices

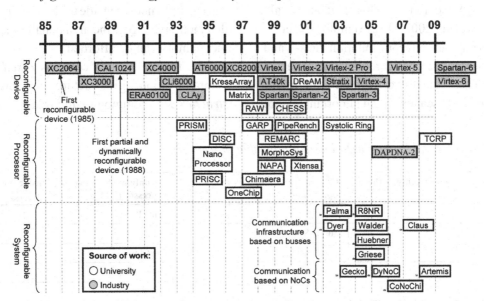

goal pursued was to accelerate software applications, migrating critical tasks to reconfigurable hardware. Table 1 presents how reconfigurable logic can be connected to the processor: (i) loosely coupled, through the system bus; (ii) dedicated bus, enabling to use the reconfigurable logic as a coprocessor; (iii) reconfigurable logic inside the processor datapath, named as RFU (reconfigurable function unit).

Loosely coupled reconfigurable logic does not require processor changes, but lead to smaller performance gains compared to RFUs. On the other side, RFUs require dedicated processors, enabling more important performance gains. An example of commercial processor with RFUs is the Xtensa platform from Tensilica (Gonzalez, 2000).

The increasing number of equivalent gates per device shifted the research effort in this century from reconfigurable processors to complete reconfigurable systems. Such systems are built around an interconnection infrastructure (point-to-point, bus, NoC), with fixed and reconfigurable IPs. Two salient features characterize reconfigurable systems: (i) a complete DRS fit in a single device, reducing communication latency; (ii) the use of coarse grain reconfigurable IPs.

Several authors proposed reconfigurable systems employing busses (Palma et al., 2002; Möller et al., 2004; Walder et al., 2004; Huebner et al., 2005; Claus et al., 2007). These employ different partial and dynamic reconfiguration flows, and display at least two reconfigurable areas connected to a bus. These fulfill the required features for DRSs, except for scalability, an important feature

for present SoCs. The Gecko platform (Marescaux et al., 2004), the DyNoC (Bobda et al., 2005) and the CoNoChi (Pionteck et al., 2006) are examples of a reconfigurable systems employing NoCs.

The work presented here, DSRS with Artemis NoC (Möller et al, 2007), has in common with Gecko the use of reconfigurable hard IPs connected to a NoC. At the *architectural level*, the contributions of this work are: (i) smaller design complexity than Gecko, since the same network is used for control and data; (ii) low area overhead NoC with explicit support for reconfiguration; (iii) reconfigurable interface components specifically designed to support partial reconfiguration.

DSRS CONCEPTUAL ARCHITECTURE

Figure 2 illustrates a conceptual DSRS architecture. Components inside the darker gray area represent the fixed part of the system, *not* requiring reconfigurable technology to be implemented. Reconfigurable technology is needed to implement the reconfigurable IPs as well as part of reconfigurable interfaces. The only external component to the SoC is the Repository, storing complete/partial configurations. The Repository can be implemented inside the SoC, using embedded memory, only if the area penalty incurred by the memory can be afforded.

The fixed SoC area contains *Fixed IPs*, the communication infrastructure, and three *infrastructure IPs*, common to any DSRS:

Table 1. Reconfigurable logic used in reconfigurable processors

Reconfigurable Logic attached to the system bus	Reconfigurable Logic used as a coprocessor	Reconfigurable Logic used inside the processor (RFU)
PRISM	Garp	Nano Processor
PipeRench	MorphoSys	DISC
	Systolic Ring	OneChip
	DAPDNA-2	Chimaera
		TCRP

Figure 2. DSRS conceptual architecture

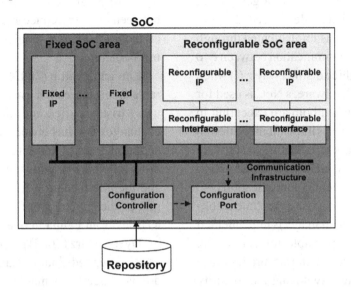

- *Configuration controller*: loads partial configurations stored in the Repository according to a schedule, choosing the reconfigurable IP to be replaced;
- *Configuration port*: transmits a partial configuration to the area assigned to it;
- *Reconfigurable interface*: defines a static interface between a reconfigurable IP and the fixed part of the DSRS.

The quality of the configuration controller and the configuration port dictate the applicability of a given DSRS. The faster the reconfiguration process, the smaller time penalty is imposed on the implemented application (Wirthlin & Hutchings, 1997). A careful design of the reconfigurable interface and the communication infrastructure is mandatory to enable reconfiguration to occur at runtime, since transient values on this interface may cause malfunction of the whole system or even damage the device.

The next Sections review related works in configuration controllers, reconfigurable interfaces, communication infrastructures, and configuration ports.

Configuration Controllers

Ullmann et al. (2004) propose a hardware task manager with real-time constraints for DSRSs. This manager is implemented in software and controls the reconfiguration process and context saving. This task manager is also responsible for controlling communication among system IPs and external devices (sensors and actuators) through a CAN bus.

Resano et al. (2004) designed a scheduler using two techniques, named Prefetch-Scheduling Technique and Replacement Technique. The Prefetch-Scheduling Technique receives as input a set of tasks and decides when these will be loaded from memory to be executed. The main goal of this technique is to hide the time penalty to load a reconfigurable IP, by analyzing task dependencies and triggering the reconfiguration process as early as possible. The Replacement Technique is responsible to increase the possibilities to reuse critical tasks to favor global system performance.

Mignolet et al. (2003) present an infrastructure for relocatable tasks management. In this system, hardware tasks can have their execution interrupted and continue its processing in software, or vice-versa. Given the nature of the target system, a unified communication model is needed. Communication among tasks is implemented through message exchange. Communication between two tasks may happen in three distinct ways: (i) when both are executing in hardware, a NoC is used for communication; (ii) when both are executing in software, communication is implemented through an API; (iii) when a task is executing in software and another in hardware, a Hardware Abstraction Layer is used to implement communication.

Griese et al. (2004) propose a run-time reconfiguration manager. This manager is implemented in hardware and is responsible for controlling and executing the reconfiguration process in a target platform. Context switching and security mechanisms were also implemented to prevent a possibly unsuccessful reconfiguration. The manager stores partial configurations in a host computer that communicates with the target platform through a PCI bus.

Carvalho et al. (2004) propose a hardware configuration controller named RSCM. It is composed by six modules: (i) a Central Configuration Control that manages the overall configuration process; (ii) a Reconfiguration Monitor, detecting situations where a reconfigurable area is available; (iii) Configuration Scheduler, statically determines which configuration is the next to be configured; (iv) a Self-Configuration module, to control the details of the reconfiguration process; (v) a Configuration Memory; (vi) a Configuration Interface.

Reconfigurable Interface

As stated earlier, a reconfigurable interface is necessary to implement the communication between a reconfigurable IP and the rest of the DSRS.

The interface proposed by Palma et al. (2002) uses two levels of tristate buffers in the input and output pins of the reconfigurable IPs. One level of tristates belongs to the reconfigurable IP and the other to the communication infrastructure. Manual routing verification and manual routing correc-

tions are required, to ensure correct connection between IPs. To reduce manual routing, Palma employs a 1-bit data serial bus as communication infrastructure.

Lim and Peattie (2002) propose a reconfigurable interface called Bus Macro. Each macro uses 8 tristate buffers and allows the simultaneous exchange of 4 bits between a reconfigurable area and another area, fixed or reconfigurable. The only difference between this reconfigurable interface and the interface proposed by Palma is that this groups the equivalent of four reconfigurable interfaces from Palma in one.

Dyer & Wirz (2002) propose a reconfigurable interface called double stage CLB macro. The double stage CLB macro uses two CLBs, one belongs to the reconfigurable IP and the other to a fixed or reconfigurable IP. Both CLBs are programmed with the identity function. Dyer & Wirz also propose the single stage CLB macro, but this macro is not a reconfigurable interface. Instead, it is a feed-through component to guide wires that are crossing an undesirable region (e.g. a wire that belongs to the fixed part of the system is crossing the boundary of a reconfigurable module).

The double stage CLB macro was originally implemented for the first family of Virtex devices.

Möller & Moraes (2005) implemented the double stage CLB macro for Virtex-II and Virtex-II Pro devices. The double stage CLB macro was also implemented by Xilinx in 2006 (Xilinx, 2008), which was still being called Bus Macro. This Bus Macro was developed for the following FPGA families: Virtex-II, Virtex-II Pro and Virtex-4. In 2008 Xilinx released the Bus Macros for Spartan-3 and Virtex-5 (Xilinx, 2008).

Huebner et al. (2005) propose a reconfigurable interface called Bus Macro (distinct from the Bus Macro proposed by Lim and Peattie, and herein renamed Huebner macro). This macro does not use tristate buffers and it is a static bus used to connect all reconfigurable IPs of the system. This reconfigurable interface is composed by two unidirectional busses, each one implemented by a set of CLBs. Each macro allows the simultaneous transmission of 8 bits from a reconfigurable area to another area, fixed or reconfigurable.

Communication Infrastructure

This Section presents some examples of reconfigurable systems presented in Figure 1. The Walder System (Walder & Platzner, 2004) employs Xilinx bus macros to construct a global bus. This global bus connects to two arbiters, BARL and BARR. The CPU implemented in the fixed part of the system manages reconfiguration at runtime. The reconfigurable areas are organized in fixed size pages, named *slots*, initially configured with dummy tasks (containing only the bus logic). The main contribution of the Walder System is to allow reconfigurable tasks to use an arbitrary number of *slots*, adapting reconfigurable resources to incoming tasks.

The Huebner System (Huebner et al., 2005) contains all IPs of the above defined DSRS conceptual architecture: (i) repository: external flash memory; (ii) configuration controller: implemented in software, using the Microblaze processor; (iii) configuration port: ICAP; (iv) reconfigurable interface: Huebner macros (Huebner et al., 2005). Partial configurations are stored compressed, to reduce the external memory size. An internal hardware module is used to decompress them on the fly. In this work, reconfigurable areas are assumed to have equal size.

One example of DRS employing NoCs is the DyNoC (Bobda et al., 2005). This work allows some routers of the NoC to be configured out of the system to give space to hardware modules. In this case, some requirements have to be fullfiled, e.g. having at least one path for all modules to communicate with each other, and the routing algorithm has to be able to find a path to connect the modules using the available channels of the NoC.

CoNoChi (Pionteck et al., 2006) is another example of DRS employing NoCs. While the previous presented work (DyNoC) uses several routers to communicate the hardware modules of the system, the CoNoChi tries to minimize the number of used routers to reduce the number of hops and area. However, minimizing too much the number of routers can make the system similar to an expensive bus in terms of area and power consumption.

The Gecko platform (Marescaux et al., 2004) is an example of a DSRS employing NoC communication. This platform contains two 3x3 mesh topology NoCs (one for data and one for reconfiguration) interconnecting the following IPs: five 16-bit RISC processors, one 2D IDCT blocks, one I/O interface to external CPU and two reconfigurable areas. It also contains a bus for operating system control. Gecko fits in a Virtex-II 6000 device, and employs the Xilinx Modular Design Flow (Lim & Peattie, 2002), based on the use of Xilinx Bus Macros.

Configuration Ports

The configuration ports used in DSRSs are usually the ones provided by the manufacturers and are available either internally or externally. As most of the DSRS works (Carvalho et al., 2004; Griese et al., 2004; Mignolet et al., 2003; Resano et al., 2004; Ullmann et al., 2004) use Virtex FPGAs, only the configuration ports of this device are discussed here. There are three available ports in Virtex FPGAs: JTAG, Select Map and ICAP (Internal Configuration Access Port). JTAG is a serial interface, while the other two can support 8-bit parallel data transmission. From these, only the ICAP is is available from inside the device and can work at 66 MHz, while Select Map and JTAG can operate at 50 MHz and 33 MHz, respectively. From the presented works, only the work (Huebner et al., 2005) and (Ullmann et al., 2004) explicitly mention the use of ICAP.

PROPOSED DSRS INFRASTRUCTURE

This Section discusses choices and trade-offs associated to these components, recommending implementation choices for each DSRS internal component. The communication infrastructure is discussed in the Artemis NoC Section.

Repositories

DSRSs need to have access to repositories able to store a potentially large number of partial configurations, often called *configuration memory*. Besides storing partial configurations, these repositories should offer fast access to its contents, to satisfy application requirements. There are basically four device types available to use as configuration memories: (i) memory internal to the reconfigurable device, usually available as RAM blocks or BRAMs; (ii) devices external to the DSRS using static RAM technology, or SRAMs; (iii) devices external to the DSRS using PROM technology, such as EPROM or flash devices called generically PROMs; (iv) devices external to the DSRS using DRAM technology, such as SDRAM and others. Table 2 presents a qualitative comparison among these devices.

Applications using BRAMs as repository may support small number of configurations and/or only small configurations, due to its limited capacity. Applications that benefit from difference-based (Lim & Peattie, 2002) reconfiguration

Table 2. Qualitative comparison of technologies applicable to build DSRS configuration memories

Memory Type	Capacity	Speed	Controller Complexity	Cost/ Bit
BRAM	+	++++	+	+++
SRAM	++	+++	++	++
PROM	++	+	++++	++
DRAM	++++	++	+++	+

techniques are among those able to employ this kind of repository.

SRAM and DRAM devices present a good compromise between access speed and storage capacity. The former imply simpler controllers added to the DSRSs, but are much more expensive per bit than DRAMs. DRAMs, on the other hand, have a low cost per storage bit, allowing storing more configurations. However, a higher area of the DSRS must be committed to implement its controller.

Contrary to the other three technologies, PROMs have the advantage of keeping configurations after turning the DSRS off. They cost more per bit than DRAMs, but imply a simpler procedure at startup of the DSRS. Also, changing the contents of the repository is more complicated than with the other technologies.

Reconfigurable Interface

The Reconfigurable Interface Subsection from the DSRS conceptual architecture Section presented three different reconfigurable interfaces. The main drawback of Palma´s work (Palma et al., 2002) is to require manual routing. The Xilinx Bus Macro (Lim & Peattie, 2002) reduces manual routing. However, it also uses tristate buffers, which are scarce resources in Xilinx FPGAs. The use of such resources overconstrains designs with complex reconfigurable interfaces.

Huebner et al. (2005) point that the routing tool sometimes does not respect the constraints set by the designer and crosses the boundary between the reconfigurable IP and the fixed part of the system. Due to this problem, these Authors developed a static bus implemented with LUTs. One advantage of this approach is to insulate computation from communication, thus preventing system interruption. However, this reconfigurable interface imposes the use of a bus-based communication infrastructure.

Configuration Ports

The external JTAG and SelectMap interfaces are alternatives for implementing configuration ports for DRSs that are not self-reconfigurable, where the configuration controller is located outside the DRS. Although these interfaces can be used for building DSRS (using external wiring connecting some of the reconfigurable device pins to them (Möller et al., 2004) most Xilinx devices have available an ICAP. The ICAP usually constitutes the best choice for supporting the construction of DSRS, since user logic can reach it from inside the reconfigurable device.

Configuration Controller

The Authors of this paper have built two versions of the RSCM (Carvalho et al., 2004) configuration controller: (i) a pure hardware version; (ii) a mostly software version (RSCM-S). Table 3 compares these two implementations qualitatively.

RSCM-S is three times slower than the RSCM hardware. This disadvantage is related to the inefficiency of the current API furnished by Xilinx to give access to ICAP. This API requires the CC to fetch 512-word blocks of each partial configuration and store these in a BRAM. Only after caching these data, the API sends configuration data to the ICAP. The RSCM sends data directly from an external memory to ICAP, leading to smaller reconfiguration time.

RSCM-S runs on an embedded 32-bit RISC designed by Xilinx, MicroBlaze. The structure of RSCM-S also includes peripheral device control-lers, memory and a communication infrastructure. If configuration control is the only task assigned to this infrastructure, the approach could hardly be justified. However, assuming that most applications today require the use of one or more processors inside the system, and assuming some of these processors have spare time to perform the configuration controller tasks, the additional hardware for configuration control requires less area than RSCM. Given the assumptions above and if the application reconfiguration time requirements are not too stringent, RSCM-S can be usefully applied.

Another important aspect regarding the design of CCs is the easiness for updating/adapting the CC to different applications. When it is necessary to include additional functionalities to the configuration controller, a software implementation is definitely more adequate. Complex tasks can be easily implemented through programming. Examples of such functionalities are configuration compression and on-the-fly decompression, on-the-fly decryption, configuration scheduling policies, and support to configuration preemption. A hardware-only implementation such as RSCM would require restructuring the CC design, realizing the CC re-synthesis and would probably increase the area overhead of the controller.

DSRS Infrastructure

Table 4 presents some recommended infrastructure choices for DSRSs. Software configuration controllers allow greater flexibility. It is possible to overcome its higher reconfiguration time disad-

Table 3. Qualitative comparison of two RSCM implementations

Characteristic	RSCM (hardware)	RSCM-S (software)
Configuration Speed	Milliseconds	Milliseconds
Area	Requires additional hardware	If processor available, small area overhead (ICAP and macro controllers)
Modification easiness	Complex / extra área	Simple / modify software

Table 4. Recommended infrastructure choices for DSRSs

Infrastructure Element	Recommended Choice
Configuration Controller	Software
Reconfigurable Interface	LUT-Macro
Repository	External SRAM
Reconfigurable Port	ICAP
Communication Infrastructure	NoC

vantage by rewriting the API to access the ICAP module, or by adding a small hardware module to directly manage ICAP.

A recommended choice for the reconfigurable interface is to use LUT-macros. Macros developed by Xilinx (Lim & Peattie, 2002) use a larger area when compared to the LUT-macros proposed in current work. Figure 3 illustrates the difference: the Xilinx Bus Macro consumes CLBs from 6 distinct CLB columns, being two in the fixed area and four in the reconfigurable area. Meanwhile, LUT-macros occupy CLBs of only two CLB columns, one at the fixed area and one at the reconfigurable area. Another difference is the number of bits transported by each macro: a Xilinx Bus Macro is 4-bit wide and LUT-macro allows 8-bit wide transfers. CLB columns used for both macros have reduced usability, due to

placement and routing restrictions imposed by the macros on both fixed and reconfigurable areas (Lim & Peattie, 2002).

Another recommendation is to use external static RAM to store partial configurations, since the controller to access these memories is very simple, present a small access time, and the capacity of such memories is sufficient to store several partial configurations. It is not advisable to waste internal FPGA memory with partial configurations, since the capacity of such memories is too small.

Table 5 compares the area required to implement different memory controllers and the typical capacity of such memories, used to store partial configurations. Internal BRAMs and external SRAMs are both static memories, consuming insignificant area to implement their controllers. SDRAMs require more complex controllers, but offer larger storage capacity.

ARTEMIS NoC

The last component of the proposed DSRS infrastructure discussed here is the communication infrastructure. As stated before, NoCs are good choices due to their scalability, increased paral-

Figure 3. Physical implementation of macros configured to communicate data from left to right: (a) Bus Macro (Lim & Peattie, 2002); (b) F2R Macro proposed in current work

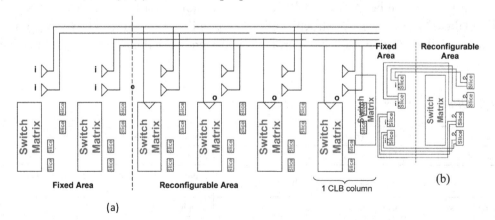

Table 5. Area required for implementing different memory controllers and the typical capacity enabled by each

Memory Type	Slices (Area)	Typical Capacity
BRAM	4	306 KB (XC2VP30)
SRAM	8	1MB
SDRAM	373	64MB

lelism and short-range wires that reduce power consumption. This work proposes Artemis, a NoC supporting specific reconfiguration services. The following Sections briefly present the Hermes NoC, used as basis for Artemis and describe the modifications carried out in Hermes to allow its use in DSRSs.

Hermes NoC

The Hermes NoC (Moraes et al., 2004) employs *packet switching*, a communication mechanism in which packets are individually routed between IPs, with no previously established communication path. The *wormhole* switching mode is used to avoid the need for large buffer spaces. The Hermes NoC supports 2D mesh topologies only, facilitating intra-chip IP placement and routing. A 2-flit header and the payload form the Hermes packet. The header contains the target address and the amount of *flits* in the payload.

The main component of the network is the Hermes router. It is assumed that each IP in the systems connects to exactly one router. This router has a centralized control logic and up to five bidirectional ports (North, South, East, West and Local). Each port has an input buffer for temporary *flit* storage. The Local port establishes the communication between the router and its local IP. The other ports connect the router to its neighbors. Credit based or handshake flow control may be used to transfer data between routers.

The centralized control logic implements the routing and arbitration algorithms. When a router receives a header flit, it performs a round-robin arbitration, and if the incoming packet request is granted, a deterministic XY routing algorithm is executed to connect an input port to an output port. If the chosen port is busy, the header flit, as well as all subsequent flits of this packet, will be blocked in intermediate input buffers. The routing request for this packet remains active until a connection is established in some future execution of the arbitration/routing in the input port. After routing all flits of the packet, connection is closed. At the operating frequency of 50 MHz, with a word size (flit) of 16 bits the theoretical peak throughput of each Hermes link is 800 Mbits/s (using credit based flow control).

Artemis NoC

The partial reconfiguration process may produce glitches in the interface between the IP under reconfiguration and the rest of the device. These glitches may introduce spurious data into the NoC, causing malfunctions or even circuit blocking. In addition, packets transmitted to an area suffering reconfiguration, must be discarded, since it is typically impossible to know if these packets are targeted to the previous configuration in this area or to the next reconfiguration. To avoid such problems, a set of services must be added to the NoC to enable its use in DSRSs.

Three services are implemented in Artemis: (i) reconfigurable area insulation; (ii) packet discarding; (iii) reconfigurable area reconnection. Hermes passed through the addition of two functionalities to support these services: (i) definition of control packets, enabling IPs to send packets to routers, not only to other IPs; (ii) capacity to disconnect/connect routers from its associated reconfigurable area. These functionalities are detailed in the next Sections.

Control Packets

The addition of two sideband signals per port to the original Hermes router serves to differentiate control packets from data packets. These signals, depicted in Figure 4, are *ctrl_in* and *ctrl_out*. For each *flit* sent by *data_out*, the *ctrl_out* is asserted together with *tx* if the *flit* is a control packet. The target router receives *flits* analogously, using *data_in*, *rx* and *ctrl_in* signals.

Figure 5(a) presents the control packet used to insulate a router from the reconfigurable IP, while Figure 5(b) presents the control packet to reconnect these. Control packets have two flits. The first is the XY target address in the network and the second is the operation code for the router. With such packet structure, the NoC can support up to 2(flit size, in bits) distinct services.

When the reconfigurable area is insulated, the router discards any data packets sent to the area under reconfiguration. Insulation also protects the network, since during reconfiguration transients can occur in the reconfigurable interface. If such signals are considered, spurious data may enter the NoC. Transients were observed in hardware by measuring the router-IP interface with a logic analyzer during reconfiguration. These events may signal a *false packet* to the router, with unpredictable outcomes. Once the new IP is configured, a control packet reconnects IP and router, enabling normal operation.

The reception and forwarding of control and data packets are similar. The major change in the router is the addition of one bit at each position of the input buffer. This is required to propagate the value of the *ctrl_out* signal to the reconfigurable IP router. When the control packet arrives at its destination router, it decodes and executes the corresponding operation.

Reconfigurable IP to Router Interface

This work proposes a new reconfigurable interface that does not impose the use of a specific communication infrastructure. This interface uses LUTs. Two unidirectional macros compose the reconfigurable interface, as depicted in Figure 6. The first one, named F2R, is responsible to send data from the fixed part of the system to a reconfigurable IP, while the second one, named R2F, implements the communication in the inverse direction. Both macros allow the simultaneous transmission of 8 data bits. The F2R macro is an identity function, while the R2F uses a special logic to avoid transient glitches during reconfiguration process from reconfigurable to fixed areas.

The complete interface between the Artemis router and a reconfigurable IP appears in Figure 7. It uses two R2F macros to connect 10 bits from right to left and two F2R macros to connect 11 bits in the reverse direction. The interface between the router and the reconfigurable IP does not contain the *ctrl_in* and *ctrl_out* signals because reconfigurable IPs neither send nor receive control packets. The *reset* is a global signal used to initialize the entire system. The router asserts

Figure 4. Interface between routers in the Artemis NoC

Figure 5. Format of control packets for insulation and reconnection of the router-IP interface in the Artemis NoC

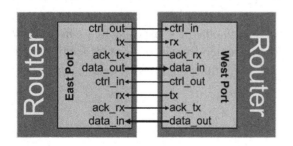

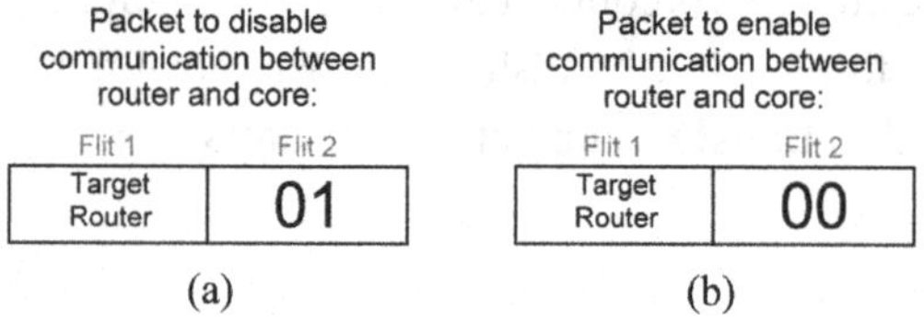

the *reconf* signal to initialize the reconfigurable core connected to the local port. The *reconf_n* signal in Figure 7 connects to the control signal in Figure 6, controlling the connection from the reconfigurable core to the router.

DESIGN FLOWS FOR DRS

The previous Sections dissertated about the infrastructure of DRSs and the interconnection to a NoC communication medium. However, no design flows were pointed out to prototype such a runtime reconfigurable infrastructure on an FPGA yet. Therefore, this Section presents not only a timeline view of design flow alternatives that can be used to create DRSs, but also presents the "straightforward" design flow, which shares the main concepts with all other flows and is used in this work.

The first sound possibility to really make use of partial and dynamic reconfiguration appeared in 1999 with the release of the Xilinx application notes 138 and 151 (Xilinx, 2009). These application notes provided important information about the structure of the Virtex bitstream, allowing one to manually manipulate the bitstream in order to create a partial bitstream. After that, different design flows for DRS emerged and will be summarized below.

Figure 6. Proposed macros for two different directions: (a) F2R; (b) R2F

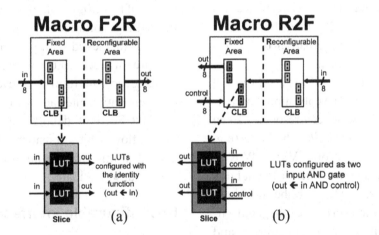

Figure 7. Router to reconfigurable core interface

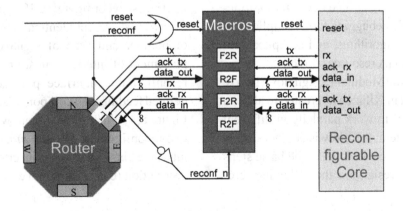

The most "straightforward" design flow for DRSs is to use the normal Xilinx tools to create the hardware project and perform three additional steps to this project: add reconfigurable interfaces; make project modules placement; and reroute wires that cross the boundaries between static and reconfigurable modules without passing through the reconfigurable interfaces. Subsequently, this project is copied and the copy is modified to create a partial configuration of the system. Finally, a tool to extract the modified logic and create the partial bitstream from a total bitstream must be used. Examples of these tools are: JBitsRipper (James-Roxby et al., 1999), JRTR (McMillan & Guccione, 2000), PARBIT (Horta et al., 2002), JBitsCopy (Dyer et al., 2002), JPEG (Raghavan & Sutton, 2002), CoreUnifier (Moraes et al., 2003), Virtex PART (Gericota et al., 2003), REPLICA (Kalte et al., 2005), and BITPOS (Krasteva et al., 2005). The BitGen (Xilinx, 2009) tool from Xilinx can also be used to generate partial bitstreams with the parameter *partialmask*.

JBits (Guccione et al., 1999) is a set of Java classes that provides an API (Application Programming Interface) to access bitstream resources. Even though JBits can be used to build other tools (e.g. JBitsRipper, JRTR, JBitsCopy, Virtex PART), JBits can also be used in two other different DRS flows (Dyer & Wirz, 2002): describe the whole circuit inside JBits and then create other configurations of this circuit to be partial bitstreams; and describe the static circuit with standard tools and the partial circuit with JBits. Both of these design flows are possible, but limited due to the simplicity of the available debugging tools, simplicity of the wire routing algorithms, and complexity of programming FPGA resources directly.

Even though the Modular Design flow (Lim & Peattie, 2002) from Xilinx was created to allow a team of engineers to work paralelly in different modules of a system, this flow is another option that can be used to create DRSs. The basic steps to use Modular Design are the following: cre-

ate a top level file, instatiate all modules (black boxes) that compose the system and connect these modules; implement, synthesize, place and route each module individually, and create the partial bitstream of each module; and finally merge all the partial bitstreams to generate the total bitstream that will initialize the system.

An upgrade of the Xilinx Modular Design flow called Early Access Partial Reconfiguration flow (EAPR) was released by Xilinx in 2005 (Xilinx, 2008). The two main modifications are: the flow allows the partial reconfiguration of modules of any rectangular size (available only for FPGAs that do support the reconfiguration of part of an FPGA column); and a significant improvement of the easy-of-use of the flow provided by the PlanAhead software.

Independently of the DRS flow in use, the layout of reconfigurable IPs shares some properties: (i) logic of a reconfigurable region must lie inside it (achieved with placement restrictions); (ii) wires of a reconfigurable region must lie inside it (achieved with routing restrictions); (iii) fixed communication interface with the rest of the DRS.

Next Sections details the "straightforward" design flow, which consider such properties and is used in this work to build two different case studies described later.

Reconfigurable Interfaces Insertion

To enable the use of reconfigurable IPs, it is necessary to impose two restrictions in reconfigurable interfaces: reconfigurable IPs sharing the same region must present identical interfaces (in terms of number and type of signals) and identical placement of interface pins. One way to define reconfigurable interface pins is to insert predefined feedthrough components, named *macros*. Figure 8(a) illustrates a system with one fixed IP, two reconfigurable IPs and *macros* defining the interface pins. *Macros* are inserted in the system description (e.g. VHDL or Verilog).

Placement Constraints

The second step is to constrain the placement of IPs and *macros*, as presented in Figure 8(b). A floorplanner tool may constrain the placement and shape of the system IPs (fixed and reconfigurable IPs), as well as the placement of *macros*. Standard place and route follows the constraints insertion.

Routing Verification / Modification

In the current generation of Xilinx physical synthesis tools, floorplanning restrictions do not have influence on the routing tool. As illustrated in Figure 8(b), some wires can still cross reconfigurable region boundaries. If this situation occurs, the associated signal can be disconnected after a reconfiguration step, possibly causing a system malfunction. In this case, the designer must either reroute the wire(s) crossing the interfaces (manually or automatically) or go back to the previous step, to try different placement constraints. The final routing must be similar to the one presented in Figure 8(c), where no wire crosses a reconfigurable interface. One noticeable exception to this rule is the global clock signals, which can safely cross the whole chip.

Partial Configurations Generation

Partial configurations, or partial *bitstreams*, are a set of bits used to configure a DRS. Partial bit-

stream generation is done by extracting a section of a total bitstream, corresponding to a reconfigurable region. This is illustrated in Figure 8(d). It is important to include part of the macro component in partial bitstreams to connect the reconfigurable core to the fixed part of the DRS. The method used here to generate partial bitstreams is straightforward, a one-phase flow. Assignment of another core to the same region requires partially repeating the flow for each core, while keeping the same placement constraints. Two tools may generate partial bitstreams. The first one is the proprietary Xilinx tool, *BitGen*, with specific commands to define the coordinates of the reconfigurable core. The second tool, compatible with all Virtex-II (Pro) devices, was developed by the authors.

In order to enable self reconfiguration, partial bitstreams have to be stored in a Repository. Each partial bitstream have an ASCII header, which have to be removed to be correctly transmitted through ICAP.

Two optional steps can be added to the flow before storing partial bitstreams in a Repository: compression, to reduce the amount of memory required in the Repository; and encryption, to increase system security. Both steps require additional hardware to perform on-the-fly decompression and decryption. The system presented by Huebner et al (2005) is an example containing an internal bitstream decompressor.

Figure 8. DRS flow proposed in this work

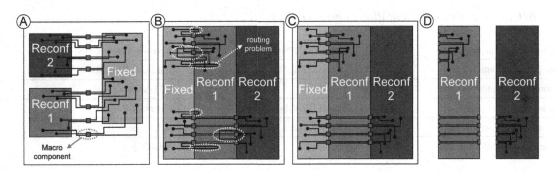

Core Relocation

Two situations require to partially repeating the DRS flow. The first one arises with the assignment of different cores to the same reconfigurable region. The second one arises with the assignment of the same core to different reconfigurable regions. It is possible to avoid the second situation if the same bitstream can be loaded at different regions. This procedure is called relocation (Krasteva et al., 2005) or placeability (Koester et al., 2009). A core originally synthesized for one reconfigurable region can be moved to another one, without re-synthesis. Core relocation also reduces the memory requirements to store partial bitstreams, diminishing system cost.

CASE STUDIES

This Section presents the implementation of two proof-of-concept DSRS case studies and their comparison. Table 6 details the characteristics of the OPB-based (Figure 9) and Artemis-based (Figure 11) case studies.

These case studies allow DSRS design space exploration, evaluating benefits, gains and limitations of each infrastructure element.

OPB-Based DSRS Description

The OPB-based DSRS, illustrated in Figure 9, contains a Microblaze processor, running an ap-

plication and the configuration controller (RSCM-S). The system also contains the following IPs connected to the OPB bus:

- An RS-232 IP connected to a host computer;
- A macro controller IP, to connect and disconnect reconfigurable IPs from the IPIF interface;
- An ICAP controller to manage the self reconfiguration process;
- A BRAM based repository to store partial configurations (Buffer in Figure 9);
- A reconfigurable IP, which can host one of several simple arithmetic modules.

The design flow to synthesize this DSRS requires additional steps w.r.t. the one previously discussed. A similar flow is also used in (Donato et al., 2005). The steps to construct the OPB-based DSRS are:

- Construct an initial system, using the Embedded Development Kit (EDK) with the Xilinx IPs and the reconfigurable IP (user function + macros + OPB wrapper);
- Insert macros to insulate the user function from the fixed part. These macros are located between the IPIF interface and the user function (the user module template generated by EDK offers to the user an interface simpler than the OPB bus, named

Table 6. Case studies implementation characteristics

Infrastructure Element	OPB-based DSRS	Artemis-based DSRS
Configuration Controller	Software (RSCM-S)	Hardware (RSCM)
Reconfigurable Interface	LUT-Macro	LUT-Macro
Repository	Internal BRAM	External SRAM
Reconfigurable Port	ICAP + Xilinx API	ICAP + dedicated hardware
Communication Infrastructure	OPB Bus	Artemis NoC

IPIF). Even if IPIF is simpler than OPB, it has 80 signals (36 from left to right, 44 from right to left), requiring 11 macros (5 R2F macros, 6 F2R macros), complicating floorplaning and routing steps;

- Generate the system netlist with EDK, exporting it to Integrated Software Environment (ISE);
- Execute the logic synthesis, followed by floorplanning and physical synthesis. The result of this step is the complete bitstream;
- Import results back to EDK for software generation. The binary code is finally added to the complete bitstream.

The above steps are repeated for each reconfigurable IP. Partial bitstreams are extracted from the obtained complete bitstreams.

The OPB-based DSRS was prototyped in a Memec Insight platform with a Virtex-II Pro XC2VP30 device. Figure 10 illustrates the physical implementation of the OPB-based DSRS. Due to routing complexity of the reconfigurable interface, macros were extended to occupy four CLB columns (extended macros). This was necessary to obtain a correct routing in the reconfigurable interface boundary.

OPB-based DSRSs have two drawbacks: bus-based communication and limited internal repository. Additionally, the design flow is quite complex, since two software environments are used: EDK and ISE. However, this simple case study allows reconfiguration time evaluation using the Xilinx API to access the ICAP module, and the area consumed to implement the reconfiguration infrastructure.

ARTEMIS-Based DSRS Description

The Artemis-based DSRS, illustrated in Figure 11, contains the following IPs:

- A 2x2 Artemis NoC;
- An RS-232 IP connected to a host computer;
- MR2, a 32-bit RISC processor, based in the load-store MIPS architecture, with 27 distinct instructions, a 32x32 register file, non-pipelined. The processor uses four internal 18 Kbits RAM blocks as instruction and data memories, providing 1K words in each memory;
- One reconfigurable IP, which can be one of three simple integer arithmetic modules:

Figure 9. The OPB-based DSRS structure

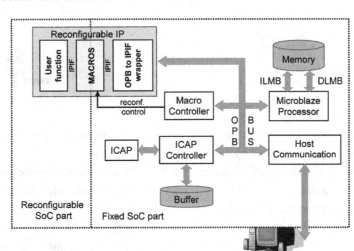

Figure 10. Physical implementation of the OPB-based DSRS in a Virtex-II Pro XC2VP30 device, with two distinct reconfigurable IPs. The OPB reconf. timer is used for debug purposes

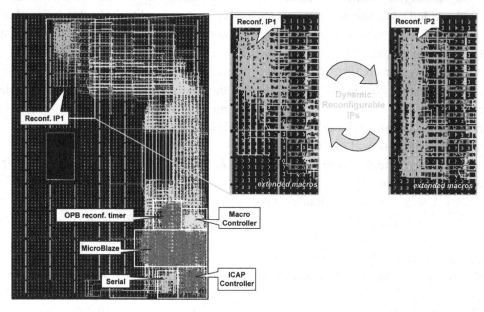

Figure 11. Artemis-based DSRS

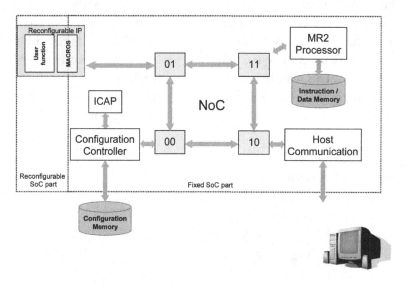

"mult" (multiplies two 16-bit operators), "div" (divides one 16-bit operator by a 16-bit operator) and "sqrt" (extracts the square root of a 32-bit operator);

- The RSCM configuration controller (CC).
- The processor is the system master. Memory mapped instructions access

reconfigurable IPs. The following protocol is used:

- The processor sends a packet to the CC, informing the identification of the desired IP.
- The CC (i) receives the reconfiguration request; (ii) selects a reconfigurable area where to configure the requested IP (if

Figure 12. Physical implementation of the Artemis-based DSRS in a Virtex-II Pro XC2VP30 device, with three distinct reconfigurable IPs

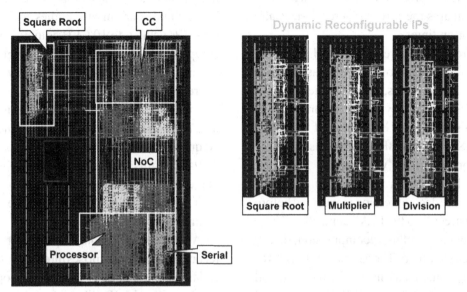

more than one reconfigurable area is available); (iii) sends a packet to disconnect communication between the router and the selected reconfigurable area; (iv) read the specific bitstream, transmitting it to ICAP.

• After reconfiguration, the CC sends a packet to reconnect communication between router and the configured IP. A second packet is sent to the processor with the networks address where the IP was configured.

If the IP is already configured in the reconfigurable area, the CC simply sends the IP network address to the processor.

The Artemis-based DSRS was also prototyped in a Memec Insight platform with a Virtex-II Pro XC2VP30 device. Figure 12 illustrates the physical implementation of the Artemis-based DSRS. The design flow to synthesize this DSRS employs the straightforward flow presented in the Design Flow for DRS Section. This is simpler than the flow used for the OPB-based DSRS, since only the ISE environment needs to be used.

Except for the configuration controller, this DSRS follows the recommended choices to implement DSRS. The configuration controller is implemented in hardware, favoring performance, but reducing flexibility.

Infrastructure Comparison

A common choice for both experiments is the use of LUT macros. LUT macros were employed in the OPB-based DSRS due to the number of bits in the reconfigurable interface (80), therefore reducing the number of CLB rows when compared to Xilinx Bus Macros. The LUT macros had to be extended to occupy 4 CLB columns each to achieve successful interface routing. The Artemis-based DSRS has a less complex interface (21 bits), using four LUT macros, exactly as presented in Figure 7, and occupying only 2 CLB columns each.

A second common choice in both experiments is the ICAP configuration port. The first case study uses the Xilinx API to access the ICAP port, while the second case study uses a dedicated module developed to access the ICAP port. As already mentioned, the Xilinx API is slower than dedicated hardware due to current buffering requirements. Table 7 compares the partial bitstream sizes and reconfiguration times.

The third column presents partial bitstream sizes. Partial bitstreams of the OPB-based DSRS occupy 10 CLB columns, while for the Artemis-based DSRS they occupy 6 CLB columns1. It is possible to store partial bitstreams of the OPB-based DSRS in internal BRAMs because a simple compression algorithm was applied to partial bitstreams, based on zeroes/ones counting. On-the-fly software decompression is executed before sending bitstreams to the ICAP controller. There is no time penalty in this decompression, due to the algorithm simplicity. The Artemis-based DSRS stores partial bitstreams in a 1 Mbyte external SRAM. The Artemis-based DSRS stores up to 10 partial bitstreams, without compression, while the OPB-based DSRS is able to store only 2 partial bitstreams using compression.

The fourth and fifth columns present the reconfiguration time using the RSCM and RSCM-S configuration controllers. The RSCM reconfiguration time is in average three times faster than RSCM-S, considering the NoC protocol. Reconfiguration times were measured using two methods: internal FPGA timers and a logic analyzer.

The sixth column presents the minimal reconfiguration time, assuming it would be possible to transmit one partial bitstream byte per clock

cycle (at 50 MHz). This column shows that it is not possible to work with reconfiguration times below 1 ms in current case studies, with reconfigurable IPs using 6 to 10 CLB columns. With more complex reconfigurable IPs, reconfigurable area is expected to increase consequently increasing the reconfiguration time.

Figure 13 details the reconfiguration time for the *divider IP*. The reconfiguration time, 9.65 ms, is equivalent to 482,500 clock cycles. Observe that 99.94% of this reconfiguration time is spent by the reconfiguration process itself (Figure 13(c)), with a very small time spent in the NoC with control packets.

After reconfiguration, the protocol to access the reconfigurable IP comprises three steps: (i) creation and transmission of a packet with the operators to the reconfigurable IP; (ii) creation and transmission of a read packet to receive results; (iii) reception of the result packet from the reconfigurable IP. Typical time spent in each step is 173, 141 and 117 clock cycles respectively. As the reconfigurable IPs are very simple in this case study, once the read request arrives at the reconfigurable IP, the packet with the results is sent immediately to the source IP, totalizing in average 439 clock cycles (sum of the time spent in each step). This protocol can be simplified by eliminating the read packet (141 cycles), with the reconfigurable IP sending the answer directly to the source IP.

At 50 MHz, 10 ms represent 500,000 clock cycles. This reconfiguration time can be hid-

Table 7. Reconfiguration time for OPB and Artemis based DSRS case studies

	Partial Bitstream Size		Reconf. Time (50 MHz)		
	Module	Size (KB)	RSCM	RSCM-S	Best†
OPB based	Arith. 1 & 2	182,180	-	63.55	3.64
Artemis based	Multiply	99,644	9,98	34.76*	1.99
	Divider	96,428	9,65	33.63*	1.93
	Square Root	101,988	10,21	35.57*	2.04

*Estimated, using data from the OPB-based system.
†Minimum reconfiguration time for the bitstream at 50MHz.

den by: (i) executing complex computations in hardware; (ii) pre-fetching reconfigurable IPs to later use; (iii) reusing the same reconfigurable IPs during a time longer than the execution in software plus the time to configure the IP into the DSRS. With such strategies, the reconfiguration time has minimal impact in DSRS performance. For example, if a given function executed in hardware is 500 clock cycles faster than an equivalent software implementation, after 1,000 consecutive executions the hardware implementation displays superior performance. This can be easily achieved with image processing algorithms, where the same operation can be repeated thousands of times.

For these proof-of-concept case studies, the average execution time for the equivalent software implementation is 26% slower (in average 600 clock cycles against 439 clock cycles). This difference in favor of the hardware implementation, 161 cycles, is not yet sufficient to demonstrate

performance gains for the proposed infrastructure, but clearly shows its viability. Some application portions (typically loops) may benefit from this approach, given they consume at least 1,000 clock cycles in the embedded processor and are repeatedly used.

Table 8 and Table 9 compare the area to implement both DSRSs The first analysis concerns the configuration controller (CC) area overhead. The RSCM uses 494 FPGA slices. The RSCM-S uses 821 slices (Microblaze, ICAP and macro controllers). However, if a processor is already available in the system (such as MicroBlaze), the area of the RSCM-S represents the area of the ICAP and macro controllers, resulting in 250 slices. As processors are ubiquitous in actual SoCs, software CC represents the implementation option with the smaller area overhead.

The area of the Artemis-NoC is 1167 slices (Table 8), representing in average 290 slices per router. For this case study, this area represents

Figure 13. Reconfiguration protocol timing, in clock cycles, for Artemis-based DSRS

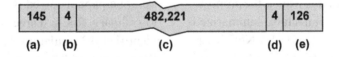

| 145 | 4 | 482,221 | 4 | 126 |

    (a)   (b)   (c)   (d) (e)

(a) packet from a source IP to the CC asking a new reconfigurable IP
(b) CC processing time and packet to the reconfigurable area to disconnect it
(c) reconfiguration time
(d) packet from the CC to the new reconfigurable IP reconnecting it
(e) packet from the CC to the source IP with the reconfigurable IP address

Table 8. FPGA area report (XC2VP30) for the Artemis-based DSRS

IP	# Slices (total: 13696)		# Flip Flops (total: 27392)	
	Total	%	Total	%
Serial	316	2.31	279	1.02
Processor	1001	7.31	555	2.03
CC (RSCM)	494	3.61	294	1.07
Artemis NoC	1167	8.52	959	3.50
DIV (reconf IP)	183	1.34	259	0.95
MULT (reconf IP)	172	1.26	259	0.95
SQRT (reconf IP)	223	1.63	269	0.98

Table 9. FPGA area report (XC2VP30) for the OPB-based DSRS

IP	# Slices (total: 13696)		# Flip Flops (total: 27392)	
	Total	%	Total	%
MicroBlaze	571	4.17	366	1.34
MicroBlaze Periph.	160	1.17	75	0.27
MicroBlaze OPB	90	0.66	11	0.04
ICAP Controller	151	1.10	155	0.57
Macro Controller	99	0.72	136	0.50
Arith 1 (reconf IP)	128	0.93	168	0.61
Arith 2 (reconf IP)	128	0.93	168	0.61

an important overhead. In practice, when using real IPs, an area overhead of 5-10% per IP is expected, justifying the use of NoCs in DSRSs. Comparing the router area to the Gecko platform (Marescaux et al., 2004), Gecko routers consume 611 slices (router plus network interfaces, data and control).

CONCLUSION AND FUTURE WORK

The main contribution of this work is the proposal of a conceptual DSRS architecture, summarized in Table 4, centered on the use of a NoC interconnection. The implementation of two proof-of-concept case studies demonstrates the viability of the proposed DSRS architecture, even if none of the case studies follow all recommendations. However, each recommendation in the Table was implemented and evaluated by at least one of the case studies. To support the development of the proposed DSRS architecture, the paper advanced two additional contributions: (i) a suggestion of a straightforward DSRS design flow; (ii) the design of a specific NoC supporting partial and dynamic hardware reconfiguration.

The ideal implementation choice for this DSRS architecture is dedicated ASICs with embedded reconfigurable areas. Nonetheless, partial and dynamic reconfigurable FPGAs were used to successfully prototype the architecture. The main ad-

vantage of the suggested flow is a reduced number of steps compared to other flows proposed in the literature, such as Modular Design. The proposed flow employs new macros, which guarantee the correct operation of the rest of the system during reconfiguration, avoiding the use of tristate buffers, components scarcely available in Virtex FPGAs. Also, the new macros enable the use of communication architectures other than busses to link reconfigurable modules to other parts of the system. To support dynamic IPs reconfiguration, the paper showed the need to add services to ordinary NoCs. Three needed services were identified: IP insulation, packets discarding and IP reconnection. These services were implemented over the existing Hermes NoC, resulting in a new NoC, Artemis, which supports DSRS.

The case studies evaluation helped to identify the area overhead incurred by the proposed infrastructure and the reconfiguration time. The addition of a Configuration Controller in a SoC represents a small area overhead (1.82 to 3.61% of the available slices for XC2VP30 device), while providing a greater flexibility to the system. The addition of hardware flexibility to a SoC enables to implement the same function both in software and in hardware. The user or the operating system may select the implementation according to performance requirements. The experiments allowed to observe that, independently of the fact that reconfiguration is controlled in software or hardware,

IP reconfiguration time is always above 2 ms for current FPGA technologies (measured times were between 9.65 ms and 63.55 ms). This represents an average value of 500,000 clock cycles. The time measured to send data to the reconfigurable IP, and to receive data from it, through the NoC is around 439 clock cycles. Performance gains can be easily obtained in loops with small/medium complexity (1,000 clock cycles) or more complex IPs.

The conceptual DSRS architecture paves the way to define a *reference architecture* to be used as a platform for future research. This platform should contain at least two embedded processors and two reconfigurable areas, characterizing a reconfigurable MPSoC.

Future work related to the software domain includes CAD tools to automatically select parts of applications that would benefit from hardware execution and integration with operating systems. Hardware/software partition and profiling techniques may be employed to select parts of applications to be accelerated by reconfigurable IPs. Scheduling is the main open issue to integrate operating systems to reconfigurable hardware. Tasks may be defined in software, in hardware or in both, hardware and software. As a reconfigurable MPSoC may contains several processors and several reconfigurable IPs, the scheduling algorithm should select where tasks will run, the task implementation, and to implement the hardware-software communication. The OS scheduler should also consider the reconfiguration time and the NoC communication protocol.

Future work related to the hardware domain includes the design of a reconfigurable MPSoC with embedded reconfigurable logic, targeting an ASIC implementation. Present FPGAs, such as Xilinx VirtexII-PRO, contain one reconfigurable logic block loosely coupled to PowerPC processors. The reconfigurable MPSoC ASIC design is a candidate to be a *future FPGA*, since it will contain several processors and embedded reconfigurable logic blocks connected to the NoC. All benefits of DSRSs could be better explored with such devices.

REFERENCES

Athanas, P., & Silverman, H. (1993). Processor reconfiguration through instruction-set metamorphosis. *Computer*, *26*(3), 11–18. doi:10.1109/2.204677

Atmel (2009). *Atmel Corporation, Inc.* Retrieved February 7, 2009, from http://www.atmel.com

Barat, F., Lauwereins, R., & Deconinck, G. (2002). Reconfigurable instruction set processors from a hardware/software perspective. *IEEE Transactions on Software Engineering*, *28*(9), 847–862. doi:10.1109/TSE.2002.1033225

Becker, J., Piontek, T., & Glesner, M. (2000). DReAM: A dynamically reconfigurable architecture for future mobile communications applications. In *Proceedings of the 10th Conference on Field Programmable Logic and Application* (pp. 312-321). Villach, Austria.

Benini, L., & De Micheli, G. (2002). Networks on chips: A new SoC paradigm. *Computer*, *35*(1), 70–78. doi:10.1109/2.976921

Bobda, C., Ahmadinia, A., Majer, M., Teich, J., Fekete, S., & Veen, J. (2005). DyNoC: A Dynamic Infrastructure for Communication in Dynamically Reconfigurable Devices. In *Proceedings of the 15th Conference on Field Programmable Logic and Application* (pp. 153-158). Tampere, Finland.

Carvalho, E., Calazans, N., Moraes, F., & Mesquita, D. (2004). Reconfiguration control for dynamically reconfigurable systems. In *Proceedings of the 19th Conference on Design Circuits and Integrated Systems* (pp. 405-410). Bordeaux, France.

Claus, C., Zeppenfeld, J., Muller, F., & Stechele, W. (2007). Using Partial-Run-Time Reconfigurable Hardware to accelerate Video Processing in Driver Assistance System. In *Proceedings of Design, Automation and Test in Europe Conference and Exposition* (pp. 1-6). Nice, France.

Dally, W., & Towles, B. (2001). Route packets, not wires: On-chip interconnection networks. In *Proceedings of the Design Automation Conference* (pp. 684-689). San Diego, CA.

Donato, A., Ferrandi, F., Redaelli, M., Santambrogio, M., & Sciuto, D. (2005). Caronte: A complete methodology for the implementation of partially dynamically self-reconfiguring systems on FPGA platforms. In *Proceedings of the IEEE Symposium on Field-Programmable Custom Computing Machines* (pp. 321-322). Napa, CA.

Dyer, M., Plessl, C., & Platzner, M. (2002). Partially Reconfigurable Cores for Xilinx Virtex. In *Proceedings of the 12th Conference on Field Programmable Logic and Application* (pp. 292-301). Montpellier, France.

Dyer, M., & Wirz, M. (2002). *Reconfigurable Systems of FPGA* (Diploma Thesis). Swiss Federal Institute of Technology Zurich (82 p.). Zurich, Switzerland.

Gericota, M., Alves, G., Silva, M., & Ferreira, J. (2003). Run-Time Management of Logic Resources on Reconfigurable Systems. In *Proceedings of the Design, Automation and Test in Europe Conference and Exposition* (pp. 974-979). Munich, Germany.

Goldstein, S., Schmit, H., Moe, M., Budiu, M., Cadambi, S., Taylor, R., & Laufer, R. (1999). PipeRench: A coprocessor for streaming multimedia acceleration. In *Proceedings of the 26th International Symposium on Computer Architecture.* (pp. 28-39). Atlanta, GA.

Gonzalez, R. (2000). Xtensa: A configurable and extensible processor. *IEEE Micro, 20*(2), 60–70. doi:10.1109/40.848473

Griese, B., Vonnahme, E., Porrmann, M., & Rückert, U. (2004). Hardware support for dynamic reconfiguration in reconfigurable SoC architectures. In *Proceedings of the 14th Conference on Field Programmable Logic and Application* (pp. 842-846). Leuven, Belgium.

Guccione, S., Levi, D., & Sundararajan, P. (1999). JBits: A Java Based Interface for Reconfigurable Computing. In *Proceedings of the 2nd Annual Military and Aerospace Applications of Programmable Devices and Technologies Conference.* Laurel, MD.

Hartenstein, R. (2001). A decade of reconfigurable computing: A visionary retrospective. In *Proceedings of Design, Automation and Test in Europe Conference and Exposition* (pp. 642-649). Munich, Germany.

Hartenstein, R., & Kress, R. (1995). A datapath synthesis system for the reconfigurable datapath architecture. In *Proceedings of the Conference on Asia Pacific Design Automation* (pp. 479-484). Makuhari, Massa, Chiba, Japan.

Hauck, S., Fry, T., Hosler, M., & Kao, J. (1997). The chimaera reconfigurable functional unit. In *Proceedings of the 5th IEEE Symposium on Field-Programmable Custom Computing Machines* (pp. 87-96). Napa, CA.

Hauser, J., & Wawrzynek, J. (1997). Garp: A MIPS processor with a reconfigurable coprocessor. In *Proceedings of the IEEE Symposium on Field-Programmable Custom Computing Machines* (pp. 12-21). Napa, CA.

Horta, E., Lockwood, J., & Kofuji, S. (2002). Using PARBIT to Implement Partial Run-Time Reconfigurable Systems. In *Proceedings of the 12th Conference on Field Programmable Logic and Application* (pp. 182-191). Montpellier, France.

Huebner, M., Paulsson, K., & Becker, J. (2005). Parallel and flexible multiprocessor system-on-chip for adaptive automotive applications based on Xilinx microblaze soft-cores. In *Proceedings of the 18th International Parallel and Distributed Processing Symposium* (pp. 149a-149a). Santa Fe, NM.

James-Roxby, P., Cerro-Prada, E., & Charlwood, S. (1999). A Core-Based Design Method for Reconfigurable Computing Applications. In *Proceedings of the IEE Colloquium on Reconfigurable Systems*. Glasgow, Scotland.

Kalte, H., Lee, G., Porrmann, M., & Rückert, U. (2005). REPLICA: A Bitstream Manipulation Filter for Module Relocation in Partial Reconfigurable Systems. In *Proceedings of the 19ᵗʰ International Parallel and Distributed Processing Symposium* (pp. 151b-151b). Denver, CO.

Keutzer, K., Newton, A. R., Rabaey, & J. M., Sangiovanni-Vincentelli, A. (2000). System-level design: Orthogonalization of concerns and platform-based design. *IEEE Transactions on CAD of Integrated Circuits and Systems, 19*(12), 1523-1543.

Koester, M., Luk, W., Hagemeyer, J., & Porrmann, M. (2009). Design Optimizations to Improve Placeability of Partial Reconfiguration Modules. In *Proceedings of Design, Automation and Test in Europe Conference and Exposition* (pp. 976-981). Nice, France.

Krasteva, Y., Jimeno, A., Torre, E., & Riesgo, T. (2005). Straight Method for Reallocation of Complex Cores by Dynamic Reconfiguration in Virtex II FPGAs. In *Proceedings of the International Symposium on Rapid System Prototyping* (pp. 77-83). Montreal, Canada.

Lim, D., & Peattie, M. (2002). Two flows for partial reconfiguration: Module Based or small bit manipulations. *Xilinx Application Note 290 (v1.0)*.

Marescaux, T., Nollet, V., Mignolet, J.-Y., Bartic, A., Moffat, W., Avasare, P., Coene, P., Verkest, D., Vernalde, S., & Lauwereins, R. (2004). Run-time support for heterogeneous multitasking on reconfigurable SoCs. *Integration, the VLSI Journal, 38*(1), 107-130.

McMillan, S., & Guccione, S. (2000). Partial Run-Time Reconfiguration Using JRTR. In *Proceedings of the 10ᵗʰ Conference on Field Programmable Logic and Application* (pp. 352-360). Villach, Austria.

Mignolet, J., Nollet, V., Coene, P., Verkest, D., Vernalde, S., & Lauwereins, R. (2003). Infrastructure for design and management of relocatable tasks in a heterogeneous reconfigurable system-on-chip. In *Proceedings of Design, Automation and Test in Europe Conference and Exposition* (pp. 986-991). Munich, Germany.

Möller, L., Calazans, N., Moraes, F., Brião, E., Carvalho, E., & Camozzato, D. (2004). FiPRe: An implementation model to enable self-reconfigurable applications. In *Proceedings of the 14ᵗʰ Conference on Field Programmable Logic and Application* (pp. 1042-1046). Leuven, Belgium.

Möller, L., Grehs, I., Carvalho, E., Soares, R., Calazans, N., & Moraes, F. (2007). A NoC-based Infrastructure to Enable Dynamic Self Reconfigurable System. In Proceedings of ReCoSoC. Montpellier, France.

Möller, L., & Moraes, F. (2005). Sistemas Dinamicamente Reconfiguráveis com Comunicação Via Redes Intra-Chip (Master Thesis). *Catholic University of Rio Grande do Sul* (142 p.). In Portuguese. Porto Alegre, Brazil.

Moraes, F., Calazans, N., Mello, A., Möller, L., & Ost, L. (2004). HERMES: An infrastructure for low area overhead packet-switching networks on chip. In *Proceedings of the 17th Symposium on Integrated Circuits and Systems Design* (pp. 69-93). Porto de Galinhas, Brazil.

Moraes, F., Mesquita, D., Palma, J., Möller, L., & Calazans, N. (2003). Development of a Tool-Set for Remote and Partial Reconfiguration of FPGAs. In *Proceedings of Design, Automation and Test in Europe Conference and Exposition* (pp. 1122-1123). Munich, Germany.

Mrabet, H., Marrakchi, T., Mehrez, H., & Tissot, A. (2006). Implementation of Scalable Embedded FPGA for SoC. In *Proceedings of the International Conference on Design and Test of Integrated Systems in Nanoscale Technology* (pp. 74-77). Gammarth, Tunisia.

Palma, J., Mello, A., Möller, L., Moraes, F., & Calazans, N. (2002). Core communication interface for FPGAs. In *Proceedings of the 15th Symposium on Integrated Circuits and Systems Design* (pp. 183-188). Porto Alegre, Brazil.

Pionteck, T., Koch, R., & Albrecht, C. (2006). Applying Partial Reconfiguration to Networks-on-Chips. In *Proceedings of the 16th International Conference on Field Programmable Logic and Applications* (155-160). Madrid, Spain.

Raghavan, A., & Sutton, P. (2002). JPG – A Partial Bitstream Generation Tool to Support Partial Reconfiguration in Virtex FPGAs. *In Proceedings of the 16th International Parallel and Distributed Processing Symposium* (pp. 155-160). Florida, USA.

Resano, J., Mozos, D., Verkest, D., Catthoor, F., & Vernalde, S. (2004). Specific scheduling support to minimize the reconfiguration overhead of dynamically reconfigurable hardware. In *Proceedings of Design Automation Conference* (pp. 119-121). San Diego, CA.

Sassatelli, G., Torres, L., Benoit, P., Gil, T., Diou, C., Cambon, G., & Galy, J. (2002). Highly scalable dynamically reconfigurable systolic ring-architecture for DSP applications. *In Proceedings of Design, Automation and Test in Europe Conference and Exposition* (pp. 553-558). Paris, France.

Singh, H., Lee, M., Lu, G., Kurdahi, F., Bagherzadeh, N., & Filho, E. (1998). MorphoSys: A reconfigurable architecture for multimedia applications. In *Proceedings of the 11th Symposium on Integrated Circuits and Systems Design* (pp. 134-139). Rio de Janeiro, Brazil.

Sullivan, C., Wilson, A., & Chappell, S. (2005). Deterministic Hardware Synthesis for Compiling High-Level Descriptions to Heterogeneous Reconfigurable Architectures. In *Proceedings of the Hawaii International Conference on System Sciences* (1-9), Big Island, HI.

Ullmann, M., Huebner, M., Grimm, B., & Becker, J. (2004). An FPGA run-time system for dynamical on-demand reconfiguration. In *Proceedings of the 18th International Parallel and Distributed Processing Symposium* (pp. 135-142). Santa Fe, NM.

Van den Branden, G., Touhafi, A., & Dirkx, E. (2005). A design methodology to generate dynamically self-reconfigurable SoCs for Virtex-II Pro FPGAs. In *Proceedings of the IEEE International Conference on Field-Programmable Technology* (pp. 325-326). Singapore.

Waingold, E., Taylor, M., Srikrishna, D., Sarkar, V., Lee, W., & Lee, V. (1997). Baring it all to software: Raw machines. *IEEE Computer*, *30*(9), 86–93.

Walder, H., & Platzner, M. (2004). A runtime environment for reconfigurable hardware operating systems. In *Proceedings of the 14th Conference on Field Programmable Logic and Application* (pp. 831-835). Leuven, Belgium.

Wirthlin, M., & Hutchings, B. (1995). A dynamic instruction set computer. In *Proceedings of the 3rd IEEE Symposium on Field-Programmable Custom Computing Machines* (pp. 99-107). Napa, CA.

Wirthlin, M., & Hutchings, B. (1997). Improving functional density through run-time constant propagation. In *Proceedings of the 5th International Symposium on Field Programmable Gate Array* (pp. 86-92). Monterey, CA.

Wirthlin, M., Hutchings, B., & Gilson, K. (1994). The Nano Processor: a Low Resource Reconfigurable Processor. In *Proceedings of the IEEE Symposium on Field-Programmable Custom Computing Machines* (pp. 23-30). Napa, CA.

Wittig, R., & Chow, P. (1995). OneChip: An FPGA processor with reconfigurable logic. In *Proceedings of the 3ʳᵈ IEEE Symposium on Field-Programmable Custom Computing Machines* (pp. 126-135). Napa, CA.

Xilinx (2009). *Xilinx Corporation, Inc.* Retrieved on March 30, 2009, from http://www.xilinx.com

Yeo, S., Lyuh, C., Roh, T., & Kim, J. (2008). High Energy Efficient Reconfigurable Processor for Mobile Multimedia. In *Proceedings of the IEEE International Conference on Circuits and Systems for Communications* (pp. 618-622). Shanghai, China.

KEY TERMS AND DEFINITIONS

Core Relocation: Process of changing the place where an IP core is implemented.

Dynamically Reconfigurable System (DRS): System that makes use of the partial reconfiguration process.

Dynamically Self Reconfigurable System (DSRS): System that makes use of the partial reconfiguration and self reconfiguration processes.

Intellectual Property (IP) Cores: Block of logic or data that is used in making a field programmable gate array (FPGA) or application-specific integrated circuit (ASIC) for a product.

Multiprocessor System-on-Chip (MPSoC): SoC which uses multiple processors, usually targeted for embedded applications.

Network-on-Chip (NoC): Approach to design the communication subsystem of SoCs. NoC brings networking theories and systematic networking methods to on-chip communication and brings notable improvements over conventional bus systems. NoC greatly improve the scalability of SoCs, and shows higher power efficiency in complex SoCs compared to buses.

Partial Reconfiguration: Process of configuring a portion of a field programmable gate array (FPGA) while the other parts of the FPGA are still operating.

Self Reconfiguration: Process of modifying a portion of a system from another part of the same system.

System-on-Chip (SoC): The electronics for a complete, working product contained on a single chip.

ENDNOTE

[1] Different bitstream sizes for the same number of CLB columns arrises because partial bitstreams are generated by *BitGen*, which uses the multi-frame write feature.

Chapter 2
Dynamically Reconfigurable Networks-on-Chip Using Runtime Adaptive Routers

Mário P. Véstias
INESC-ID/ISEL/IPL, Portugal

Horácio C. Neto
INESC-ID/IST/UTL, Portugal

ABSTRACT

The recent advances in IC technology have made it possible to implement systems with dozens or even hundreds of cores in a single chip. With such a large number of cores communicating with each other there is a strong pressure over the communication infrastructure to deliver high bandwidth, low latency, low power consumption and quality of service to guarantee real-time functionality. Networks-on-Chip are definitely becoming the only acceptable interconnection structure for today's multiprocessor systems-on-chip (MPSoC). The first generation of NoC solutions considers a regular topology, typically a 2D mesh. Routers and network interfaces are mainly homogeneous so that they can be easily scaled up and modular design is facilitated. All advantages of a NoC infrastructure were proved with this first generation of NoC solutions. However, NoCs have a relative area and speed overhead. Application specific systems can benefit from heterogeneous communication infrastructures providing high bandwidth in a localized fashion where it is needed with improved area. The efficiency of both homogeneous and heterogeneous solutions can be improved if runtime changes are considered. Dynamically or runtime reconfigurable NoCs are the second generation of NoCs since they represent a new set of benefits in terms of area overhead, performance, power consumption, fault tolerance and quality of service compared to the previous generation where the architecture is decided at design time. This chapter discusses the static and runtime customization of routers and presents results with networks-on-chip with static and adaptive routers. Runtime adaptive techniques are analyzed and compared to each other in terms of area occupation and performance. The results and the discussion presented in this chapter show that dynamically adaptive routers are fundamental in the design of NoCs to satisfy the requirements of today's systems-on-chip.

DOI: 10.4018/978-1-61520-807-4.ch002

INTRODUCTION

The increasing density of integrated circuits allows the implementation of systems-on-chip (SoC) with hundreds of homogeneous or heterogeneous processing and memory units. The communication between these units may become intensive requiring high bandwidth, low latency, quality of service to guarantee real-time functionality, and low power consumption. These requirements and constraints must be guaranteed by the communication infrastructure.

Traditional interconnection architectures, such as single buses or a hierarchy of buses, are no longer able to support the increasing interconnection complexity and the bandwidth demands of such platforms due to their poor scalability and shared bandwidth. The Network-on-Chip (NoC) communication structure has been introduced as a new interconnection paradigm able to integrate a significant number of IP cores while keeping a high communication bandwidth between them (Hemani et al., 2000; Dally & Towles, 2001).

Networks-on-Chip are becoming definitely the only acceptable interconnection structure for today's multiprocessor systems-on-chip (MPSoC) for several reasons: (1) hundreds of IP cores can be "easily" integrated in a single device since NoC structures are scalable and do not have the problems associated with hierarchical bus structures which are usually irregular and harder to route; (2) lower design effort since a specific technological NoC solution can be easily scaled up, that is, designers do it once for each technology; (3) permits modular design with a hierarchy of communication layers that hide the inherent complexity of lower levels; and (4) robust designs can be achieved with fault tolerance using multipaths for the same source-destination pair.

The first generation of NoC solutions considers regular topologies, typically 2D meshes under the assumption that the wires' layout is well structured in such topologies. Routers and network interfaces between IP cores and routers are mainly homogeneous so that they can be easily scaled up and facilitate modular design. All advantages of a NoC infrastructure were proven with this first generation of NoC solutions.

However, soon, the designers started to be worried about the two main disadvantages associated with NoCs, namely, area and speed overhead. Routers of a NoC need space for buffers, routing tables, switching circuit and controllers. On the other side, direct bus connection is always faster than pipelined connections through one or more routers since these introduce latency due to packaging, routing, switching and buffering.

In a first attempt to consider area and latency in the design process, designers considered that regular NoC structures may probably be adequate for general-purpose computing where processing and data communication are relatively equally distributed among all processing units and traffic characteristics cannot be predicted at design time. But, many systems developed for a specific class of applications exhibit an intrinsic heterogeneous traffic behavior. Since routers introduce a relative area overhead and increase the average communication latency, considering a homogenous structure for a specific traffic scenario is definitely a waste of resources, a communication performance degradation and an excessive power consumption.

Application specific systems can benefit from heterogeneous communication infrastructures providing high bandwidth in a localized fashion where it is needed to eliminate bottlenecks (Benini & De Micheli, 2002), with sized communication resources to reduce area utilization, and low latency wherever this is a concern. More, a specific component of the infrastructure can be complemented with some capabilities to deal with, for example, quality of service.

Homogeneous and heterogeneous solutions of first generation NoCs follow different design methodologies but have one thing in common: their architectures are found at design time and are kept fixed at running time, i.e., the topology and the architecture of the routers are fixed at design time. Apparently, this is not a problem, but since several applications may be running with the same NoC, the same topology and router will generally not be equally efficient in terms of area, performance and power consumption for all different applications. The efficiency of both homogeneous and heterogeneous solutions can be improved if runtime changes are considered. A system running a set of applications can benefit from the runtime reconfiguration of the topology and of the routers to improve performance, area and power consumption considering a particular data communication pattern. Customization of the number of ports, the size of buffers, the switching techniques, the routing algorithms, the quality of service techniques, the switch matrix configuration, etc. should be considered in a reconfigurable NoC. Both general purpose and application-specific SoCs will benefit from using dynamically reconfigurable NoCs since the performance and power consumption of data communication can be optimized for each specific application.

The second generation of NoCs will be dynamically or runtime reconfigurable providing a new set of benefits in terms of area overhead, performance, power consumption, fault tolerance and quality of service compared to the previous generation where the architecture is decided at design time.

There is no doubt that the best NoC structure for an application specific multi-core system will have in most cases an irregular topology with heterogeneous routers. The problem now is to decide about the reconfigurable structures of these routers, which components should be considered in the configuration process, and how they should dynamically adapt to the traffic.

In the following sections, we are concerned with the design of a dynamically reconfigurable router. In particular, the chapter will address the following:

- **Analysis of a typical static NoC** – Understand the limitations of a static structure, how application specific NoCs benefit from design time customization of routers and topology and what improvements do we expect when using dynamically adaptive NoCs;
- **Customization of a router** – Analyze and discuss methods to customize a router with the objective to improve its area and performance;
- **Analyze the tradeoffs between the customization methods** – There are several customization methods to improve the architecture of the router for specific communication scenarios. Naturally, the improvements obtained using all methods are not cumulative, i.e., after using one type of customization, the next one will not be so effective compared to what would be obtained if it was utilized at first place. For example, changing the routing or the switching algorithm determines the size of the buffers, or using a specific routing algorithm will determine the connectivity demands of the switch matrix, etc. We will analyze and discuss the effectiveness of some customization methods and how they influence each other;
- **Dynamic customizations** – Propose an adaptive router based on the identified customizations. See how the customizations methods can be adapted for a dynamically reconfigurable router.

BACKGROUND

Almost one hundred proposals of NoC architectures can be found in the literature (Salminen, Kulmala & Hämäläinen, 2008). These NoC proposals differ in the used topology, the routing and the switching schemes, the design metrics and the target application. Routers have been also extensively studied, designed and implemented with different flit widths, buffer sizes, switching mechanisms, routing mechanisms considering latency, area, power consumption, fault-tolerance, quality-of-service, etc.

Some of these architectures are the basis of a number of methodologies dedicated to help in the design of application specific NoCs, where topology, routing, buffer size, switch matrix, link size and number of ports are customized to efficiently run an application in terms of performance, area and power consumption. These design methodologies for application-specific NoCs vary in the way they customize the NoCs. Some consider the mapping of tasks to IP cores as a way to improve traffic latency. Others consider the customization of the topology or the router, while others only use an adaptive routing algorithm to achieve better traffic distribution.

A few approaches consider a homogeneous infrastructure and just change the mapping of tasks. The idea is that different mappings of tasks influence the traffic distribution which will balance the bandwidth demand over each router turning the homogeneous solution a better approach. This way it is possible to minimize the effects of a homogenous infrastructure over the average communication delay, the area overhead or the power consumption of a NoC. For example, Murali & Micheli (2004) presented a mapping algorithm aiming to minimize the energy consumption and the average communication delay. Lei & Kumar (2003) use a two step genetic algorithm to map tasks in order to maximize timing performances. Hansson et al. (2005) developed an algorithm to map cores onto a NoC topology and at the same time statically route the communications.

A careful mapping of tasks improves traffic latency but it is somehow limited by the fixed topology. Application-specific topology customization has a major influence in the final NoC performance and area. Topology customization considers regular and irregular topologies where

the number of links and routers are optimized for specific applications or set of applications with a traffic behavior known before execution. This design exploration has been considered in several works. Pinto, Carloni & Sangiovanni-Vincentelli (2003) proposed a design process to optimize the topology for a specific application without any floorplanning considerations, while Ahonen (2004) considered a floorplanner during topology design to improve power consumption. Using a floorplanner is important to estimate the design area and wire-lengths. These estimates can be used to estimate the target frequency and the power consumption. Benini (2006) proposed a methodology for NoC design that customizes the topology, the frequency of operation and link-width. Floorplaning is also considered to place cores and NoC components. Recently, Elmiligi, Morgan, El-Kharashi & Gebali (2008) proposed a topology generation methodology to minimize the power consumption of application specific NoCs. A network partitioning technique is employed to optimize the topology and, therefore, to reduce the number of links. Also, the number of ports of a router can be reduced based on the topology connections, which will reduce the size of the routers.

Application-specific routing algorithms can also be successfully used to improve the NoC in terms of performance by equally distributing the traffic throughout the network. Routing algorithms are classified as deterministic or adaptive. Deterministic routing always uses the same route for a particular destination without considering any information about the state of the network. Adaptive routing considers the state of the network, such as the status of a link or buffer, to route data.

Compared to adaptive, deterministic routing requires fewer resources while guaranteeing an ordered packet arrival. On the other hand, adaptive routing provides better throughput and lower latency by allowing alternate paths. Deterministic routing is more appropriate if the traffic generated by the application is predictable, while adaptive deals better with irregular networks and/or stochastic traffic.

Deterministic routing usually has poor capacity to equally distribute the traffic along all links of the network since the routes are statically assigned independently of the traffic requirements. On the other hand, highly adaptive algorithms have the potential to reach a uniform utilization of resources and provide fault tolerance. These algorithms distribute the traffic through all links to reduce congestion. However, the efficiency of highly adaptive algorithms is compromised by the necessity to guarantee deadlock free scenarios. Generally, to keep the high adaptiveness of these algorithms a number of virtual channels are required (Bjerregaard & Mahadevan, 2006) increasing the cost of the solution compared to that using deterministic routing.

Some researchers have improved the performance of routing algorithms using adaptive techniques (Duato, 2003; Chiu, 2000; Glass, 1994). Hu & Marculescu (2004) have proposed a smart routing for networks-on-chip which is a combination of static and dynamic routing. The router switches between Odd-Even and XY routing using a simple congestion monitoring. XY routing was also improved to load balance the traffic (Dehyadgari et al., 2005). Pseudo adaptive XY-routing uses the deterministic XY-routing algorithm under low congestion and adaptive routing under heavy traffic conditions. A few solutions use the load state of input or output buffers of neighbor routers to determine the best path (Ascia et al., 2006). In (Faruque et al., 2007), the best path is determined at runtime considering the distance to the destination and link bandwidth usage.

The topology of the network and the routing mechanism to be followed are two important design optimizations to be considered in the development of a NoC. However, router optimizations are also fundamental to reduce area overhead and latency. Several aspects may be considered in the optimization process of the router: number and size of buffers, switch matrix structure and

number and size of links. The number and size of links are consequences of the topology structure. The optimization of the switch matrix is also very important since for specific technologies (e.g., FPGA – *Field Programmable Gate Array*) this is one of the most area consuming resource. Buffers also take a significant portion of the area and power consumption of a NoC (Michelli & Benini, 2006) and so their size must be carefully minimized. Reducing the size of buffers has a negative impact over the performance of the NoC, especially when the network becomes congested. However, recent works (Martini et al., 2007; Medardoni et al., 2007) show that as the switch buffer size is decreased, the clock frequency can be increased.

Only a few works have considered router customization. Kreutz et al. (Kreutzh et al., 2005) proposed three different routers with different performance, energy and area occupation and an algorithm to find the right combination of router architectures and the optimal placements of cores that produce a NoC complying with the latency and the energy requirements. While important to improve NoC area overhead, this work uses three fixed router configurations, which may produce a solution still far from the optimal area consumption. Recently, Véstias and Neto (Véstias & Neto, 2007) proposed a generic router that can be used to design a NoC with a specific area/performance tradeoff according to the traffic patterns. The work considers the optimization of the size of buffers and of the switch matrix given a specific traffic behavior. Buffer size optimization is also a major concern for several authors. In (Hu, Ogras & Marculescu, 2006) a system-level buffer planning algorithm is proposed. Given the traffic characteristics and the buffer space budget, the algorithm automatically finds the buffer depth for each and every buffer of all routers to optimize performance.

The first generation of NoCs already provided some kind of reconfigurability or adaptability through the use of adaptive routing and reconfigurable computing. Adaptive routing algorithms are utilized to uniformly distribute the traffic throughout all links. This permits the NoC to improve the utilization of the resources at runtime, and potentially improve the performance and power consumption of the NoC. However, the routing algorithms run under the same topology and router configurations, and thus the optimization that can be gained with the adaptive routing is limited by the structure of the NoC and their network components. Besides, highly adaptive routing algorithms are expensive in terms of virtual channels and, consequently, increase the area overhead. Reducing the number of virtual channels usually means that the adaptiveness of the algorithms and hence the efficiency of the approach are reduced. Changing structures can be implemented with reconfigurable technology. Given a particular application or traffic behavior, it is possible to dynamically reconfigure the NoC to better fit the communication needs of the application. A few approaches have been developed. For example, Pionteck (Pionteck et al., 2006) proposed an approach to dynamically add or remove routers from the network allowing the NoC to adapt to the running application. Reconfiguration is done without stopping the operation of the circuit since dynamic routing tables are used. This is the typical approach with reconfigurable technology of the first generation of NoCs, i.e., architectures dynamically adapt to successive applications.

Recently, state-of-the-art proposals of dynamically reconfigurable NoCs have focused their attention into the reconfigurable topology. Stensgaard & Sparsø (2008) present a NoC architecture with a reconfigurable topology. The logical topology is configured at design time based on the communication requirements of the application. Topology switches are introduced between the network and the routers allowing links to be connected to a port of a router or to another link, whether using packet switching or circuit switching, respectively. Ahmad, Erdogan & Khawam (2006) present a dynamically reconfigurable NoC, where routing, switching and data packet size can be dynamically reconfigured, but resources are

fixed at design time. Kumar et al. (Kumar et al., 2007) follow a similar approach where runtime configuration reduces to a simple change of the contents of a memory. Topology, buffers and port connections are determined at design time. In the CuNoC approach (Jovanvic et al., 2007) the reconfigurable device approach is filled with small units that can establish a communication between two cores. This approach suffers from huge power consumptions and high latency due to the significant number of units in a communication path. Furthermore, a custom topology cannot be defined. CoNoChi (Pionteck, Koch & Albrecht, 2006) is an adaptable NoC. The reconfigurable device is divided in a matrix where each cell can hold a computational module, and a switch or a point-to-point link. This approach has major drawbacks, including high area overhead and latency. In (Rana et al., 2008), a runtime reconfigurable NoC is presented using an FPGA. The NoC can add or remove express lines, and perform run-time NoC topology and routing table reconfiguration to deal with traffic congestion.

Runtime switching was considered by Wang et al. (Wang, Gu & Wang, 2007) that proposed a hybrid switch mechanism to guarantee quality of service for real-time applications. In this approach, circuit switching is used for real-time traffic and virtual cut through switching is used for best effort traffic. At runtime, data is identified according to the type of switching to be used for their commutation within each router.

According to this state-of-the-art description, second generation NoCs are still in its infancy. Basic runtime topology configuration and dynamic switching together with adaptive routing algorithms were the first attempts towards a dynamically reconfigurable NoC. Runtime router configuration with an adaptive structure is still an open issue that must be mitigated so that an integrated runtime NoC will be designed.

STATIC NOC DESIGN

As we are aiming at a discussion of runtime features of NoC routers, we start with an analysis of a static NoC and its routers. The objective is to understand the limitations of a static structure, how application specific NoCs benefit from design time customization of routers and topology and what improvements do we expect by using adaptive NoCs.

In a typical homogeneous NoC, each router has the same structure and is connected to at most four neighbor routers and to a local core. Each core connects to the communication network through only one router, and links between routers have the same bandwidth (see figure 1a).

Other homogeneous topologies are possible as long as all IP cores have a connection to the communication infrastructure. Also, routers can connect to more than one IP core (see figure 1b). In this case, each router is connected to four cores and to two neighbor routers. The links between routers can be customized to support a higher bandwidth than the links between a router and a local core.

For a $n \times n$ mesh with bidirectional links there are $n-1$ bidirectional links in each line or column of routers. Therefore, we have $n \times (n-1)$ bidirectional horizontal links and $n \times (n-1)$ bidirectional vertical links. So, the total bandwidth, TB, of the mesh is given by:

TB $= 4 \times n \times (n - 1) \times bl$, where bl is the raw bandwidth of a link.

Assuming that each router is connected to a single IP core, the maximum traffic that the cores sent to the network, CT, is given by:

CT $= n^2 \times bc$, where bc is the raw bandwidth of injected traffic of one core.

Given the maximum injected traffic sent to the network and the maximum available aggregate bandwidth of the network, we can estimate the maximum percentage utilization (MPU) of the network as a function of the average routing distance, ard, that is, the average distance a packet travels from the source to the destination. Basically, a single packet consumes the bandwidth of ard links. Formally,

Figure 1. 2D-mesh homogeneous NoC Architectures: a)1-local connection; b) 4-local connections

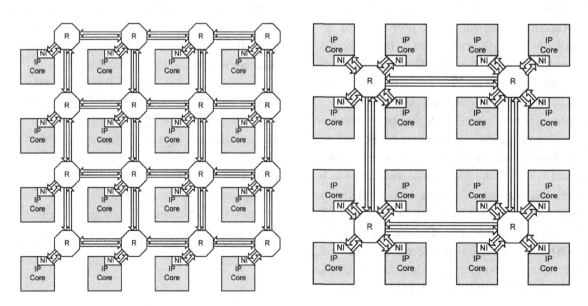

MPU (%) = CT / TB = 100 × ard × bc/bl × n/ (4n - 4))

The average routing distance depends on traffic pattern, stochastic variations in traffic and on the routing algorithm. Assuming an average routing distance of $n - 1$, and bc = bl, we have a maximum network utilization (MNU) given by $(n - 1) \times n/(4n - 4)) = n/4$. For example, for n = 4, MNU = 100%, and for higher values of n the network would have saturated, that is, it would not have enough bandwidth to support the communication and the IP cores could not inject data at the desired rate.

In practice, these utilization values are too high for a uniform 2D-mesh network due to traffic pattern and stochastic variations. During traffic routing there are conflicts in the utilization of some links and, consequently, packets have to be buffered. Buffering packets increases the average latency and so the delivered bandwidth is lower than the offered bandwidth. When we increase the injected rate of traffic, the average latency also increases until the network saturates, that is, until the delivered bandwidth has been totally utilized and the average latency goes to infinity (see figure 2).

Typical saturation points range from 40 to 80% depending on traffic pattern, stochastic variations in traffic and on the routing algorithm. To improve the bandwidth utilization of a NoC, we can use better routing algorithms capable to uniformly redistribute the traffic, which will reduce traffic congestion, and consequently improve the saturation point. For application specific NoCs, it is also possible to customize the topology so that hot spots of the network can be attenuated with more bandwidth to reduce traffic congestion. Low used links can be removed and highly utilized links can be improved with bigger buffers and link sizes. These customizations improve the utilization of the network and so the average latency and the saturation points are enhanced. Basically, what NoC customization does is to redistribute resources according to traffic patterns to improve average latency. In certain cases, customization can be used to reduce the area overhead with small or no degradation of the average latency.

Customization of a NoC

To give a better insight of what can be achieved with topology customization, we are going to use the methodology of (Véstias & Neto, 2006) to optimize the topology of a NoC. The approach starts with a fully interconnected NoC designated *Generic NoC* (GNoC) (see figure 3).

In the generic NoC architecture, a router can have up to four local connections with neighbor cores. Therefore, neighbors can exchange data

Figure 2. Relation between the total bandwidth of a NoC, the delivered bandwidth and the latency

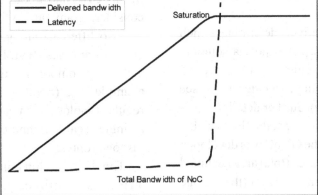

Figure 3. A generic network-on-chip

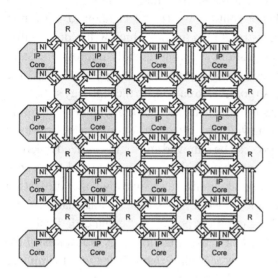

through a single router, reducing communication latency. From the generic architecture, it is possible to obtain irregular architectures by removing links and routers.

With the proposed methodology it is possible to customize the NoC structure with the appropriate number of routers and the proper number of connections and bandwidth, so that the communication structure uses less area and exhibits lower average communication latency than the typical NoC structure.

To customize the GNoC for a specific application or set of applications, a design space exploration (DSE) process was proposed to explore the design space, which consists on improving the area of the architecture. The DSE process assumes a given mapped application on the set of IP cores of the GNoC.

The DSE process iteratively determines which routers and links should be part of the final solution. The iterative process of the algorithm is controlled with a simulated annealing (SA) algorithm (see Véstias & Neto, 2006, for further details).

A discrete-event simulator at the flit level that mimics the behavior of the GNoC was developed and used to measure the area (total area occupied by the GNoC) and the average latency (the average time elapsed from the moment a flit is created until the moment it is received at the destination node for all flits). In each NoC simulation experiment, 1,000 packets were delivered using the round-robin arbitration mechanism and the store and forward switching mechanism.

Packets are generated at each node with a user defined injection rate, *irate*, and contain 4 flits. The injection rate is a fraction of the maximum channel bandwidth that in our experiments varies between 5% and 40% to avoid saturation. The injection rate is an ideal value that may not be respected if the network entrance is congested. In this case, the network is saturated.

In the experiments, we considered uniform traffic, local traffic and two real applications. With the uniform traffic, a node sends packets to all other nodes with the same probability. With the local traffic, a node sends packets to its nearer nodes with higher probability. One of the real applications is a video/audio system (multimedia application) from [Hu & Marculescu, 2005] with 40 tasks assigned onto 16 cores. We have mapped the 16 cores onto a 4×4 NoC according to figure 4 with the main objective of mapping cores with communicating tasks in neighbor tiles of the NoC, as far as possible.

Also, we have considered an execution of 30 frames/s.

The other real application is a radio system [Jantsch & Lu, 2009] already mapped onto a 4×4 NoC where the 26 traffic flows are clearly identified ranging from 64 upto 4096 Mbits/s. In this case, and in our simulations, we are not considering any quality-of-service restrictions.

Two different routing algorithms were considered: XY and west-first. The results obtained with GNoC are compared to the results obtained with a simple NoC (SNoC – a NoC with a 2D mesh regular topology with each router connecting to a single local core and each core connecting to just one router).

The NoC architectures were described in VHDL, synthesized, placed and routed with

Figure 4. Mapping of cores of the multimedia application into tiles of the NoC

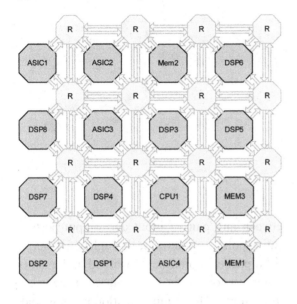

ISE 9.2 from Xilinx using a Virtex-5 FPGA to determine the occupied area and the operating frequency.

A set of experiments with different injection rates for the random application were considered. For all applications 64-bit links were assumed (see figures 5, 6 and 7).

The results show the area improvement achieved when topology optimization with GNoC is followed. The relative area of the optimized NoC goes from almost 85% down to 60% compared to the SNoC architecture.

When the west-first routing algorithm is used, we observe that better results are obtained because this algorithm is partially adaptive.

The latency follows a similar behavior. The SNoC can only achieve latencies above 11.5, while the optimized topology can achieve better latencies starting with 7. Similar results were obtained with west-first routing algorithm.

We applied the methodology in the design of the NoC for the real applications. The results are also very promising. In the case of the multimedia application, we obtain 40% improvement in the area with both routing techniques. In fact, the communications of the multimedia application are mostly to neighbor cores. Therefore, the final solution reduces to routers with multiple local ports that greatly improve the communication between neighbors while the links between routers have enough bandwidth to support non-local communications. In the case of the system radio application, the communications are much more demanding and so a NoC solution with the same area of the one used for the multimedia application cannot achieve the required data rates. Still, some improvement in the area was obtained, about 9% with XY and 20% with west-first. In this case, the west-first routing algorithm achieved better results than those obtained with the XY algorithm since it allows different paths for the same source-destination pair and therefore the communication paths can be better redistributed.

Figure 5. Irate/area ratio for a) uniform traffic, b) local traffic

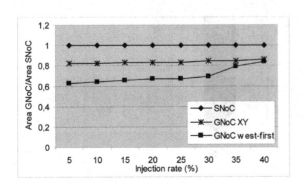
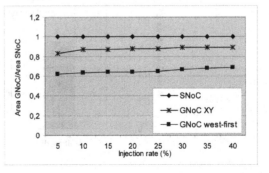

Figure 6. Latency/area ratio for a) uniform traffic, b) local traffic

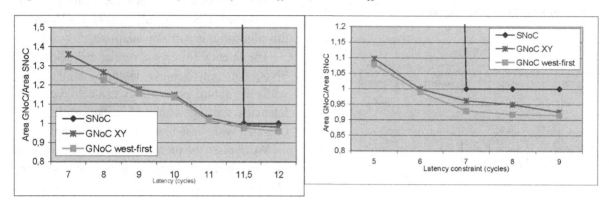

This discussion shows how a simple optimization methodology can easily improve the area and the average latency of a NoC customized for a specific application or traffic pattern.

ROUTER DESIGN

A router is used to forward packets between processing IP cores. After receiving a data packet, the router reads the packet destination address and forwards it to the correct output port. To achieve this, a router has several input and output ports connected to neighbor routers and to local IP cores, a switch that establishes a connection between any pair of input and output ports, a routing policy to determine through which output port a packet should be forwarded and a set of arbiters to control the simultaneous accesses to output ports. The communication ports include buffers to temporarily store packets (see a router with four input/output ports in figure 8).

The routing policy is implemented in each input port using a route block, which may be a hardwired implementation of a routing algorithm or a simple routing table.

Having determined the destination port, the route block sends a request to the arbiter associated with the destination output port. Since the arbiter may receive more than one request from

Figure 7. Area ratio for both real applications

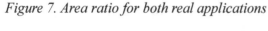

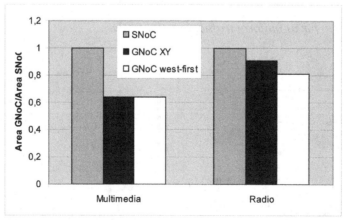

Figure 8. Router architecture with four input and four output ports

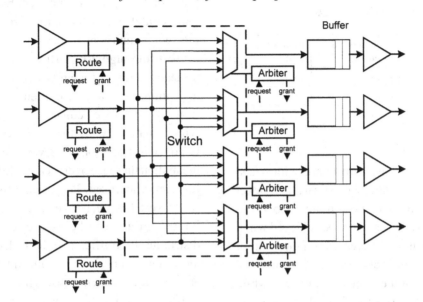

different input ports, it uses an arbitration policy to choose among the requesting input ports, usually round-robin, which is a fair mechanism that avoids starvation of packets. The packet is finally switched from the input port to the output port through the switch, which is nothing more than a set of multiplexers (or a crossbar for ASIC technology).

To better understand the tradeoffs involved in the customization of a router it is important to quantify the area occupied by each component of a router and the bandwidth availability.

Tables 1 and 2 show the values for the area occupation, for a Virtex-5 target device.

From the tables, the most area consuming components of a router are the buffers, the switches and the arbiters. For example, a typical router with 5 ports with size 16 (4 ports connected to neighbor routers and 1 local port connected to the network interface) uses 55 ($5 \times \text{arbiter}_5$) + 80 (switch matrix$_{5 \times 16}$) + 120 ($5 \times \text{FIFO}_{16 \times 16}$) + 30 ($5 \times \text{router}_5$) = 285 LUTs.

For a router with n bidirectional ports the total bandwidth, TB, is given by:

TB = $n \times$ br, where *br* is the raw bandwidth of a single port.

Table 1. Logic area (Virtex-5 LUTs) utilized by the arbiters, the switch and the routing blocks

# Ports	Arbiters	Switch (8 bits)	Switch (16 bits)	Routes
2	$2 \times 2 = 4$	$2 \times 8 = 16$	$2 \times 16 = 32$	$2 \times [3, 8] = [6, 16]$
3	$3 \times 5 = 15$	$3 \times 8 = 24$	$3 \times 16 = 48$	$3 \times [3, 8] = [9, 24]$
4	$4 \times 8 = 32$	$4 \times 8 = 32$	$4 \times 16 = 64$	$4 \times [3, 8] = [12, 32]$
5	$5 \times 11 = 55$	$5 \times 8 = 40$	$5 \times 16 = 80$	$5 \times [3, 8] = [15, 40]$
6	$6 \times 14 = 84$	$6 \times 16 = 96$	$6 \times 32 = 192$	$6 \times [3, 8] = [18, 48]$
7	$7 \times 18 = 126$	$7 \times 16 = 112$	$7 \times 32 = 224$	$7 \times [3, 8] = [21, 56]$

Table 2. Logic area (Virtex-5 LUTs) utilized by the buffers implemented as FIFOs

# Depth	Buffers (8 bits)	Buffers (16 bits)	Buffers (32 bits)
16	16	24	40
32	18	26	42
64	26	44	76

For example, a router with 4 ports with size *m* bits and a latency of one cycle to forward a packet from an input to an output can route up to *4 × m* bits/cycle. This result can only be achieved when all input packets have different destination output ports. Otherwise, the packets share a common output port and one of them must be buffered while the other is being forwarded. In this case, one output port is overused and another is underused and, consequently, the bandwidth utilization of the router decreases. The worst scenario is when all input packets compete for the same output port.

Once again, to improve the bandwidth utilization of a router, we can use better routing algorithms capable to uniformly redistribute the traffic, which will reduce the traffic congestion, and consequently improve the ports utilization and reduce the buffer requirements. For application specific NoCs, it is also possible to customize the router by increasing the bandwidth for heavily utilized ports and reducing it for ports with low utilization. These customizations improve the utilization of the router and so the average latency is enhanced. Also, in certain cases, customization can be used to reduce the area overhead with small or no degradation of the average latency, that is, when the average latency has no tight constraints it is possible to share resources of the router to reduce its area (Véstias & Neto, 2006).

Customization Methods for a Router

Each router of a network-on-chip can be customized to improve area or performance, or even both metrics. Customization can be done by changing the routing algorithm, the switching policy, the number and size of ports, the size of buffers and the switch matrix as follows.

Routing algorithm. Routing algorithms can be static, partially adaptive or adaptive. Static algorithms are easy to implement but are poor in terms of traffic redistribution. For example, with a static XY routing algorithm the traffic is concentrated along the vertical lines, which means that these are overused while the horizontal lines are underused. Partial adaptive algorithms (e.g., west-first, negative-first) are also easy to implement and can be used to dynamically redistribute the traffic. The main disadvantage of these algorithms is that some traffic directions are forbidden, which restricts the optimization of traffic redistribution. However, this is a necessity to avoid deadlocks and virtual channels. Finally, adaptive algorithms are able to achieve the best redistribution of traffic. However, they are harder to implement and need virtual channels to avoid deadlocks. The efficiency of adaptive routing depends on the ability to keep real-time information about the network utilization, which is a complex task. Generally, these routing algorithms rely only on neighbor information.

Switching policy. Several switching techniques have been considered in NoCs: circuit switching, store and forward, wormhole, virtual cut through and pipelined circuit switching. Circuit switching establishes a path before transmitting a message. Once the transmission completes, the path is turned down. Circuit switching can be used to guarantee latency constraints at the cost of low link utilization. With store and forward, messages are divided in packets to be routed. The entire packet is stored before being forwarded; therefore the latency increases with the number of hops of the path. In wormhole, packets are sub-divided into smaller units called flits. The first is used to determine the route and the remaining follow the first. Virtual cut through behaves like wormhole switching except when the outgoing channel is busy. In this case, virtual cut through buffers the entire packet in a single node while in wormhole the packet is stored in all nodes along the path. Pipelined circuit switching is like circuit switching except that the packet waits in the source until the path from source to destination is established. It has the advantage that after the path is established it is possible to guarantee a particular latency.

In the customization of a router, we can use wormhole or virtual cut through (store and forward is rarely used) to switch best effort traffic and circuit switching or pipelined circuit switching for delay sensitive traffic. For example, (Wang, Gu & Wang, 2007) consider an hybrid switching mechanism with virtual cut through and circuit switching.

Number and size of ports. The ports of a router may have different sizes to account for different utilizations. For example, the size of a heavily used port can be doubled and hence support twice the bandwidth of another port (see figure 9).

Besides increasing the number of lines of the link connected to the larger port, this solution also requires a larger output buffer and some additional hardware at the receiver, since there may be two flits at a single input port to be routed independently. Although more expensive in terms of area, this solution will reduce the average latency.

Another approach is to increase the number of ports in the same direction (see figure 10).

This solution provides the same bandwidth of the former but with independent links. The advantage is that both links can be customized and managed independently. For example, the buffers may have different sizes or one link may

be used for real-time traffic and the other for best-effort traffic. Also, one link may used with virtual cut through switching and the other with circuit switching.

Like the previous solution, it allows improving the latency at the cost of more links, extra buffering and more complex control and arbitration.

Size of buffers. Buffer resizing has proven to be an effective technique to improve latency (Hu, Ogras & Marculescu, 2006). An empirical rule for resizing the buffers is to increase the size of buffers associated with heavily used ports and decrease the size of the others.

For FPGAs buffer resizing is easier to design since the resources used to implement the buffers do not scale linearly with the size of buffer (See section "Router Design"). For example, FIFOs with up to 32 words of space have nearly the same size when implemented in a Virtex-5 device[1]. Therefore, for this technology resizing only makes sense with multiples of 32 words.

Switch matrix. With an FPGA, a switch matrix is implemented with a set of multiplexers. In the case of a fully connected switch, there will be one multiplexer for each output and each multiplexer has an input for each input of the router. Therefore, the switch has as many internal buses as the number of input and output ports. For example, a fully connected router with five input and output ports has a switch matrix with five internal buses that allow five simultaneous communications. A fully connected switch matrix with m ports can switch up to m flits at the same time, as long as

Figure 9. Router with ports of different sizes

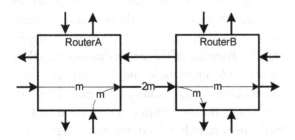

Figure 10. Router with multiple ports in the same direction

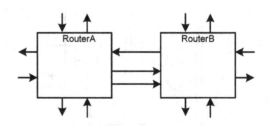

they all have different destinations. This is the best performance that can be achieved with a fully connected switch matrix. Therefore, the performance of the switch cannot be improved, but it is still possible to reduce its area.

Reducing the area of the switch can be accomplished by sharing internal buses. For example, consider the implementation shown in figure 11.

In this case, two pairs of input ports share buses of the switch and two pairs of output ports share outputs. Hence, instead of using 4 multiplexers with 4 inputs each, the solution uses 4 multiplexers with 2 inputs each. Instead of four arbiters with four input requests, it only requires four arbiters with only two input requests.

Many configurations can be obtained by changing the number of buses inside the switch. For example, for a router with 8 ports, there are 21 possible configurations for sharing the input ports (see figure 12).

As can be seen from the figure, there can be different configurations with different areas for the same number of internal buses. For example, a switch with two internal buses can have 4 different configurations: 1+7, 2+6, 3+5, 4+4. Each configuration has a different hardware implementation and consequently a different hardware size.

Figure 11. Switch matrix with shared resources

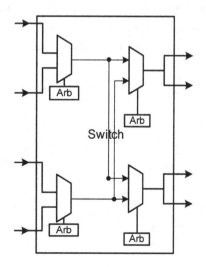

Similar conclusions may be obtained for routers with different number and size of ports. With this technique it is possible to save up to more than 80% of the area of a switch at the cost of bandwidth reduction, since the total bandwidth of the switch is proportional to the number of internal buses of the switch matrix.

Sharing buses of the switch permits to shrink the router but some care has to be taken to avoid a deadlock situation. Deadlocks can be avoided using virtual cut through switching where a packet is forward only if the output FIFO has enough space to accommodate all flits of the packet in case the header flit gets blocked due to congestion. See (Véstias & Neto, 2007) for further details.

Analysis of the Customization Methods

There are several customization methods to improve the architecture of the router for specific communication scenarios. Naturally, the improvements obtained using all methods is not cumulative, i.e., after using one type of customization, the next one will not be so effective compared to what would be obtained with its application at the first place. In the following sections we analyze the effectiveness of some customization methods (routing, number and size of ports, size of buffers, and the switch matrix) and how they influence each other.

Routing Algorithm

Adaptive routing algorithms can be used in the customization process of a router. In this case, the same algorithm is used for all applications with the advantage that it dynamically changes the routes according to the traffic requirements. The disadvantage of these algorithms is that they are harder to implement and real-time information about the congestion of links must be available so that the algorithm can take good routing decisions. Partially adaptive routing algorithms are easier to implement but also depend on good traffic congestion information to achieve good routing decisions. Usually, this type of algorithms improves the delivery time for some packets but increases it for others. Typical implementations of the partially adaptive algorithms are on average worst than the deterministic XY algorithm (see some results in Vieira, Ost, Moraes & Calazans, 2004).

Adaptive algorithms are being more extensively used to achieve fault tolerance in NoC fabrics (Zhu, Pande & Grecu, 2007). Due to technological defects, the final circuit may have defected links. To overcome the communication problems associated with a defective link, adaptive algorithms are used. These algorithms can easily find other paths around a defective path.

Instead of relying on the real-time behavior of adaptive algorithms to improve the average communication latency of a NoC communication infrastructure, it is helpful to customize the routes of a NoC considering the communication requirements of the application(s). In these cases, we know at compile time all pairs of communications and so we can set the paths for all communications to optimize the performance. For this approach to be effective, it is important that multiple different paths exist for the same source-destination pair.

This is where adaptive routing algorithms can be used efficiently (Palesi, Holsmark, Kumar & Catania, 2009).

To observe this behavior, we have measured the average latency for a 6×6 NoC with (1) uniform traffic and different injection rates, and (2) the two real applications utilizing a deterministic algorithm, XY, and a partially adaptive algorithm, west-first, using buffers with depth 16. For XY routing, there are no customizations since there is only one path for each communication source-destination pair. On the other hand, west-first allows different paths for communications with the same source-destination. In this case, to take advantage of the partial adaptability of west-first, we have implemented a very simple approach where all set of communications sharing the same source and destination are uniformly distributed through the best alternative paths (see results in figure 13).

With this simple approach it was possible to achieve around 10% average packet latency (the average time elapsed from the moment the first flit of the packet is created until the moment the last flit of the packet is received at the destination node. From now on, we will refer to average packet latency as simply average latency) reduction with the west-first algorithm compared to the XY in the

Figure 12. Ratio between area and number of buses of the switch for a router with 8 ports

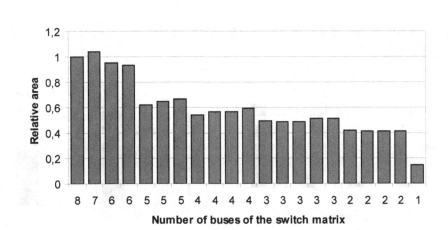

case of uniform traffic. At 50% injection rate both cases are saturated. For the two real applications, we achieved 3% improvement in the multimedia application and 20% in the radio application. The multimedia application is mostly neighbor traffic, so the expected improvement is low. For the radio application, the traffic is less localized and so the improvement is higher.

Number and Size of Ports

Overused links increase the average latency since packets must be buffered a number of cycles proportional to the utilization of the link. As already mentioned, one way to reduce the utilization of a link is to increase the number of ports in the same direction. It is possible to include any reasonable number of ports. However, each link needs a port an arbiter and a multiplexer, which increases the cost of the router.

Any direction of a router can be augmented with any number of links. However, the gains obtained do not increase linearly with the extra links because the next router may not be able to route the extra injected data unless it has also extra ports. We can add extra ports to all the routers. However, with such extra bandwidth added to the network at the cost of more area the bottlenecks will soon be the extremes of the communications: the processing units.

Instead of using independent extra links in the same direction, we can use a wider link. Compared to the previous solutions this one is less expensive in terms of area since it uses a single arbiter and a single FIFO. With FPGA technology a single FIFO is smaller than two half sized independent FIFOs (see table 2).

A 6×6 NoC was tested considering uniform traffic with injection rates of 40% and 50% and with both real applications utilizing the XY routing algorithm. The original network uses a single link in each direction, for a total of 120 links. Then, an extra link is added to the directions with an utilization higher than a specific value. Independent links and wider links were tested.

As shown in figure 14, independent links achieve better performance at the cost of higher areas. With independent links and an injection rate of 40% it is possible to achieve up to 70% of reduction in the average latency at the cost of more than 50% increase in the area. With single wider links the reduction in the average latency is about 60% with an increase of at most 35% increase in the area. We also observe, for example, that to achieve a latency of 20 cycles with two independent links we need an area of 5700 LUTs that is higher than the area achieved using the single wider link (5476 LUTs) to get the same 20 cycles of latency. So, the faster solution is not necessarily the better. In many cases the single-wider link is the best

Figure 13. Average latency for different routing algorithms with a) uniform traffic with different injection rates b) Real traffic

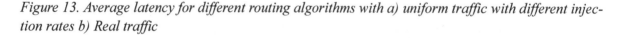

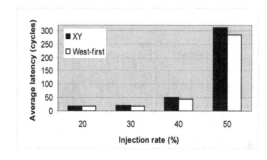
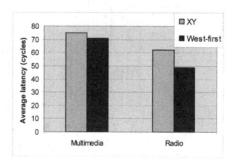

option. One other important point is the number of independent links doubled. For the case with latency of 20 cycles half of the links were doubled. Also, for this example, achieving 20 cycles of latency corresponds to an improvement of about 85% of the total possible reduction (35 cycles) with an additional 50% overhead in the area. The remaining 15% of latency improvement can be achieved at the cost of the same area overhead.

For an injection rate of 50% the results are even more interesting (see figure 15).

As shown, in this case independent links can achieve a large latency reduction from 310 cycles

(saturation) down to 18 at the cost of the same 50% increase in the area. With single wider links the reduction is from 310 cycles (saturation) down to 25 with an increase of at most 35% increase in the area. We also observe, for example, that with 50% injection rate, the two independent links technique achieves better latencies at the same area cost.

From this test we also observed that to achieve a reduction of about 80% in the average latency (corresponds to an average latency of 50) only 35% of the links had to be doubled.

When applied in the design of the NoC for the multimedia application (see figure 16), both

Figure 14. Average latency for uniform traffic with an injection rate of 40% with different link number and sizes

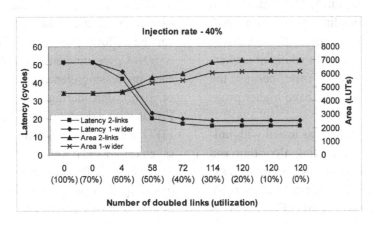

Figure 15. Average latency for uniform traffic with an injection rate of 50% with different link number and sizes

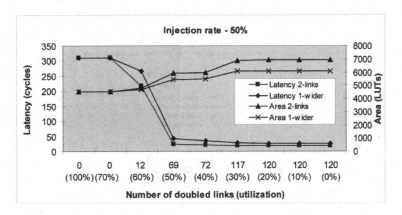

techniques achieved improvements in the latency (8% and 4% for double links and double sized buffers, respectively) and the cost of an increase of 50 and 30% increase in the area. Once again, since the link utilizations are low, only marginal improvements are achieved since there are only a few points of congestion.

For the radio application (see figure 17) the area improvements are close to 10%. In this case, the utilization of the links is very high. However, the main bottleneck is in the destination core and so we have reduced improvements.

Size of Buffers

Buffer resizing can also be used to improve latency or minimize area. For custom NoCs the buffers can have different sizes depending on the utilization of the links. For heavy loaded networks increasing the buffer size will decrease blocking of packets since buffers can be emptied faster because the following buffers in the path have higher probability of not being full.

Once again, a 6×6 NoC was tested with uniform traffic and the XY routing algorithm and with buffers with different sizes (see figure 18).

The average latency grows faster with increase injection rate for NoCs with smaller FIFOs. FIFOs with a depth of 16, 32, and 64 provide results closer to each other. For example, with 50% network loading and FIFOs with depth of 16 the average latency is about 67 cycles (already saturated). Increasing the FIFOs to 32 words will improve the latency by about 37% and increasing it to 64 words will improve the latency by about 43%. The main disadvantage of this approach is the area overhead associated with the buffers. From 32 to 64 words there is an increase of about 70% in the area of a FIFO implemented in a Virtex-5 and the best performance gain achieved is just 1% for an injection rate of 40%.

Therefore, in the customization process of the router, the size of the FIFOs must be carefully chosen to avoid using FIFOs deeper than what is needed to achieve the system requirements while optimizing area utilization. A simple experiment can be followed to show this tradeoff. With a 6×6 NoC, we injected a burst of 50 packets from each processing element (with an injection rate of 30%), considering 2-hotspot traffic and all FIFOs with a depth of 16. Then, the size of each FIFO with an average utilization higher than a given threshold value was doubled. For each configuration, the area and average latency were determined through simulation. The results are as expected (see figure 19).

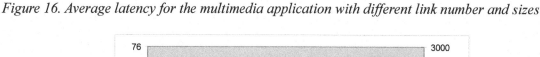

Figure 16. Average latency for the multimedia application with different link number and sizes

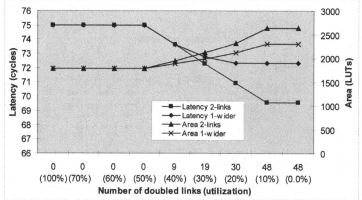

Figure 17. Average latency for the radio application with different link number and sizes

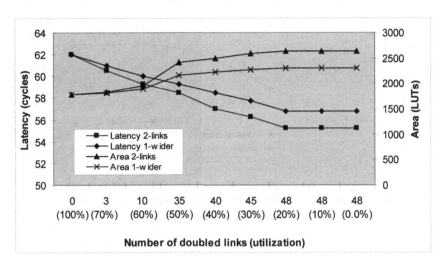

As we can observe from figure 15, we can reduce the latency from 73 cycles down to 58 cycles (20% improvement) by increasing the size of the FIFOs. The best average latency reduction is achieved with an increase in the area from 4540 to 5500 LUTs (21% increase). The latency decreases rapidly when the depth of the most utilized FIFOs are doubled. For example, when FIFOs with an utilization higher than 40% are increased, the latency is reduced about 15% (remember that the best achievable latency reduction is 20%).

The results are not so expressive with uniform traffic since the traffic is more distributed and the FIFOs utilization is less unbalanced. In this case, tests have shown an average latency of 47 cycles when all FIFOs have a depth of 16 and an average latency of 40 with all FIFOs with depth 32. This is about 15% reduction in the average latency, less than the achievable 20% latency reduction observed with 2-hotspot traffic.

We have also applied the technique considering the real applications and the results were even less expressive than with the uniform traffic. Improvements from 3 to 9% were achieved. Also, for example, increasing the buffers from 32 to 64 words had no effect on the average latency.

Figure 18. Average latency for different FIFO sizes and injection rate for uniform traffic

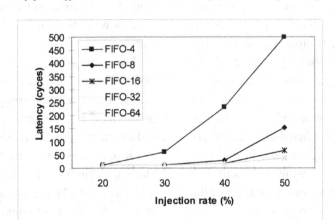

Figure 19. Average latency for number of resized FIFO with 2-hotspot traffic

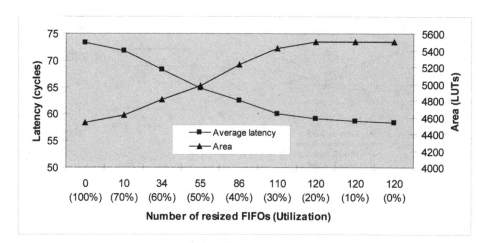

Switch Matrix

The customization of the switch matrix can be used to reduce the area overhead of a router. This approach will not improve the performance since it shares internal resources of the router.

As explained above, the area of the switch matrix can be reduced by sharing internal buses. Consequently, the input ports must be multiplexed in the access to the switch matrix. The outputs can also be shared to reduce the number of multiplexers establishing the paths from the internal buses of the switch matrix to the outputs. A single router can be configured in multiple ways, each with its area and performance cost. To analyze how the customization of the switch matrix affects the performance and the area of the router, a simple heuristic was developed based on the percentage of utilization of each input and output. The algorithm uses a control sharing variable (*share*) starting at 0, meaning there is no switch sharing. Then, at each step it increases the *share* and joins input and output ports whose added utilization is lower than *share*. It stops when there are no more ports to share. The sharing variable increases at each step by a factor given by the user (step). The higher this factor, the faster the algorithm but the final solutions may not be so good.

In each NoC simulation, 1000 packets with 8 flits each were delivered using XY, round-robin arbitration and virtual cut-through switching to avoid deadlocks (see Véstias & Neto for further details). Packets are generated at each node with an injection rate of 30%. The traffic models considered in our evaluation were uniform (see figure 16) and 2-hotspot (see figure 20). The hotspot model realistically mimics the behavior of systems where some of the cores (hotspots), such as memories, are accessed by most of the others.

As shown in figure 16, if an average latency increase of 10% is acceptable, then the area can be reduced around 10%, and for a 50% latency increase the area can be reduced 32%. Since the traffic is uniform the routers have similar utilizations. So, all routers suffer similar area reductions. That is why the graphic shows an almost linear performance/area ratio.

For 2-hotspot traffic, a different picture is obtained (see figure 21). With this traffic type, the traffic is not uniformly distributed through all routers. So, we observe a steep initial area reduction with small average delay degradation because many routers are over dimensioned. As shown, an (initial) area reduction of 37% is possible without any performance degradation. For only a 10% average latency degradation, up to

50% reduction in the area is possible. These are very good results, better than the ones obtained for the more regular traffic models.

Tradeoffs with the Customization Methods

The previous analysis gave us the performance and area improvements that can be achieved with four customization techniques:

- Up to 10% reduction with a partially adaptive algorithm. This result is relatively low possibly due to the necessity to base its routing decisions on good real-time analysis techniques to help in the routing process;
- With independent links it was possible to achieve near 70% of reduction in the average latency at the cost of almost 85% increase in the area. With single wider links the reduction in the average latency is only about 60% with an increase of at most 40%. For certain values of average latency, the NoC using wider links occupies a lower area than what would be obtained using double independent links;
- Increasing the FIFOs from 16 to 32 words will improve the latency by about 20% and

increasing it to 64 words will improve it by about 30%;
- Switch matrix sharing achieves about 30% area reduction with 50% latency degradation.

The studied customization techniques can be used alone or together in the design of a specific NoC. Designing a NoC with two or more of these techniques will provide a solution with better performance compared to that obtained using only one technique. However, it is not expected a final solution with the best of both worlds, because using one technique will reduce the design space available for optimization using another technique. The following tradeoffs are involved in the customization of NoCs with these techniques:

- A partially adaptive algorithm reduces the bottlenecks of the network by improving the distribution of the traffic. This reduces the opportunities for improvement achieved with link or buffer resizing since there are fewer congested links;
- Increasing the number of links in a single direction or increasing their size will reduce the effects of link congestion, which consequently reduces the optimizations achieved with adaptive routing and buffer

Figure 20. Average latency and area for different configurations of the switch matrix (uniform)

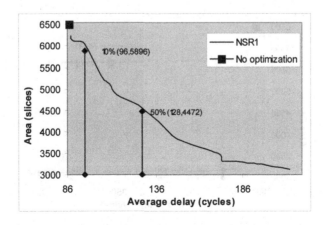

resizing. However, this technique achieves better performance than the adaptive routing technique since it does not increase the number of hops of a path. Also, while buffer resizing reduces congestion it does not increase the bandwidth of links. So, its latency improvements are worst than those from link customization. The problem of this technique has to do with the higher area overhead;

- Buffer resizing reduces the congestion in the links. Adaptive routing will be less efficient after buffer resizing, but link resizing will still provide a considerable design space for improvement since buffer resizing does not improve link bandwidth;

- Switch matrix sharing can be used for area improvement. The technique can be used with any other of the previous techniques to improve the area of each routing. For example, high congested links can be improved by resizing the link while the other less utilized links can be shared to reduce the area. The final average latency will probably be better.

To quantify the above discussion we performed a set of experiments using different combinations of the customization techniques, namely:

- **Case 1** – XY routing with FIFOs with 16 positions and using one or two independent links;
- **Case 2** – XY routing with FIFOs with 32 positions and using one or two independent links;
- **Case 3** – West-first routing with FIFOs with 16 positions and using one or two independent links;
- **Case 4** – West-first routing with FIFOs with 32 positions and using one or two independent links.

In all cases, we have used uniform traffic with an injection rate of 50%.

In cases 1 and 2, we have used XY routing and fixed sized FIFOs (16 positions) and then incrementally doubled the links starting with the most utilized. The same simulation was done with deeper FIFOs (32 positions) (see results in figure 22 and 23).

According to the results illustrated in the figure, doubling the FIFO size has a major impact at design points where the link doubling technique is still not effective. After doubling the first 40 most utilized links, the FIFO doubling is less effective until being completely inefficient (after 72 doubled links). At this point, we are very near to the minimum achievable average latency at the cost of an area increase of about 32%. The remaining

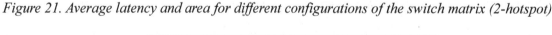

Figure 21. Average latency and area for different configurations of the switch matrix (2-hotspot)

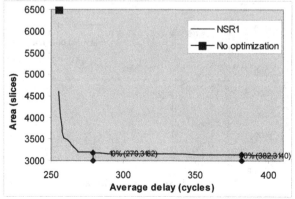

20% increase in the area is very expensive since it only contributes to an improvement of the latency of about 0,64%.

With the west-first algorithm (cases 3 and 4) (see results in figure 23) the conclusions are similar except that the results achieved in terms of latency are a bit better. However, this solution is more expensive in terms of area occupation since the implementation of the west-first algorithm uses more resources than those used by the XY.

As seen, the design space is considerably large and hard to analyze. However, in a first approximation to the design of a customized NoC with these techniques it would be valuable to know which seems to be the best approach to achieve a specific latency and which combination of customizations occupies less area. To give an insight about this, we have determined solutions with all cases mentioned above to achieve an average latency of 20 and 30 (these latencies represent the situation in which the area is in between the lowest and the highest figures).

As shown in figure 24, in both cases the best solution is obtained using small FIFOs and XY routing. The main reasons for this are that the efficacy of the adaptive west-first routing is difficult to maintain, that increasing the number of links

in a single direction compensates somehow the static nature of the XY algorithm, and that FIFOs are very expensive to implement compared to the improvements that can be achieved by increasing their size. West-first and increased buffer sized should only be used if the main objective is to improve latency without considering the area overhead.

All solutions studied so far are basically concerned with the improvement of the average latency. However, if we need to improve the area then the customization of the switch matrix must also be considered in the design process. As stated above, this technique improves the area at the cost of increasing the average latency in most cases. Therefore, when the customization of the switch matrix is considered together with the other customization methods there will several tradeoffs between performance and area. It is up to the designer deciding about the best trade-off between area and performance.

RUNTIME ADAPTIVE ROUTER

In the discussion above, a set of customization techniques that can be used to tailor the router

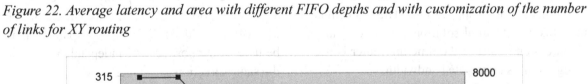

Figure 22. Average latency and area with different FIFO depths and with customization of the number of links for XY routing

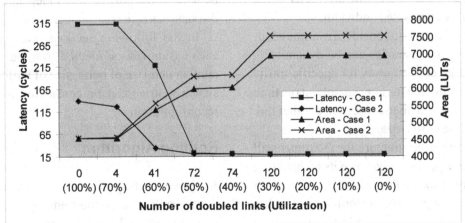

Figure 23. Average latency and area with different FIFO depths and with customization of the number of links for west-first routing

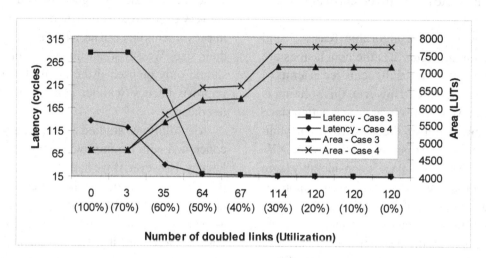

for specific traffic requirements at design time were discussed. However, for general purpose architectures used to run applications where traffic cannot be predicted at design time or for executing applications with diverse traffic patterns, these customization techniques may be applied at design time but without any warranties since there is not traffic information to support the design decisions. Instead, a full featured router with two or more links in each direction, each with its own buffer with high depth, and a complete switch matrix must be considered to achieve a good performance for all applications. The problem is that this architectural solution consumes many resources and most of the time the router is not using all of its aggregate bandwidth.

A better solution is to consider a runtime adaptive router whose resources can be dynamically reorganized given the quests for specific traffic bandwidth requirements. For example, at runtime a router may decide to use a second parallel link or increase the buffer size.

Using a dynamic structure for the router will provide better performance since its architecture dynamically adapts to the traffic requirements to achieve the best possible performance. Another perspective is that a runtime adaptive router can achieve the same performance of a static router using fewer resources since the reduction in resources for packet routing is compensated by the runtime adaptability of the existing resources.

Routing adaptability may be complemented with topology reconfiguration at runtime. Besides changing the router configuration it would also be possible to change the links between routers. In this chapter, we are only concerned with router adaptability where the size of the links connecting the routers is fixed. State-of-the-art chips have high wiring capability. Some authors already proposed NoCs with link widths of 256 bits (Mullins, 2006). This wiring capability may be used to support several independent links in the same direction.

In the following sections we analyze how each customization method (routing, switching, number and size of ports, size of buffers, and the switch matrix) can be considered for dynamic reconfiguration.

Routing Algorithm

Adaptive routing algorithms can be used to dynamically change the routes given the traffic requirements and the links' congestion. The ap-

Figure 24. Relation between average latency and area for different combinations of customization techniques

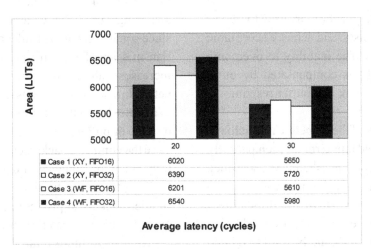

	Average latency (cycles) 20	Average latency (cycles) 30
■ Case 1 (XY, FIFO16)	6020	5650
□ Case 2 (XY, FIFO32)	6390	5720
□ Case 3 (WF, FIFO16)	6201	5610
■ Case 4 (WF, FIFO32)	6540	5980

proach is very attractive from the point of view of the compiler since the routing process is managed by the network controller. The problem, as already stated, is that it needs a well designed protocol to maintain information about the utilization of the links for the router to take good routing decisions.

An alternative is to determine the routes of each application at compile time and then change the routing tables before starting the execution of each particular application. This option is attractive for the network since it just has to support the reload of new routes, while it is harder for the compiler that must determine the best routes for each application. Compared to the approach using adaptive routing algorithms this is easier to achieve and possibly will offer better performances, but only applies to applications whose traffic pattern is known at compile time. On the other side, adaptive routing algorithms have a much broader usage since they can be used in all types of applications, whether specific or generic.

Many adaptive routing algorithms have been recently proposed. Solutions exist that base their routing decisions on the load state of the buffers of neighbor routers (Ascia, et al.,2006 & Don et al., 2006). Another one dynamically determines

its routes using the distance to destination and current link bandwidth utilization (Faruque, Ebi, Henkel, 2007). Others (Hu & Marculescu, 2004) use a combination of static and partially dynamic routing. A fully adaptive algorithm was proposed in (Bartic et al., 2005) using an operating system to program the routing tables.

According to the results presented in the previous section around 10% average latency reduction can be achieved with the west-first algorithm compared to the XY algorithm.

Considering the costs associated with implementing adaptive or partially adaptive routing, the gains obtained with it, and knowing that larger improvements can be achieved with other techniques we conclude that:

- Adaptive routing algorithms are still a very expensive design option for NoCs;
- Partially adaptive routing algorithms are a good design option for static NoCs or when other adaptive techniques are unacceptable since it is possible to achieve some latency improvements with low area overhead. However, they are less important for dynamically adaptive routers where other

dynamic techniques can be used to improve the latency;

- Static routing algorithms are easy to implement, deterministic and achieve acceptable routing results. The inability to overcome congested links is compensated by other techniques like link and buffer resizing.

Design remarks: If the objective is to achieve a low cost runtime adaptive router then partially adaptive routing algorithms should be used since the other customization techniques are more expensive in terms of area. However, if the objective is to achieve an efficient runtime adaptive router then XY routing should be used to reduce implementation costs. The loss in performance is compensated with the other customization techniques at the cost of some area.

Switching Algorithm

Switching can be dynamically changed from wormhole or virtual cut through to circuit switching to guarantee the requirements of some delay sensitive traffic. The decision about the type of traffic is taken at the source of the data, i.e., the processing unit. Then, the network interface adds one bit to the header of the packet to indicate the type of traffic. When the router receives the packet it checks the type of traffic. In the case of best effort traffic, the packet is sent to the appropriate output using wormhole or virtual cut through switching. Otherwise, the router initiates a process to reserve an output port for circuit switching, namely: (1) requests a dedicated link, (2) the associated output arbiter grants the link as soon as the link is free and (3) the input releases the link at the end of the message. All the routers belonging to the path from the source to the destination processing units (circuit) do the same port reservation. The circuit is kept reserved until the end of the transmission of the message or set of messages associated with a specific flow.

The flits of a packet may be buffered in one or more FIFOs of several routers whenever wormhole switching is used. This generates traffic congestion since some ports of the router are reserved for the transmission of a specific packet. The congestion increases with the injection rate of traffic into the network, since more packets require a specific port of a router for switching, and with the decrease of the size of buffers since more routers are used to store the complete packet. One possible approach to balance the congestion due to an increase of traffic is to reduce the size of packets. However, smaller packets imply more packets and consequently a higher header overhead.

To analyze these tradeoffs we have injected uniform traffic at a rate up to 50% into a 6×6 NoC with all buffers with a depth of 32 and using wormhole switching. Then we measured the total latency (see figure 25).

The results show that for high injection rates the best total latency is achieved with small packets. For example, with an injection rate of 50% the transmission of packets with 64 flits takes on average about 50% more cycles to transmit all flits, than with packets of size 4. This number reduces to less than 40% with an injection rate of 40% and to about 10% when the injection rate reduces to 30%. The scenario inverts for injection rates less than 20%, although the difference is small (less than 4%). In these cases, the best alternative is to use large packets.

A single iteration of the multimedia application runs faster for bigger packets. This was expected since it has low injection rate and neighbor traffic. Therefore, the extra latency when smaller packets are used is mainly due to the extra header overhead. In the case of the radio application, we run for a pre-determined number of flits and obtained a similar behavior compared to that of the multimedia application. This is possibly due to the fact that some links are heavily loaded and the extra header data increases this load.

If the buffer's depth is changed to half the size, similar conclusions can be obtained, but this time the percentage improvements are smaller (see figure 26).

For example, with 50% injection rate, the communication with packets of size 64 takes about 45% more cycles, than with packets of size 4. This increase is lower than what was obtained with FIFOs of size 16. This was expected, since for the same injection rate and using wormhole switching the congestion increases when the size of buffers is reduced. The results for the real applications are not shown since they are almost identical to those of the previous figure.

The size of the packets is decided at the network interface based, for example, on link congestion. To support this runtime configuration, the input controllers of the router must be able to route packets with different sizes reading this information from the header.

Design remarks: When the NoC must support real-time traffic, the runtime adaptive router has to support dynamic switching between wormhole or virtual-cut through and circuit switching. Also, to improve traffic congestion, the router has to able to route packets with different sizes. However, this

is only used when the network interface generates packets with different sizes.

Number and Size of Ports

As stated above, the size of the links between the routers is assumed to be fixed. The router may use a link for a single port or for multiple independent ports. The number and size of ports in a single link is decided at design time. The tradeoff involved in this design step is that independent links achieve better performance than a single link with the same number of lines but occupy more area. A good design option is to consider at least two independent links, which can achieve performance improvements compared to a single link and can be used to implement dynamic switching, as explained above.

An adaptive router with independent links in each direction is a costly solution in terms of area since each independent link needs an arbiter and a controller. Furthermore and recalling the results presented in figures 14 and 15, for typical traffic scenarios only half of the links need to be resized

Figure 25. Relation between total latency and injection rate for different packet sizes using FIFOs of depth 32

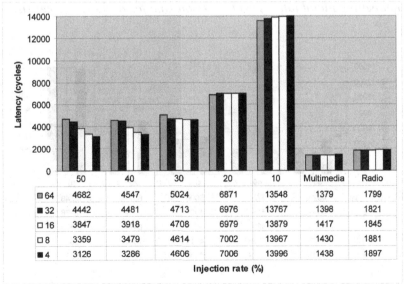

to obtain very good performance improvements (up to 85%). Therefore, instead of including buffers in all independent links, a runtime adaptive router can share some buffers, that is, the router can include floating buffers that are dynamically associated with a link depending on the traffic congestion of the associated port (see figure 27) – *port resize*.

In the example of figure 27, the adaptive router includes one floating buffer that can be associated with any output. This configuration saves three buffers. The buffer is dynamically associated with, for example, the most utilized output. Other configurations with more floating buffers and with different possible associations between floating buffers and output links can be considered.

The assignment of each floating buffer is determined by a floating buffer controller that reassigns the floating buffer whenever it is empty. It basically sends a signal to the arbiter associated with an output port indicating that the floating buffer is assigned to it. Several policies can be followed to assign the floating buffer. For example, it can be reassigned to a port whose fixed buffer is full. For a fair assignment, all eligible ports for assignment should be chosen in a round-robin manner.

The arbiter associated with an output port receives the requests from all input ports and grants access to its buffer and to the floating buffer case this is assigned to it. Therefore, two independent requests can be assigned at the same time.

Compared to a static router, this architecture permits to reduce the size of the buffers to half (basically, instead of a buffer with size m, the port has an associated buffer of size $m/2$ (fixed buffer) + $m/2$ (floating buffer)) but the arbiters are more complex since they receive up to seven requests instead of four. Additionally, it also needs one extra buffer and the controller of the floating buffer. The switch matrix must also consider almost two times the number of inputs and outputs but the data inputs of the multiplexers are half the size.

A NoC with adaptive routers was tested with different numbers of floating FIFOs using XY routing and wormhole switching for uniform traffic and for the two real applications (see figure 28).

From the results presented in the figure, we observe that using a single floating buffer the latency reduces about 50% of the achievable latency reduction, except for the lowest injection rate (30%) where it achieves a reduction of only 1 cycle from a total of 6. With 2 floating buffers

Figure 26. Relation between total latency and injection rate for different packet sizes using FIFOs of depth 16

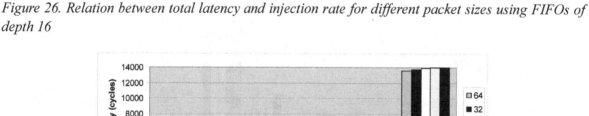

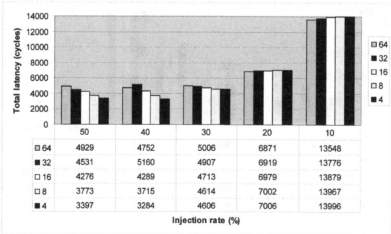

Injection rate (%)	50	40	30	20	10
□ 64	4929	4752	5006	6871	13548
■ 32	4531	5160	4907	6919	13776
□ 16	4276	4289	4713	6979	13879
□ 8	3773	3715	4614	7002	13967
■ 4	3397	3284	4606	7006	13996

Figure 27. An adaptive router with floating buffers for port resize

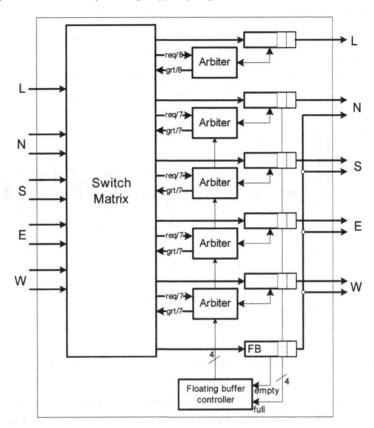

the reductions go from about 66% up to about 95%. So, a single floating buffer already achieves very good results and it is probably enough for most design cases. The most effective solution for the real applications is an adaptive router with a single floating buffer.

Design remarks: The runtime adaptive router should use independent links in each direction to improve performance. Whenever this customization technique it is too expensive it can be partially considered in directions were traffic congestion is more probable (for example, Y directions in XY routing are more utilized than X directions). The number of floating buffers will depend on the necessary performance. The design option will be in most of the cases between a quarter and half of the number of ports of the router. Additional floating FIFOs are inefficient.

Size of Buffers

Buffer resizing can be used to reduce latency. However, the previous analysis about using simultaneously link doubling and FIFO resize has shown that the latter is relatively inefficient when half (or more) of the ports of a router use two independent links. Hence, if we decide to use two independent floating buffers, then FIFO resize is ineffective. If the design option is to consider a single floating FIFO, then it can be used for buffering independent links or to resize a fixed output buffer (see figure 29).

The results show, as expected, that using two floating buffers is more efficient than using a single floating buffer for resizing fixed buffers or to buffer independent links. Also, using the single floating buffer to resize the fixed buffers

Figure 28. Relation between average latency and injection rate for different number of floating buffers with depth 16 considering port resize

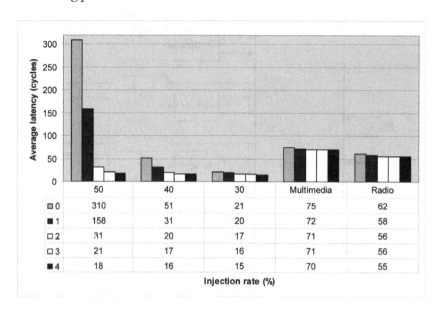

	50	40	30	Multimedia	Radio
□ 0	310	51	21	75	62
■ 1	158	31	20	72	58
□ 2	31	20	17	71	56
□ 3	21	17	16	71	56
■ 4	18	16	15	70	55

is worst than using it to buffer independent links. So, whenever possible it is better to use floating buffers for independent links.

A runtime adaptive router with floating buffers for independent links consumes more area than a router with floating buffers for resizing. Therefore, when cost is a constraint, the designer should consider the runtime adaptive router with a floating buffer to resize the fixed buffers instead of independent links (see figure 30).

The floating buffer of the figure can be used with any other buffer to dynamically increase the size of the output buffers. This configuration reduces the total buffering of the router. The buffer is dynamically associated with the most utilized output. Other configurations with a different number of floating buffers and with different possible associations between floating buffers and output links can be considered.

Reading from a port with two associated buffers (one fixed and one floating) is made according to the arrival time, that is, older packets are dispatched first. This is implemented using a simple bit-width FIFO that stores the sequence of accesses to be followed.

Compared to a static router, this architecture needs four additional two-input multiplexers at the outputs, one extra buffer and the controller of the floating buffer. The arbiter and the controller are also slightly more complex. These extra resources are somehow compensated by the reduction in the size of the fixed buffers.

To analyze the influence of the number of floating FIFOs over the performance, a NoC with adaptive routers was tested with different numbers of floating FIFOs using XY routing and wormhole switching. The fixed and the floating FIFOs are all of size 16. Therefore, when a floating FIFO is associated with a particular fixed FIFO, is like using a fixed FIFO with 32 positions (see figure 31).

FIFO resizing achieves considerable reductions in the average latency (for 50% the NoC is saturated). However, these results are worst than those achieved using independent links.

In the case of the real applications, the technique achieves only a marginal improvement, less than that obtained with link resize.

Design remarks: The adaptive router should use the floating buffers for independent links in-

Figure 29. Relation between average latency and injection rate for different customizations of the router

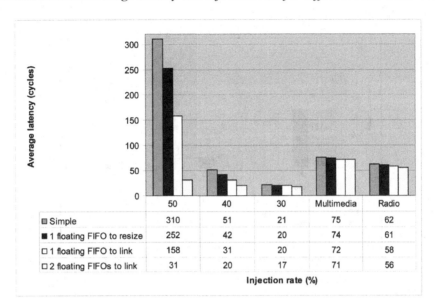

	50	40	30	Multimedia	Radio
▣ Simple	310	51	21	75	62
■ 1 floating FIFO to resize	252	42	20	74	61
▢ 1 floating FIFO to link	158	31	20	72	58
▢ 2 floating FIFOs to link	31	20	17	71	56

Injection rate (%)

Figure 30. An adaptive router with one floating buffer for buffer resizing

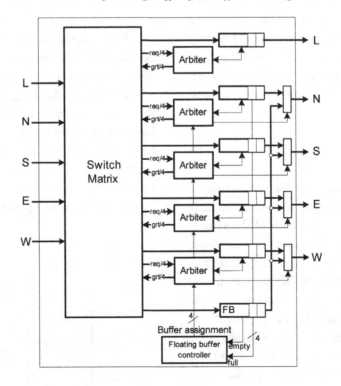

Figure 31. Relation between average latency and injection rate for different numbers of floating FIFOs used to resize fixed FIFOs

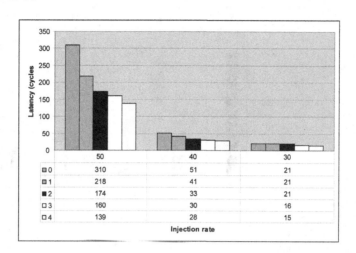

	50	40	30
0	310	51	21
1	218	41	21
2	174	33	21
3	160	30	16
4	139	28	15

Injection rate

stead of buffer resizing, as long as the cost is not a major constraint. Otherwise, buffer resizing should be considered with one or two floating buffers.

An additional design technique (not analyzed in this chapter) is to use both techniques together to achieve improved performance but following a different configuration where the floating buffers have a larger size than that of the fixed buffers. The cost is marginal and the improvements are expectedly better than those achieved with equally sized buffers.

Switch Matrix

The customization process of the switch matrix determines the set of input ports connected to each internal bus and the set of outputs of the switch matrix. The customization supports the decision based on the utilization of the ports avoiding sharing resources between highly utilized ports. Statically, it determines the utilization of each port when running a particular set of applications and then share less used resources. Dynamically, the router determines resource sharing at runtime based on the real-time utilization of the ports (adaptive switch matrix).

To implement this adaptability, the router needs an interconnection matrix between the inputs and the internal buses of the switch matrix (input switch matrix) and an interconnection matrix and multiplexers between the switch matrix and the outputs (output switch matrix) (see figure 32).

The cost of this adaptive configuration depends on the number of internal buses of both switches and the technology to implement it. For example, 8 multiplexers with 8 inputs each are needed to implement a complete switch matrix (8 internal buses) with 8 input/output ports. With a Virtex 5 FPGA, we need $8 \times 16 \times bw$ (bitwidth) LUTs to implement it. With four internal buses, we need $(4 \times 16 + 8 \times 8) \times bw$ LUTs, that is, an identical area without counting that some extra control

Figure 32. Adaptive switch matrix

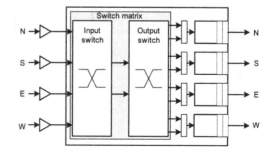

is needed. With two internal buses, this number reduces to $(2 \times 16 + 8 \times 4) \times$ bw LUTs, which is half the size of the previous configurations. This reduction in the size of the area of the switch matrix comes at the cost of a reduction in the aggregate bandwidth to a quarter of that of a complete switch matrix since there are only 2 available buses (out of 8 of the complete switch) to forward packets.

Considering a runtime adaptive router with one floating FIFO, the switch matrix configuration requires 8 input multiplexers each with 8 inputs and 5 output multiplexers each with 5 inputs, which corresponds to $120 \times$ bw LUTs. When compared to the $128 \times$ bw LUTs of the complete switch matrix, this is a very small area reduction that results in a 5/8 reduction in the aggregate bandwidth.

To conclude, unless the technology allows implementing a low cost adaptive switch matrix compared to that of the static switch matrix, the reduction in the aggregate bandwidth due to the customization of the switch matrix dictates that this dynamic customization should not be included in the adaptive router.

Concluding Remarks

A router consists of input and output ports, a switch matrix, a routing algorithm, a switch mechanism and buffers. For a runtime adaptive router each port may consist of multiple independent links or a single link with a size multiple of the width of the flits. Ports have associated buffers implemented as FIFOs which are fixed (statically associated with a specific port) or float (dynamically associated with a specific port). Buffers may be of fixed size or dynamically resizable using one or more floating FIFOs. The routing algorithm can be static or partially adaptive, while the switching technique can be dynamically changed between wormhole or virtual cut-through and circuit switching. With so many design variables the design space is considerably large.

In the previous sections we have analyzed and discussed how dynamic customization techniques improve the performance of the router and consequently the NoC. Designing NoCs with runtime adaptive routers is not a one time job even for a specific technology. The architecture of the adaptive router should be designed specifically for each target application, requirements and technology.

The design of NoCs for application specific systems, for domain specific systems or for general-purpose systems can greatly benefit from optimized runtime adaptive routers as follows:

- **Application specific systems** – Since the applications are known at design time, any configuration of the router is possible depending on the communication requirements of the application. A simple adaptive and/or customized routing algorithm may be enough to fulfill the requirements. If not, multiple links, link resize, buffer resizing or switch matrix customization are design options with different area/performance tradeoffs that can considered in the design of the router;
- **Domain specific applications** – In this case the designer only knows some specificities of the communications: size of messages, burst or periodic, etc. For example, the architecture of the system may target scientific computing applications, digital image or video processing, etc. In these cases, for example, data communication is intensive with long packets to be transmitted. Hence, the runtime adaptive routers should consider using independent links with floating buffers, possibly with bigger sizes;
- **General-purpose computing** – In this category the features of the traffic are not known at design time. Therefore, the router should be designed with some features

to be associated with the characteristics of the system architecture. For example, if the system must support real-time traffic then it is a good design option to include hybrid switching including circuit switching, or if the system must support burst traffic, then independent links with floating FIFOs should be included.

FUTURE RESEARCH DIRECTIONS

This chapter has identified a number of static and dynamic customizations that should be utilized in the design of routers. Each of these techniques achieves different tradeoffs between performance, area and power consumption. Also, the methods are correlated, that is, using one method reduces the effectiveness of another one and so the appropriate combination of customization techniques must be subjected to a careful analysis in order to optimize the final network-on-chip solution. The analysis and discussions elaborated in this chapter are the basis for further research directions, namely:

- **Data gathering** – All adaptive customization methods: adaptive routing, redirection of floating FIFOs, buffer resize, switch matrix customization, etc.; rely on timely information about the congestion of the network. Typically, adaptive methods rely on neighbor information about the state of buffers and the utilization of links. While this is a simple solution it is important to research other data gathering methods relying on information beyond local information; what can be achieved in terms of performance, area and power consumption considering the costs associated with the data gathering system;

- **Quality-of-Service** – Many applications have real-time tasks with real-time traffic associated. Several works have proposed

techniques to deal with such real-time requirements, namely: reservation of virtual channels, prioritization of traffic and circuit switching for real-time traffic. Additional aspects must be considered in the design of NoC for real-time applications: the influence of the discussed customization methods and how they can be used in the design of routers with support for real-time traffic; and the best router design options when hard, soft and best effort deadlines are present (should we reserve separate resources for real time traffic or can they be also used by the other types of traffic without compromising real-time traffic);

- **Power consumption** – Power consumption is an important aspect of many on-chip systems. Several works have analyzed the power consumption of NoC and how to reduce the power consumption. A similar analysis must be performed for runtime adaptive NoC with adaptive routers to see how the runtime techniques affect power consumption and hw efficiently they are when power consumption is a concern;

- **Heterogeneous routers** – The design of an application specific NoC with statically customized routers will probably generate an irregular network with heterogeneous routers with different number of resources and connections. However, for dynamically reconfigurable NoCs the adaptive routers used are homogeneous in the sense that all use the same architecture. The discussion and results obtained throughout the text were for homogeneous adaptive routers. These are able to improve the NoC solution, but due to the heterogeneity of the processing elements (for example, some elements are generic purpose processors, others are memories and others may be dedicated processing units) or to the asymmetric aspects of a NoC (routers closer to the border of the NoC are probably less used than those in the middle) heterogeneity can and should also be considered in the design of adaptive routers in the same NoC;

- **Design methodology** – The design space of NoC systems with static or adaptive routers is huge with many options available. The development of efficient design methodologies and tools to help the designer explore such huge space is therefore a very important topic of research. One possible approach is to iteratively apply the customization techniques starting with those considered more efficient to improve the performance or the area.

CONCLUSION

Dealing efficiently with today's traffic requirements on chip needs an advanced intercommunication network capable to dynamically adapt to the traffic requirements using information about the congestion of the links. This communication approach where information about the state of the network is collected and processed to support routing decisions can be identified as some sort of "intelligent" communication. Therefore, we designate this second generation of networks-on-chip *Intelligent Networks-on-Chip (INoC)*. INoCs can dynamically adapt its topology as well as their resources to best fit the demands of processing units for communications. Dynamic adaptability of the topology as well as the network resources are recent topics of research.

This chapter addressed the design of adaptive routers: which mechanisms can be dynamically used, what do we expect from them and how do they relate to each other. From the several analyses presented in the chapter we conclude that adaptive routers are fundamental in the design of an INoC since they improve the bandwidth capacity of the network, as well as the quality-of-service while reducing the area overhead associated with these structures.

REFERENCES

Ahmad, B., Erdogan, A., & Khawam, S. (2006). Architecture of a dynamically reconfigurable noc for adaptive reconfigurable mpsoc. Adaptive Hardware and Systems, 2006. In *Proceedings of the 1st NASA/ESA Conference on AHS*, (pp. 405–411).

Ahonen, T., Sigüenza-Tortosa, D., Bin, H., & Nurmi, J. (2004). Topology Optimization for Application Specific Networks on Chip. In *Proceedings of the 2004 international workshop on System level interconnect prediction* (pp. 53-60).

Al Fanique, M. Ebi. T., & Henkel, H. (2007), Run-Time Adaptive On-chip Communication Scheme, In *proceedings of the International Conference on Computer Aided Design*, (pp. 26-31).

Ascia, G., Catania, V., Palesi, M., & Patti, D. (2006). Neighbors-on-path: A new selection strategy for on-chip networks. In *Proceedings of the 4th IEEE Workshop on Embedded Systems for Real Time Multimedia*, (pp. 79-84).

Ascia, G., Catania, V., Palesi, M., & Patti, D. (2006). Neighbors-on-Path: A new selection strategy for On-Chip Networks. In *the Proceedings of the 4th IEEE/ACM/IFIP Workshop on on Embedded Systems for Real Time Multimedia*, (pp. 79-84).

Bartic, T. (2005). Topology Adaptive Network-on-Chip Design and Implementation. *IEE Computers and Digital Techniques*, *152*(4), 467–472. doi:10.1049/ip-cdt:20045016

Benini, L. (2006). Application Specific NoC Design. In *Proceedings of the Conference on Design, Automation and Test in Europe*. (pp. 491-495).

Benini, L., & De Micheli, G. (2002). Networks on Chips: a New SoC Paradigm. *Computer*, *35*(1), 70–78. doi:10.1109/2.976921

Bjerregaard, T., & Mahadevan, S. (2006). A Survey of Research and Practices of Network-on-Chip. *ACM Computing Surveys, 38*(1), 1–51. doi:10.1145/1132952.1132953

Chiu, G. (2000). The odd-even turn model for adaptive routing. *IEEE Transactions on Parallel and Distributed Systems, 11*(7), 729–728. doi:10.1109/71.877831

Dally, W., & Towles, B. (2001). Route Packets, Not Wires: On-Chip Interconnection Networks. In *Proceedings of the 38ᵗʰ Design Automation Conference* (pp. 684-689).

Dally, W. J., & Seitz, C. L. (1987). Deadlock-free message routing in multiprocessor interconnection networks. *IEEE Transactions on Computers, 36*(5), 547–553. doi:10.1109/TC.1987.1676939

De Micheli, G., & Benini, L. (2006). *Network on chips*. San Francisco: Morgan Kaufmann.

Dehyadgari, M., Nickray, M., Afzali-kusha, A., & Navabi, Z. (2005). Evaluation of pseudo adaptive XY routing using an object oriented model for NOC. In *Proceedings of the International Conference on Microelectronics*. (pp. 204-208).

Dong, W., Al-Hashimi, B., & Schmitz, M. (2006). Improving Routing Efficiency for Network-on-Chip through Contention-Aware Input Selection. In *the Proceedings of the Asia and South Pacific Design Automation Conference* (pp. 36-41).

Duato, J. (1993). A new theory of deadlock-free adaptive routing in wormhole networks. *IEEE Transactions on Parallel and Distributed Systems, 4*(12), 1320–1331. doi:10.1109/71.250114

Elmiligi, H., Morgan, A. A., El-Kharashi, M. W., & Gebali, F. (2008) Power-Aware topology optimization for networks-on-chips. In *Proceedings of the IEEE International Symposium on Circuits and Systems*, (pp. 360–363).

Faruque, M., Ebi, T., & Henkel, J. (2007). Run-time Adaptive on-chip Communication Scheme. In *Proceedings of the International Conference on Computer Aided Design*, (pp. 26-31).

Glass, C., & Ni, L. (1992). Maximally fully adaptive routing in 2D meshes. In *Proceedings of International Conference Parallel Processing*, (pp. 101–104).

Glass, C., & Ni, L. (1994). The turn model for adaptive routing. *Journal of the ACM, 31*(5), 874–902. doi:10.1145/185675.185682

Hansson, A., Goossens, K., & Radulescu, A. (2005). A unified approach to constrained mapping and routing on network-on-chip architectures. In *Proceedings of the 3rd IEEE/ACM/IFIP International Conference on Hardware/Software Codesign and System Synthesis*, (pp. 75-80).

Hemani, J. A., Kumar, S., Postula, A., Oberg, J., Millberg, M., & Lindqvist, D. (2000). Network on Chip: An Architecture for Billion Transistor Era. In *Proceedings of the IEEE NorChip Conference*. Turku, Finland.

Hu, J., & Marculescu, R. (2004). DyAD – smart routing for networks-on-chip. In *proceedings of the 41st Annual Conf. Design and Automation*, (pp. 260–263).

Hu, J., & Marculescu, R. (2005). Energy- and Peformance-Aware Mapping for Regular NoC Architectures. *IEEE Transactions on Computer-Aided Design of Integrated Circuits and Systems, 24*(4), 551–562. doi:10.1109/TCAD.2005.844106

Hu, J., Ogras, U., & Marculescu, R. (2006). System-Level Buffer Allocation for Application-Specific Networks-on-Chip router Design. *IEEE Transactions on Computer-Aided Design of Integrated Circuits and Systems, 25*(12), 2919–2933. doi:10.1109/TCAD.2006.882474

Jantsch, A., & Lu, Z. (2009). Resource Allocation for QoS On-Chip Communication. In Gebali, F., Elmiligi, H., & Watheq el-Kharashi, H. (Eds.), *Networks-on-Chip, Theory and Practice*. Boca Raton, FL: CRC Press.

Jovanovic, S., Tanougast, C., Weber, S., & Bobda, C. (2007) Cunoc: A scalable dynamic noc for dynamically reconfigurable fpgas. In *International Conference on Field Programmable Logic and Applications*, (pp. 753–756).

Kreutz, M., Marcon, C., Carro, L., Wagner, F., & Altamiro, A. (2005). Design Space Exploration Comparing Homogeneous and Heterogeneous Network-on-Chip Architectures. In *Proceedings of the 18th Symposium on Integrated Circuits and Systems Design* (pp. 190-195).

Kumar, A., Hansson, A., Huisken, J., & Corporaal, H. (2007). An fpga design flow for reconfigurable network-based multi-processor systems on chip. In *Proceedings of the Design, Automation and Test in Europe Conference and Exhibition*, (pp. 1–6).

Lei, T., & Kumar, S. (2003). Algorithms and Tools for Network-on-Chip Based System Design. In *Proceedings of the 16th Symposium on Integrated Circuits and Systems Design* (pp. 163-168).

Lei, T., & Kumar, S. (2003). A Two-step Genetic Algorithm for Mapping Task Graphs to a NoC Architecture. In *Euromicro Symposium on Digital System Design* (pp.180-187).

Martini, F., Bertozzi, D., & Benini, L. (2007) Assessing the Impact of Flow Control and Switching Techniques on Switch Performance for Low Latency NoC Design. In *the First Workshop on Interconnection Network Architectures On-Chip*.

Medardoni, S., Bertozzi, D., Benini, L., & Macii, E. (2007). Control and datapath decoupling in the design of a NoC switch: area, power and performance implications. In *Proceedings of the International Symposium on System-on-Chip* (pp. 1-4).

Mullins, R., et al. (2006). The Design and Implementation of Low-Latency on-Chip Network. In *Proceedings of the Asia and South Pacific Design Automation Conference* (pp. 164-169).

Murali, S., & De Micheli, G. (2004). Bandwidth-Constrained Mapping of Cores onto NoC Architectures. In *Proceedings of Design* (pp. 896–901). Automation and Test in Europe.

Palesi, M., Holsmark, R., Kumar, S., & Catania, V. (2009). Application Specific Routing Algorithms for Networks on Chip. *IEEE Transactions on Parallel and Distributed Systems, 20*(3), 316–330. doi:10.1109/TPDS.2008.106

Pinto, A., Carloni, L., & Sangiovanni-Vincentelli, A. (2003). Efficient Synthesis of Networks on Chip. In *Proceedings of the 21st International Conference on Computer Design*, (pp. 146-150).

Pionteck, T., Albrecht, C., & Koch, R. (2006). A Dynamically Reconfigurable Packet-Switched Network-on-Chip. In *Proceedings of the conference on Design, automation and test in Europe*, (pp. 136-137).

Pionteck, T., Koch, R., & Albrecht, C. (2006). Applying partial reconfiguration to networks-on-chips. In *proceedings of International Conference on Field Programmable Logic and Applications*, (pp. 1-6)

Rana, V., Atienza, D., Santambrogio, M., Sciuto, D., & De Micheli, G. (2008). A Reconfigurable Network-on-Chip Architecture for Optimal Multi-Processor SoC Communication. In *the 16th IFIP/IEEE International Conference on Very Large Scale Integration*, (pp.).

Salminen, E., Kulmala, A., & Hämäläinen, T. (2008). *Survey of Network-on-Chip Proposals. (WHITE PAPER, OCP-IP)*. Finland: Tampere University of Technology.

Stensgaard, M., & Sparsø, J. (2008) ReNoC: A Network-on-Chip Architecture with Reconfigurable Topology. In *the 2ⁿᵈ ACM/IEEE International Symposium on Networks-on-Chip* (pp. 55-64).

Véstias, M., & Neto, H. (2006). Area and Performance Optimization of a Generic Network-on-Chip Architecture, In *Proceedings of the 19th Symposium on Integrated Circuits and Systems Design* (pp.68-73).

Véstias, M., & Neto, H. (2007). Router design for application specific networks-on-chip on reconfigurable systems. In *Proceedings of the International Conference on Field Programmable Logic and Applications* (pp.389.394).

Vieira, A., Ost, L., Moraes, F., & Calazans, N. (2004). Evaluation of routing algorithms on Mesh Based NoCs. (Technical Report).Porto Alegre, Brasil.

Wang, K., Gu, H., & Wang, C. (2007). Study on Hybrid Switching Mechanism in Network on Chip. In *Proceedings of the 7ᵗʰ International Conference on ASIC.* (pp. 914-917).

Zhu, H., Pande, P., & Grecu, C. (2007). Peformance Evaluation of Adaptive Routing Algorithms for Achieving Fault Tolerant in NoC Fabrics. In *IEEE International Conference on Application-Specific Systems, Architectures and Processors,* (pp. 42-47).

ENDNOTE

[1] A LUT in a Virtex-5 can be configured as a shift register with up to 32-bits without using the available flip-flops. Therefore, the data storage part of a FIFO can be efficiently implemented with several LUTs.

Chapter 3
Keys for Administration of Reconfigurable NoC:
Self–Adaptive Network Interface Case Study

Rachid Dafali
European University of Brittany, France

Jean-Philippe Diguet
CNRS, Lab-STICC (UMR 3192), France

ABSTRACT

This chapter presents an analysis of current needs in the domain of Reconfigurable Network on chip. We first detail our motivations for NoC reconfiguration, which is followed by a description of our model for Reconfigurable Network on chip in relation with the usual OSI network layers. Then, we propose a study of outstanding research issues of current work and open issues organized into three topics: dynamic reconfiguration administration, network infrastructure reconfiguration and network protocol reconfiguration. To finish, we present our strategy for reconfiguration and introduce a self-adaptive Network Interface architecture as a part of the configuration manager.

INTRODUCTION

The design of recent Systems-on-Chip (SoC) for a large scope of applications, such as telecom and multi-media domains, and in various platform types ranging from dedicated platforms to fully programmable platforms, is mainly constrained by communications. Networks-on-Chip (NoC) have recently emerged as a promising concept to support communication on SoCs providing a solution to connect different IP-cores through an effective, modular, and scalable communication network.

In recent years, the introduction of hard and soft processor cores into reconfigurable circuits, exploiting the flexibility of FPGA, has changed the job and horizon of designers by providing them with complete Reconfigurable System-on-Chip (RSoC). Key industry issues such as design productivity and embedded system reliability have turned RSoC into viable solutions for embedded systems design.

However, we observe that recurrent drawbacks are usually stressed by industrial partners regarding

DOI: 10.4018/978-1-61520-807-4.ch003

academic innovative NoCs and their associated CAD tools. The first one is the final cost in terms of area (thus, static power), which remains the number one concern in embedded systems. For example, the priority that has driven the Arteris industrial solution is the optimization of the NoC area according to market constraints. This means that strong efforts may be made to reduce features that have the most impact, such as memory.

The second usual criticism is the inadequacy of current tools and methodologies with real applications. The main drawback is the lack of flexibility to support dynamic systems, where communication features greatly change at run-time. This aspect is obvious, for instance, when architectures with multiple general purpose processors are considered. In such cases, the number of processes communicating over the network usually cannot be predicted.

Actually, in such a context, current NoC CAD approaches, based on path and time-slot allocations (e.g Evain & Diguet, 2007; Radulescu et al., 2004) or on simulations (e.g. Arteris 2005), can lead to extra cost, since worst-cases must be considered to support communication demand fluctuations including peaks. Thus, a new NoC design methodology, which provides adaptivity of infrastructure resources and of network protocols, emerges as an alternative solution to face current constraints. It appears that learning-based methods relying on observations are necessary to adapt the NoC to application needs mainly in terms of bandwidth.

In this chapter, we explore the different dimensions of NoC reconfigurations. Then, we propose a methodology that we first apply to Network Interfaces (NI) according to our NoC model and CAD tool. It is organized as follows; part 1 provides the NoC model we consider for exploring the reconfiguration space and classifies current work. In part 2, we give an overview of our configuration strategy and present a new model of self-adaptive NI illustrated with results that clearly show how optimizations can be obtained. Finally we conclude.

MODELING OF RECONFIGURABLE NETWORK-ON-CHIP (RNOC)

The development of embedded systems is based on hardware and software resources that are physically separated, but cooperate to achieve various tasks. The architecture is, therefore, composed of a set of components collaborating to execute application(s). At run-time, communication requirements can change according to data dependency, to application needs, to user choices in terms of concurrent application choices, to architecture hazards (cache coherency, for instance) and to network access conditions (data-rate, protocols, standards). It can also depend on architecture reconfiguration itself to implement or remove given functional blocks. Moreover, it is increasingly needed to be able to change the system architecture to add new services, which were not defined or known at design time.

To cope with these changes caused by 'natural' consequences of the system evolution, the designer must include a control mechanism that transforms and adapts HW/SW environmental parameters to the new process and execution conditions, while avoiding or minimizing system downtime or interruption. This mechanism is called Dynamic Reconfiguration (DR).

In this section, we present the dynamic reconfiguration model and define the layered abstraction-based view. The model is derived from standard network abstraction OSI (Open Systems Interconnection) models and is adapted to RNoC.

Dynamic Reconfiguration Model

The approach of the DR model proposed by Kramer & Magee (1985; 1990) is organized around three stages:

Figure 1. Illustration of the dynamic reconfiguration process and design method for reconfigurable NoC according to network layers

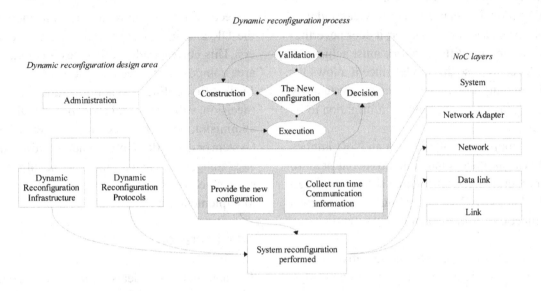

- *The Configuration Validation*: The system can be considered as an implementation of many specifications, which consists of logical parameters and a physical structure. Hence, its reconfiguration is presented in the form of changes in specifications, such as adding, modifying or removing components from the environment. The validation step ensures that changes made by new specifications are compatible with the logical and physical architecture of the system.
- *The Configuration Management*: The configuration manager translates the valid changes expressed for configuration specification into executable commands to the operating system to change the current system. This translation requires knowledge of the state of the system. The required information can be obtained from a database managed by the configuration manager, and fed continuously by the updates from the different components of the system.
- *The Configuration Execution*: The operating system assembles, builds up and

installs the new configuration on the system, according to the instructions sent by the configuration manager.

This DR model is based on an approach, which does not include a decision-making process, because it considers that the decision depends on the designer strategy for the implementation of the DR. So, to extend the above model to have a complete design methodology, we are adding to the DR model a new level of abstraction, which is the Configuration Decision-Making.

Figure 1 illustrates our proposed DR model for NoC, this representation is based on an approach where the configuration manager (System) decides, validates, and builds a new configuration, according to information content collected by the network adapter. Then the system provides the new configuration to be performed on the NoC infrastructure and protocols through the network adapter.

Bjerregaard and Mahadevan (2006) represent the operating process of the NoC by the simplified OSI model of computer networks. In Benini & De Micheli (2001) and Arteris (2005), the OSI model of layered network communication was shown to be easily adapted to NoC cases.

This model does not define services and protocols to be used for each layer; it does nevertheless describe what the network must achieve at each level. We use this model since it allows for the interconnection of heterogeneous systems by adapting the changing flow of information to be processed.

The application of a DR on an NoC implies modifying protocols and standards of one or more layers of the OSI model. Figure 1 shows our design methodology of RNoC according to the properties and functions of individual NoC layers:

- *System:* it represents the operative part of the reconfiguration, its global knowledge of the application and network architecture provides the parameters to decide, validate, and implement a new configuration.
- *Network adapter:* it introduces adaptation protocol between IP and the network. Its knowledge of the number of messages transmitted, consumed, blocked or rejected allows it to collect the necessary information needed by the system to decide the new implementation of reconfiguration.
- *Network:* it defines the routing technique used for delivery of packages through the network and topology. Thus, the system can reconfigure and modify protocols in this layer amending:
  ○ The network topology.
  ○ The routing algorithm.
  ○ The switching technique.
- *The data link*: it manages network resources to deliver a packet; reconfiguration at this layer can handle data exchange protocol between routers, multiplexing methods (e.g. TDMA or SDMA) and techniques of detection and correction of errors.

RNoC Model Dimensions

In section 1.1, we have defined the model of the DR and its correspondence with NoC layers. This correspondence induces a new design area composed of three axes (left side of Figure 1). The first defines the administration methods (decision, validation, and execution) of DR. This administration can be integrated at different levels (system or network adapter). The second and third axes analyze the changes that may be introduced respectively on the structure and network protocols to implement the DR.

Administration

The administration deals with the management of each component of the network, and also the management of the global configuration. Its ultimate goal is to maintain the overall behavior according to the quality of service constraints, which can be the throughput and/or the latency depending on the communication type and priority. The configuration manager is responsible for the administration, and its deployment must adhere to the following points to ensure a dynamic and runtime reconfiguration:

- The system must ensure the management of *coherency* and maintain the *quality of service* of the network. By ensuring that all changes with a new configuration have no inconsistency or negative impact on network performances.
- The system must enable a *delegation* of decision-making, validation and implementation of one or more components of the network. Because the centralized management requires more infrastructure and increases the time for implementation.
- The system must manage the *evolution* of hardware or software components. It ensures an exact knowledge of the infrastructure and elements connected to the

network in case a new component is added dynamically.

- The system should minimize *disruption*. The deployment of a new configuration should monopolize the minimum elements of the network to ensure continuity in transfers not affected by the new configuration.

Most solutions proposed for the RNoC design do not integrate any administration strategy. With the exception of the approach defined by Nollet et al. (2004; 2005), which present a method that incorporates an Operating System (OS). The role of the OS is to manage the allocation of resources and to minimize the number of blocked packets in the NoC. For this reason, it uses a management model based on the collection, analysis, and interpretation of statistical blocked and consumed packets.

These statistics of network traffic enables the OS to regulate the time slot during which each network element will be allowed to send packets. In addition, the OS dynamically adjusts the routing in the NoC.

However, the dynamic routing adaptation is not performed at run-time, and requires a complete stop of traffic, which makes the system unworkable throughout the period of reconfiguration. In addition, the OS administration doubles the cost of NoC in terms of surface and energy consumption because it must use an independent network to collect the control messages from the NI. Finally no details are given about the policies implemented for deciding the size of network access windows.

The administration of DR is effective when it respects the rules of coherency, quality of services, delegation, and disruption. To realize this class of administration, it must be shared between the manager (e.g. OS) and network components (e.g. network adapter). Consequently, the manager will have the role of supervisor, monitor and will react only to totally reconfigure the network. While the components can manage the reconfiguration

of local mechanisms (e.g. buffer sizing, priority, TDMA table).

This technique of DR administration includes a distribution and a delegation of management. The delegate manager decides, validates and achieves partial reconfiguration according to the rules established by the supervisor and it stands to inform the supervisor of the state of the network. The configuration manager assembles the administration rules for each delegate manager according to the overall network state. Moreover, it is able to configure directly a part or the full network. Such an approach has been presented in the very specific domain of security management in (Diguet et al., 2007).

This method reflects the suggestion of integrating an efficient DR administration for RNoC design. However, the efficient and low cost management of RNoC configurations remains an open problem.

Infrastructure

The NoC infrastructure is the combination of various elements (routers, network interfaces and links) that determine the communication architecture. At run-time, adding or removing configured hardware modules in an RSoC structure and changing constraints for communication require a dynamic communication infrastructure, which provides the features of adaptivity in topology, dynamic buffer sizing, and dynamic link bitwidth distribution.

Previous works for adapting NoC architecture target network topology reconfiguration. The first solution addressed by Braun et al. (2007), is called circuit switched. This approach allows modules, which are willing to communicate, to establish a physical connection by setting some multiplexers on the communication links. Other methods addressed in (Bobda et al., 2005; Pionteck et al., 2006) adapt the number of switches and their location by a partial DR when hardware modules are inserted or removed. The realization of these approaches strongly depends on the characteristics of the underlying hardware. To make use of dynamic topology reconfiguration, a homogeneous FPGA, which is dynamically reconfigurable at logic block level, is required.

The NoC structure reconfiguration at run-time offers the ability to effectively establish communication according to the new constraints. Thus far, the solutions proposed to adapt the network structure by removing or adding switches require the use of a reconfigurable device and demand a significant overhead for implementation. Some studies, such as (Stensgaard et al., 2008) use an intelligent switch where packet-switching and physical circuit-switching is combined. This arrangement offers more flexibility in topology reconfiguration. However, there is still some work to be done to build the switch, which consumes the minimum amount of power and area.

The network topology is not the only point of investigation one can explore to achieve a reconfiguration of the NoC infrastructure. Indeed, the reconfiguration of the size of the buffer represents an opportunity to reduce the surface and the energy consumption of NoC. Several solutions exist to scale the size of buffers at design time. But so far, no research has been conducted to change dynamically the size of the buffer.

Protocols

The main goal of the NoC is to support the end-to-end communication between the modules with a specified Quality-of-Service (QoS). To support the QoS requirements, the NoCs must define and include specific network protocols, which determine the Data switching technique, the addressing and routing policies, the multiplexing technique, the end-to-end congestion, and flow control schemes.

The DR of an NoC implies a change in the quality of service requirements. In terms of protocols, this change imposes the exploration and

the calculation of new routing paths and a new multiplexing technique for time slot distribution.

Several studies have suggested algorithms and methods for dynamically calculating either paths or slot tables. Hansson, Coenen and Goossens (2007) present a model that enables the spatial reconfiguration of NoC and an algorithm that uses the model to map multiple applications onto an NoC, delivering undisrupted QoS. This solution presents several limitations. First, it only considers static applications, defined at design time. Secondly, the run-time choices are restricted to a limited choice of precomputed NoC configurations.

Moussa, Baghdadi and Jezequel (2008) propose a dynamic and fast routing algorithm based on the self-routing algorithm for multiprocessor systems with shuffle interconnections (Francalanci & Giacomazzi, 2006). This algorithm takes only one clock cycle to analytically compute the shortest path to destination. However, this approach is quite restrictive since it imposes a De Bruijn network topology and packet length limited to a single one physical unit (PHIT) that, moreover, eliminate deadlock analysis. It is, nevertheless, particularly well adapted to turbo-communication applications.

Marescaux, Bricke, Debacker, Nollet and Corporaal (2005) developed a fast heuristic to perform dynamic time-slot allocation on NoC that provides hard-guaranteed QoS with TDMA techniques, but this approach is applied only for regular networks based on mesh topology.

These solutions are developed in a way that requires knowledge of the various changes that are applied on RNoC for a given configuration. Actually, we can determine two kinds of situations. The first one corresponds to mode changes, in such a case the transition from a set of applications or standards to another usually offers opportunities, for instance based on anticipation and initialization phases, to launch runtime heuristics for computing new path and time-slot allocation. The second case is intrinsically dynamic and corresponds to communication natural variations for a given set of applications. To optimally adapt RNoC to unforeseen changes and guarantied communications, we must develop algorithms providing a safe and smooth adaptation scheme to obtain low cost solutions for run-time spatio-temporal exploration and assignment of paths and time slots.

State of the Art Synthesis

To complete our investigation work on reconfigurable NoC, a survey of some research and design methods for RNoC is presented in Figure 2. The comparison of these methods is made according to dynamic reconfiguration administration, infrastructure and protocols. We found interesting methods, but none of them really covers all three areas, which have been defined in the NoC dynamic reconfiguration model. The solution addressed in (Pionteck et al., 2006) presents a methodology, which is based on the three axes except that the administration is not performed at run-time, it relies on a global control instance of the system.

Figure 2. Survey of research and design methods for reconfigurable NoC and comparison of these methods according to dynamic reconfiguration administration, infrastructure and protocols

Research and design methods	Concept	Dynamic Reconfiguration Administration	Dynamic Reconfiguration Infrastructure	Dynamic Reconfiguration Protocols
Operating system controlled NoC (Nollet et al., 2004)	Includes an OS that can manage communication on a NoC. The OS optimizes communication resource allocation and minimizes interaction between concurrent applications	Operating System	Not defined	Dynamic Injection rate control and routing adaptation
Dynamic Time-Slot Allocation (Marescaux et al.,2005)	Proposes an algorithm to dynamically perform routing and allocation of guarantied communication resources on NoC that provides QoS with TDMA techniques	Extended Iterative deepening algorithm (IDA)	Not defined	Dynamic Time-slot allocation and routing adaptation
DyNoC: Dynamic Network on Chip (Bobda et al., 2005)	Consists in processing elements surrounded by a huge number of switches. At run-time, switches car be disabled and their hardware resources can be reused for dynamically inserted hardware modules. Routing is realized as an extension of the XY algorithm which is capable of surrounding obstacles	Not defined	Network topology reconfiguration	Update the routing protocol
CoNoChi: Configurable NoC (Pionteck et al., 2006)	Adapts network structure to the location, number and size of currently configured hardware modules. Switches can be added or removed from the network by a global instance at runtime.	Global control instance of the system	Network topology reconfiguration	Update the routing protocol
ViChaR: Virtual Channel Regulator (Nicopoulos et al., 2006)	Introduces a centralized buffer architecture, which dynamically allocates virtual channels and buffer slots in real-time, depending on traffic conditions	Unified Control Logic (UCL)	Not defined	Dispensing dynamically a variable of Vcs on demand
Undisrupted QoS during NoC reconfiguration (Hansson et al.,2007)	This work presents a model that enables partial reconfiguration of NoC and an algorithm that uses the model to map multiple applications onto a NoC, delivering undisrupted QoS during reconfiguration	Unified mapping and configuration algorithm	Not defined	Dynamic selection of QoS constrained paths
ReNoC (Stensgaard et al., 2008)	NoC architecture viewed by the application as a logical topology built on top of the real physical architecture. To create these application-specific topologies, ReNoC combines packet-switching and circuit-switching in the same topology switch	Not defined	Network topology reconfiguration	Dynamic combination of packet switching and circuit switching

Although, the solution presented in (Nollet et al., 2004) offers a complete administrative system based on an operating system, no OS policies are really proposed and the area and power overhead must be dramatically decreased to obtain viable and acceptable solutions. Our conclusion is that we must focus on specific interesting aspects of reconfiguration. One promising research direction is for instance an NoC reconfiguration solution where bandwidth can be adapted at run-time, by considering low cost solutions and a hierarchy of controllers for decoupling local and global decisions with different sampling rates.

RECONFIGURABLE NETWORK-ON-CHIP ADMINISTRATION STRATEGY

Global / Local Strategy

The configuration manager is composed of three main components, which are observation, decision, and configuration. An important parameter is the tradeoff between the complexity of the manager and its impact in terms of performance gains. The last point is related to the sampling and adaptation rates compared to application time frames. These three points may be considered to solve the NoC online reconfiguration problem.

Regarding the different tradeoffs, we propose an approach based on two ideas. We use source routing, which is the more efficient technique in our case and consequently we consider that observation of the NoC can be based only on information provided by NIs.

Secondly, we propose a configuration model based on a hierarchy of two kinds of managers, which are a single Global Manager (GM) and multiple Delegate Managers (DM) in each NI as illustrated in Figure 3. The objective is to get a balance between fast and simple local decisions and slower but more optimized ones.

A DM is implemented as a hardware module in each NI, which is in charge of local configuration

decisions. Here the objective is to take fast configuration decisions according to observations limited to a single NI. The kind of decisions at this level deal with memory and bandwidth allocations and sharing regarding communications crossing this particular NI. The DM has a limited configuration range and the strategy is to consider that the GM can decide a new configuration only when DMs have reached their limit. Here the main challenge is to design DM with a very short response time and a low footprint compared to expected gains in performance and area.

On the contrary, the global manager has a complete view of the NoC. It is based on the initial solution, namely the previous complete configuration decided by the GM and updated with short single-packet messages issued from NI. These messages contain some information about the FIFO status including size decided by DMs and use rates due to real communications. Different policies may be implemented at the GM level, since it can be implemented as a software daemon. It is comparable to a new OS service. A balance has to be found between the decision complexity for computing new paths and TDMA (Time Division Multiple Access) tables, for instance, and the expected response time. Another important point is the issue of transition between configurations.

Figure 3. Global / local management

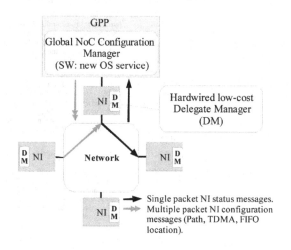

Single packet NI status messages.
Multiple packet NI configuration messages (Path, TDMA, FIFO location).

Two possibilities can be explored, the first one is the equivalent to a change of modes and requires freezing new communications, waiting for all the NI to achieve ongoing communications, reconfiguring all NI, and then, restarting. This solution is safe, but can be unacceptable if fast reconfigurations are needed. The second solution consists in overlapping configuration schemes. This approach is fast, but requires some strict conditions to avoid deadlocks and packet reordering.

We have currently implemented and tested a solution for the DM presented hereafter in section 2.3 and current ongoing work is focused on the GM algorithms.

Evolution of NoC Design Methodology

The design of NoCs implies a complex set of optimization tasks organized as a suite of design flow stages. Moreover, the ordering of these stages, their interactions and any border effect may impact a large number of parameters. To address the question of design complexity and also to improve productivity, we developed μSpider (Evain et al., 2007) tool that automates critical design steps. Various arbiter and routing policies may be defined and a large set of system parameters is available. However, the network generated by μSpider uses packet switching protocol for the flexibility it offers. Furthermore, it is based on the wormhole memorization strategy, which provides

a low latency and a moderate memorization cost in routers. Moreover, it integrates an end-to-end flow-control using credit-based flow-control. This flow-control guarantees, before emission, that the network interface of destination has enough space to store the message in the associated buffer. In addition, there is no packet-loss in the network because the packets which meet a contention are not destroyed, but are stored in routers. Finally, it can implement virtual channels in order to furnish channels to provide best effort (BE) and guaranteed throughput (GT) by using the technique of bandwidth reservation via TDMA.

This CAD tool performs design space exploration and code generation. Design space exploration is implemented in an interactive way based on designer choices. Automatic procedures are then available for time-consuming and error prone tasks such as TDMA organization, buffer sizing and path allocation for guaranteed traffic management.

As shown in Figure 4, the CAD environment flow design is composed of four main parts:

- The first one is based on simulation results from which communication features such as bandwidth, latency, synchronization points and communication ordering are extracted. The objective is to obtain the maximum benefit from this set of information that represents the input data of the NoC design flow. They are captured within a

Figure 4. μSpider NoC design flow

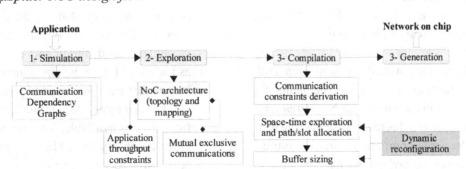

CDG (communication dependency graph) to provide a fine description of the communication constraints to be respected by the final NoC. A front end has been implemented on the top of the CAD tool to remove a tedious design task. These graphs can be, for instance, generated after parsing a simulation input file specified with cadence {E} language. The CDG describes the communications ordering constraints, the direction (W/R), ID of emitters and receivers, expected size of message. The ordering constraints enable automatic extraction of information such as:

- ◦ Parallel communications.
- ◦ Mutual exclusion between communications.
- ◦ Synchronization constraints.

- Thanks to the ordering of communications in the CDGs, the second step explores these graphs to calculate the real bandwidth and latency and provide parameters to generate a specific topology and determine mutual exclusive communications.

- The third step deals with derivation of local latency and bandwidth constraints for each unidirectional communication, and computes the minimum TDMA table size required for implementing guaranteed traffic communications and a minimum bandwidth for all best effort communications. Then, it computes and allocates time slots to each guaranteed traffic communication.

- The last step is the VHDL code generator targeting synthesis tools, some additional C APIs are also provided for interfacing NoC components with IP cores.

When designing an NoC, silicon area and power consumption are two key elements to optimize. A dominant part of the NoC area and power consumption is due to the buffers in the NIs needed to decouple computation from communication. The main drawback we have observed,

regarding our NoC model, is the cost of the NI buffers. This cost is mainly due to the buffer size required to provide guaranteed throughput services. The computation of buffer sizes is based on the knowledge of the application communication constraints. In the case of concurrent applications with variable communications, this estimation becomes difficult, if not impossible. This is the case, for instance, when multiple GPPs (General Purpose Processor) are considered.

Thus, we have integrated programmable and self-adaptive components to be able to modify, at runtime, NoC features as depicted in Figure 4. Then, the idea is to no longer consider worst cases, but average values for constraints and finally reduce the cost of an NoC that can be adapted dynamically to support peak communications.

Local Management: Self-Adaptive Network Interface (SANI)

SANI Description

For previously given reasons, we have developed an RNoC based on the DR model presented in section 1. The RNoC is based on a self-adaptive network interface (SANI), which relies on the observation that only one memory access operation can occur per cycle on NI/Processor link and simultaneously on NI/NoC link. This means that only one FIFO can be modified per cycle on both reception and emission sides. Thus, as shown in Figure 5, the new Network Interface on both emission and reception sides, is composed of:

- A *shared memory* that replaces the fixed FIFOs of the previous NI. A shared memory implements different soft-coded buffers as shown in Figure 6. The main advantage here is that each buffer can spread over its neighbors if more space is needed and when free space is available. This approach provides flexibility to configure dynamically buffer size according to variable processor

Figure 5. The self-adaptive network interface architecture is composed by a forward/reverse shared memory controller, an input/output port, a network input/output port, and a local reconfiguration manager

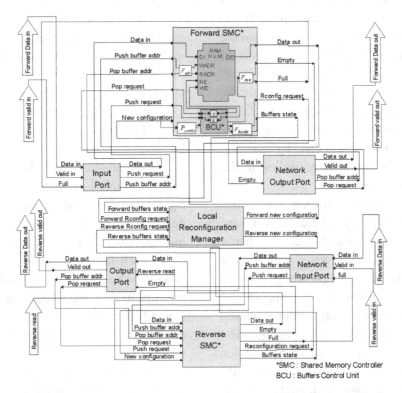

Figure 6. Buffers management

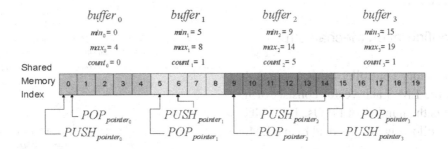

needs in terms of bandwidth and burst size. It presents two main interests. First, it helps to reduce buffer over-sizing based on worst case analysis. Basically, it enables memory-use optimization by adapting buffer size and by minimizing unused free space. A key point is the relative position of buffers since a FIFO can use the free space of its

neighbors. These positions and the initial FIFO size are, like paths and TDMA slots, configurable parameters that can be modified by the NoC GM according to statistics based on DM feedback. Secondly, for a given memory size, it enables communication performance optimization by improving TDMA slots use.

- A *shared memory controller*. In a traditional NI, each buffer has its own FSM that controls its own signals (Read, Write, Full, Empty, Counter status). By moving to a shared memory, which is a cluster of different configurable FIFOs, a global controller is required. The latter controls the Push, Pop, Begin, and End Pointers of each FIFO as shown in Figure 6. Thus, a modification of a buffer size is done by changing Begin and End pointers according to the mobility of neighbors.

- A *Delegate Manager* observes and configures buffers in a very simple and fast way; it runs as follows. At each cycle, the shared memory controller updates a status vector that indicates how FIFOs are used (Full, Nearly Full, Empty, Nearly Empty) and which are current reconfiguration requests. Moreover, the DM checks configuration possibilities and when necessary modifies FIFO parameters as soon as an opportunity occurs. The complete configuration operation is done in one cycle, which means that no additional delay is introduced compared to the usual NI.

SANI Reconfiguration Mechanism

The shared memory controller (SMC) shares 4 vectors with the LM. Each vector consists of n bits, where n is the number of FIFOs. The SMC updates the 3 following vectors read by the LM:

- *FIFO reconfiguration request vector*: the SMC calculates asynchronously the filling ratio of each FIFO using the comparison given by Eq.2.1. When the FIFO filling ratio reaches the threshold fixed by the global manager, the SMC sets the corresponding bit of the FIFO to 1.

- *Right (resp. Left) FIFO spreading possibility vectors*: in our architecture, a FIFO

reconfiguration means increasing or decreasing FIFO size. This modification implies a right or a left move of FIFO boundaries. The right (resp. Left) extension of the *fifo(n)* involves the removing of one location at the beginning (Resp. end) of the *fifo(n+1)* (Resp. *fifo(n−1)*), and adding this location at the end (Resp. beginning) of the *fifo(n)*.

For no data loss, the SMC guarantees that the FIFOs are able to be extended. If a FIFO meets right (Resp. left) spreading conditions briefly described in Eq.2.2 (Resp. 2.3), the corresponding right (Resp. left) bit of this FIFO is set to 1.

At each leading edge of the clock, the LM processes the information issued from the 3 vectors updated by the SMC. Then, it creates the new FIFO configuration by updating the new configuration vector read by the SMC that modifies FIFO sizes. The choice is made according to the reconfiguration requests and the reconfiguration priority fixed by the global manager.

$$count\left(fifo_n\right) > max\left(fifo_n\right) - min\left(fifo_n\right) \quad (1)$$

$$
\begin{pmatrix} pop\ pointer\left(fifo_n\right) < push\ pointer\left(fifo_n\right) \end{pmatrix} \quad and \\
\begin{pmatrix} max\left(fifo_{n+1}\right) - min\left(fifo_{n+1}\right) > threshold\left(fifo_{n+1}\right) \end{pmatrix} \quad and \\
\begin{pmatrix} \left(pop\ pointer\left(fifo_{n+1}\right) < push\ pointer\left(fifo_{n+1}\right)\right) and \left(pop\ pointer\left(fifo_{n+1}\right) > min\left(fifo_{n+1}\right)\right) \end{pmatrix}
$$

$$(2)$$

$$
\begin{pmatrix} pop\ pointer\left(fifo_n\right) < push\ pointer\left(fifo_n\right) \end{pmatrix} \quad and \\
\begin{pmatrix} max\left(fifo_{n-1}\right) - min\left(fifo_{n-1}\right) > threshold\left(fifo_{n-1}\right) \end{pmatrix} \quad and \\
\begin{pmatrix} \left(pop\ pointer\left(fifo_{n-1}\right) < push\ pointer\left(fifo_{n-1}\right)\right) and \left(push\ pointer\left(fifo_{n-1}\right) < max\left(fifo_{n-1}\right)\right) \end{pmatrix}
$$

$$(3)$$

Hardware Test-Bench Conditions

To measure performances of NoC enhanced with SANI interfaces, an emulation system has been implemented on a Chipit platform designed by ProDesign. This platform embeds two Xilinx Virtex 5 FPGA and provides an Ad Hoc Communication Bus (UMR), which allows a C/C++ application to interact with a Hardware Design implemented on the FPGAs. A ProDesign PCI card makes the physical interface.

A UMR Bus is hardwired on the board, allowing, for example, to preload some on-board memories or to interact with an LCD screen with native functionalities. The UMR bus can be extended transparently within the FPGA design. Thus, we have implemented our own ad hoc protocol to control read and write operations on chained on-chip memories.

The emulation system uses all these interfacing functionalities to evaluate functional and performance aspects of the NI. Figure 7 shows the structure of the emulation system, which is done in 5 steps:

- On the Software part, C++/SystemC threads use some API to communicate some data to the Hardware part that is to be sent to NI FIFOs with specific headers containing information required for emulation such as emission date, data-block size and destination ID (equivalent to FIFO numbers).
- On the Hardware part, a module called sequencer interprets, with a given protocol, all communication from C++ applications

and stores data in an On-Chip RAM memory.

- Then, when all data is stored, the software application sends RUN signals to the sequencer, which starts sending data blocks into the NI, after decoding each data block header. To avoid loss of data, the sequencer considers NI control signals, which indicate some full FIFO situations.
- During the emulation, a hardware module receives (on the fly) data provided by NI outputs and stores it in a second On-Chip RAM memory.
- The Software application gets back all data stored by the receiver, to enable comparisons with sent data and check the NI right behavior and some additional information, which allow for performance evaluation.

Results

To demonstrate the effectiveness of our approach, we have realized different experiments comparing SANI and NI implementations and performances.

Shared memory vs. fixed FIFO costs: The cost is given as a number of LUTs, no block RAM memories have been used to get comparable conditions. BRAM have not been selected since

Figure 7. Emulation system

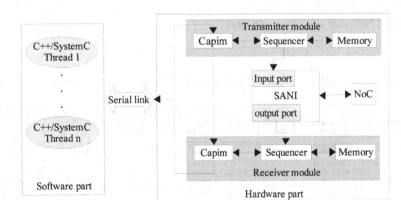

the result, in that case, depends on the BRAM size. Our objective was to remain as independent as possible from device architectures, which is also why we have selected a solution based on our own VHDL FIFO component. Figure 8-a) and b) present the evolution of logic and memory costs respectively for two FIFOs, while increasing the memory size. Figure 8-c) and d) present the evolution of logic and memory costs respectively for a fixed global memory size, while increasing the number of FIFOs. First, we observe that the logic cost remains equivalent for both solutions. Secondly, for memory cost, we notice an average overhead of 13% for fixed FIFOs compared to shared memory. This is due to threshold effects, the main lesson we can draw is that the shared memory solutions offer programmable FIFOs and thus flexibility without any impact on the cost, on the contrary a slight benefit can be obtained.

Bandwidth and cost evaluations: To compare performances, both interfaces have been implemented in network where they can have four possible destinations, namely four FIFOs. The global clock cycle is equal to 70MHz and 32bit links are used. The same TDMA table is used for all cases, it is composed of 12 time slots, each destination has one slot per round. Then, six scenarios of data transfers were tested with various burst sizes and various destination orderings. In the following output bandwidth means the bandwidth observed on the (SA)NI/ Network link, is actually limited to the number of slots offered by the TDMA table, which represent an upper bound equal to 560 Mbits/s. The input bandwidth is obtained between the processor and

the (SA)NI; it is limited by the status of the FIFOs, since data can be written every clock cycle, thus the upper bound would be 2.24 Gbits/s, therefore, in practice, it is also limited to the output bound.

An important issue in this experiment, is the whole cost of the DM. In our first implementation, we obtain an overhead around 18%. Nevertheless, we are quite confident with the fact that this value can now be optimized. In any case, the following experiments give interesting results and show the benefits of self-adaptivity at the NI level for both cost and performance objectives.

Bandwidth performances with a fixed memory size: Figure 8 presents the results we obtain for the six scenarios, when a fixed memory size is considered. Here the idea is to observe how a SANI can reduce the NoC access time by dynamically adapting FIFOs, thus by reducing idle time due to full FIFO. We observe improvements for both output and input bandwidths, which go from 7 to 28% with an average around 18%. This benefit is provided by the flexibility offered by the SANI and the DM reconfiguration decisions.

Memory size optimization with a fixed bandwidth: The second experiment is set up to show how memory space can be saved with self-adaptivity. Thus, we measure the increase of memory required with NI to reach the bandwidth offered by the equivalent. We can observe in Figure 9, that important memory savings can be obtained with SANI. The average value is around 30%, which is larger than the cost of the first DM version.

Figure 8. Bandwidth optimization with SANI compared to NI for a given global memory size

		Scenario 1	Scenario 2	Scenario 4	Scenario 4	Scenario 5	Scenario 6
NI	Output bandwidth (Mbits/s)	205.63	237.74	345.85	347,32	389.02	281.54
	Input bandwidth (Mbits/s)	212.94	247.57	367.03	368,69	416.02	295.42
SANI	Output bandwidth (Mbits/s)	221.32	275.57	420.26	450.48	486.36	354.4
	Input bandwidth (Mbits/s)	241.9	291.67	456.79	509.53	517.4	363.38
	The output bandwidth gain (%)	7.09	13.72	17.71	22.9	20.02	20.56
	The input bandwidth gain (%)	11.97	15.12	19.65	27.64	19.59	18.7

CONCLUSION

Given the current limitations of NoCs regarding flexibility and cost, we have first focused our study on the definition of a global model for RNoC management based on three dimensions: administration, infrastructure and protocols. For this model, we provide a comprehensive state of the art. We made two main observations; first, the Network Interfaces that represent an important part of NoC costs in general are not investigated here, as far as reconfiguration opportunities are concerned. Secondly, there is few work that address the question of configuration administration. Therefore,

Figure 9. Memory size optimization with SANI compared to NI for a given bandwidth constraint

		Scenario 1	Scenario 2	Scenario 3	Scenario 4	Scenario 5	Scenario 6
Conditions	Output bandwidth (Mbits/s)	221.32	275.57	420.26	450.48	486.36	354.4
	Input bandwidth (Mbits/s)	241.9	291.67	456.79	509.53	517.4	363.38
Total memory size	SANI (kbits)	6.4	6.4	6.4	6.4	6.4	6.4
	NI (kbits)	11.5	9.6	7.68	7.94	7.27	8.16
Gain (%)		79.69	50	20	24	13.5	27.5

Figure 10. Comparison of buffer costs based on shared memories and fixed FIFOs

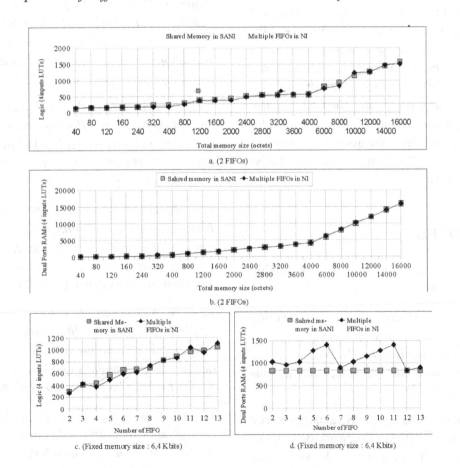

we have proposed a hierarchy of configuration managers to implement configuration decisions and, thus, self-adaptivity within our NoC model.

Our approach also deals with infrastructure (FIFO sizing) and protocol (TDMA, path) dimensions. The preliminary results we present in this chapter are focused on the local configuration manager, namely the Delegate Manager that provides NI with a local decision on buffer allocations. These first results show how self-adaptivity can optimize performances or cost at the NI level. These gains are mainly related to the management of unpredictable burst transfers, which are usual in graphics or multimedia applications. However, we also note some restrictions. Actually the impact of such an approach is positive if applications and architectures present some interesting features, which are typically the number, the size and the ordering of FIFOs.

Based on these promising results, two kinds of work are today undeveloped. The first one is the optimization of the DM architecture to decrease its footprint. The second one is the implementation of the global manager as a software task with a slower adaptation rate. The GM will deal with other parameters that will improve our solution by benefiting from the association between TDMA and path allocation, and between SANI and FIFO ordering.

REFERENCES

Arteris. (2005). A comparison of network-on-chip and buses. *White paper*.

Benini, L., & De Micheli, G. (2001). Powering networks on chips: energy-efficient and reliable interconnect design for SoCs. In *Proceedings of the 14th international symposium on Systems synthesis*, (pp. 33–38).

Bjerregaard, T., & Mahadevan, S. (2006). A survey of research and practices of network-on-chip. In *Proceedings of ACM Computing Surveys*, (vol. 38).

Bobda, C., Ahmadinia, A., Majer, M., Teich, J., Fekete, S., & van der Veen, J. (2005, August). *Dynoc: A dynamic infrastructure for communication in dynamically reconfugurable devices*. In *Proceedings of the Field Programmable Logic and Applications International Conference*.

Braun, L., Hubner, M., Becker, J., Perschke, T., Schatz, V., & Bach, S. (2007, August). *Circuit switched run-time adaptive network-on-chip for image processing applications*. In *Proceedings of the Field Programmable Logic and Applications International Conference*.

Diguet, J.-Ph., Evain, S., Vaslin, R., Gogniat, G., & Juin, E. (2007, May). *Noc-centric security of reconfigurable soc*. In *Proceedings of the 1st ACM/IEEE International Symposium on Networks-on-Chips*. Princeton, NJ.

Evain, S., Dafali, R., Diguet, J.-Ph., & Juin, E. (2007). *μspider CAD tool: Case Study of NoC IP Generation for FPGA*. In *Proceedings of the Workshop on Design and Architectures for Signal and Image Processing*. Grenoble, France.

Evain, S., & Diguet, J.-Ph. (2007 March). *Efficient space-time NoC path allocation based on mutual exclusion and pre-reservation*. In *Proceedings of the 17th ACM Great Lakes Symposium on VLSI (GLSVLSI)*. Italy.

Francalanci, C., & Giacomazzi, P. (2006, January). *High-performance self-routing algorithm for multiprocessor systems with shuffle interconnections*. In Proceedings of IEEE Transactions on Parallel and Distributed Systems, (vol. 17).

Hansson, A., Coenen, M., & Goossens, K. (2007, April). *Undisrupted quality-of-service during reconfiguration of multiple applications in networks on chip*. In *Proceedings of Design, Automation, Test in Europe Conference and Exhibition*.

Kramer, J., & Magee, J. (1985 April). Dynamic configuration for distributed systems. In *Proceedings of IEEE Transactions on Software Engineering*, (vol. 11).

Kramer, J., & Magee, J. (1990). The evolving philosophers problem: Dynamic change management. In Proceedings of IEEE Transactions on Software Engineering, (vol. 16).

Marescaux, T., Bricke, B., Debacker, P., Nollet, V., & Corporaal, H. (2005,September). Dynamic time-slot allocation for qos enabled networks on chip. In Proceedings of Embedded Systems for Real-Time Multimedia Workshop.

Moussa, H., Baghdadi, A., & Jezequel, M. (2008, June). Binary de bruijn on-chip network for a flexible multiprocessor ldpc decoder. In *Proceedings of the 45th ACM/IEEE Conference on Design Automation.*

Nicopoulos, C. A. Park. D., Kim, J., Vijaykrishnan, N., Yousif, M. S., & Das, C. R. (2006 December). Vichar: A dynamic virtual channel regulator for network-on-chip routers. In *Proceedings of MICRO-39, the 39th Annual IEEE/ACM International Symposium*, (pp. 333–346).

Nollet, V., Marescaux, T., Avasare, P., Verkest, D., & Mignolet, J.-Y. (2005 March). Centralized run-time resource management in a network-on-chip containing reconfigurable hardware tiles. In *Proceedings of Design*. Automation and Test in Europe.

Nollet, V., Marescaux, T., Verkest, D., Mignolet, J. Y., & Vernalde, S. (2004). Operating-system controlled network on chip. In *Proceedings of DAC '04, the 41st annual conference on Design automation*. New York.

Pionteck, T., Koch, R., & Albrecht, C. (2006). *Applying partial reconfiguration to networks-on-chips*. FPL.

Radulescu, A., Dielissen, J., Goossens, K., Rijpkema, E., & Wielage, P. (2004). An efficient on-chip NI offering guaranteed services, shared-memory abstraction, and flexible network configuration. In *Proceedings of IEEE TCAD*.

Stensgaard, M. B., & Sparso, J. (2008, April). *ReNoC: A network-on-chip architecture with reconfigurable topology*. In *Proceedings of NoCs 2008, the Second ACM/IEEE International Symposium*, (pp. 55–64).

KEY TERMS AND DEFINITIONS

Global Decision: Configuration decision for parameters impacting the whole system.

Local Decision: Configuration decision limited to the configuration of a sub-system.

Reconfigurable Network on Chip (RNoC): NoC where at least one of the following capabilities can be modified at run-time: topology, bandwidth between two nodes, source routing paths, memory buffer size, communication priorities.

Reconfigurable System on Chip (RSoC): Multi-processor system on a single chip with flexible software and hardware features such that computation and/or memory and/or communication capabilities can be modified at run-time.

RNoC Administration: Organization of local and global decision capabilities.

Self Adaptive Network Interface: Interface with reconfiguration and decision-making capabilities regarding network access including bandwidth, latency and buffer size.

Self-Adaptive Embedded System: Embedded system with observation and decision-making capabilities such that it can select and implement a new configuration according to goal-settings without any designer intervention.

Chapter 4
An Efficient Hardware/Software Communication Mechanism for Reconfigurable NoC

Wei-Wen Lin
National Chung Cheng University, Taiwan, R.O.C.

Jih-Sheng Shen
National Chung Cheng University, Taiwan, R.O.C.

Pao-Ann Hsiung
National Chung Cheng University, Taiwan, R.O.C.

ABSTRACT

With the progress of technology, more and more intellectual properties (IPs) can be integrated into one single chip. The performance bottleneck has shifted from the computation in individual IPs to the communication among IPs. A Network-on-Chip (NoC) was proposed to provide high scalability and parallel communication. An ASIC-implemented NoC lacks flexibility and has a high non-recurring engineering (NRE) cost. As an alternative, we can implement an NoC in a Field Programmable Gate Arrays (FPGA). In addition, FPGA devices can support dynamic partial reconfiguration such that the hardware circuits can be configured into an FPGA at run time when necessary, without interfering hardware circuits that are already running. Such an FPGA-based NoC, namely reconfigurable NoC (RNoC), is more flexible and the NRE cost of FPGA-based NoC is also much lower than that of an ASIC-based NoC. Because of dynamic partial reconfiguration, there are several issues in the RNoC design. We focus on how communication between hardware and software can be made efficient for RNoC. We implement three communication architectures for RNoC namely single output FIFO-based architecture, multiple output FIFO-based architecture, and shared memory-based architecture. The average communication memory overhead is less on the single output FIFO-based architecture and the shared memory-based architecture than on the multiple output FIFO-based architecture when the lifetime interval is smaller than 0.5. In the performance analysis, some real applications are applied. Real application examples

DOI: 10.4018/978-1-61520-807-4.ch004

show that performance of the multiple output FIFO-based architecture is more efficient by as much as 1.789 times than the performance of the single output FIFO-based architecture. The performance of the shared memory-based architecture is more efficient by as much as 1.748 times than the performance of the single output FIFO-based architecture.

INTRODUCTION

With the progress of technology, we are able to integrate several *intellectual properties* (IPs) into one single chip. The performance bottleneck has shifted from the computation in individual IPs to the communication among IPs. Common on-chip communication architectures include point-to-point dedicated wire-based and shared bus-based architectures. Although the dedicated wire-based architecture can guarantee the required communication bandwidth between IPs, it suffers from low link utilization. The shared bus-based architecture provides higher link utilization than the dedicated wire-based architecture. But, when the number of IPs increases, the shared bus-based architecture faces the drastic contention problem caused by IPs. A feasible solution to the link utilization and the contention problem has been proposed called the *Network-on-Chip* (NoC). In an NoC, routers are responsible for data transmissions among IPs. NoC provides high link utilization and alleviates the contention problem through the parallel execution of routers. However, an ASIC-implemented NoC lacks flexibility and has a high non-recurring engineering (NRE) cost. As an alternative, we can implement an NoC in a *Field Programmable Gate Arrays* (FPGA). Such an FPGA-based NoC is more flexible and the NRE cost of FPGA-based NoC is also much lower than that of an ASIC-based NoC. In addition, FPGA devices can support dynamic partial reconfiguration such that the hardware circuits can be configured into an FPGA at run time when necessary, without interfering hardware circuits that are already running. Because of dynamic partial reconfiguration, an FPGA-based reconfigurable

NoC system is able to accommodate the execution of more hardware applications than an ASIC-implemented NoC system given the same amount of system resources. Nevertheless, an FPGA-based reconfigurable NoC system has many problems to be solved such as hardware module placement, hardware applications scheduling, reconfigurable topologies, reconfigurable IP design and hardware/software (HW/SW) communication.

In general, a system is composed of many components, such as processors, memory, and I/O peripherals. With the improvement of semi-conductor manufacturing technology, an entire system can be integrated into a single chip, often called a *System-on-Chip* (SoC). As summarized by the *International Technology Roadmap for Semiconductors* (ITRS) [2008], the semiconductor manufacturing technology is progressing from 180nm in year 2000 to 22nm in the future year 2016. This trend reveals that future SoCs will include more and more IP cores. Consequently, with the increasing number of IP cores in SoCs, the communication architecture plays a more important role in system performance.

A common communication architecture consists of dedicated wires as shown in Figure 1. The dedicated wire-based architecture has a high communication efficiency due to the point-to-point routing resources between two IP cores. However, when the number of IP cores increases, the dedicated wire-based architecture suffers from the increasing complexity of placing and routing. In addition, it also suffers from low link utilization because the dedicated wires can be utilized only by the connected IP cores. Figure 2 illustrates another general communication architecture, namely the shared bus. As implied by the name, a bus is shared

Figure 1. Dedicated wire-based communication architecture

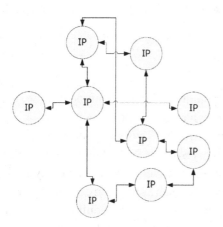

by several IP cores which are connected to the bus. Therefore, the shared bus-based architecture has higher link utilization than the dedicated wire-based architecture. However, it needs a global arbiter to grant bus access to an IP core. The shared bus-based architecture must resolve bus access contentions among IP cores, which ensures that two or more IP cores cannot access the bus concurrently. When we integrate more and more IP cores in a shared bus-based architecture, IP cores face a long waiting period to access the bus due to severe contentions. From the above observation, not only the dedicated wired-based architecture but the shared bus-based architecture also suffers from the scalability issue. Dedicated wires cause low link utilization and the shared bus lacks the capability for parallel communication. Hence, a

novel interconnect communication architecture called the *Network-on-Chip* (NoC) was proposed to provide a feasible solution [Benini & Micheli, 2002; Duato et al., 1997].

Figure 3 illustrates that an NoC is composed of several routers and IP cores. Each IP is attached to a router on the NoC. Each router has a unique address and is responsible for routing packets to their destinations and for receiving packets destined for its attached IP. In NoC, since links are shared by packets destined for different IPs, the link utilization in an NoC is higher than that in a dedicated wire-based architecture. In an NoC, the data transmission authority has shifted from a global arbiter to the local judgements of routers, because routers are in charge of transmitting packets of attached IPs. Compared to the shared bus-based architecture, an NoC avoids the communication resource contentions through distributed arbitrations of routers. Since NoC allows routers to work concurrently, thus parallel communication is supported. In summary, NoC allows high link utilization, eases the contention issue among IPs, supports parallel communication and high scalability.

Though an ASIC-implemented NoC provides high system performance, it also incurs high area overhead and long time-to-market. Due to the high *non-recurring engineering* (NRE) cost, an ASIC-based NoC also has a high production cost. Nowadays, not only the gate count capacity of FPGA has been increasing rapidly, but it is also possible to

Figure 2. Shared bus-based communication architecture

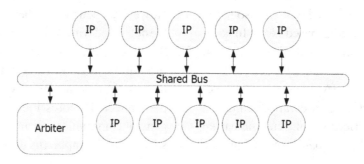

Figure 3. Network-on-chip communication architecture

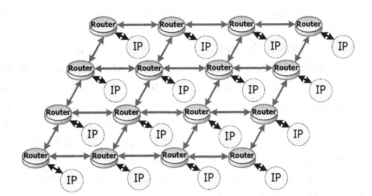

configure an FPGA chip at run-time and partially. Now hardware circuits can be configured into an FPGA, without interfering other hardware circuits that are running. *Dynamically Partially Reconfigurable System* (DPRS) is a system composed of processors, memory, communication architectures, peripherals, and FPGAs to support reconfigurability. Compared to an ASIC-based system, DPRS has less production cost and is more flexible. With DPRS, we can run more hardware applications with less hardware area through time overlap than an ASIC-based system. Because of reconfigurability, it is easy to upgrade hardware circuits such that the production lifetime is extended.

To utilize the limited reconfigurable resources efficiently, DPRS needs an operating system to manage the reconfigurable resources, namely *Operating System for Reconfigurable System* (OS4RS). Traditionally, an operating system should manage several system functions such as program execution, memory usage, file systems, I/O and peripheral devices, and *inter-process communication* (IPC). In an OS4RS, additional typical functionalities include scheduling both the software tasks and hardware tasks, placing the hardware tasks by allocating reconfigurable resources, supporting the communication between the software and hardware tasks, and ensuring the correct reconfiguration operation are typical additional functionalities.

To prototype a system with scalability in a short time and to save the hardware area overhead, we will construct a DPRS with a *Reconfigurable NoC* (RNoC). From the point of view of a reconfigurable architecture, there are several issues including the design of RNoC topology reconfiguration, reconfigurable buffer size of routers, the design of reconfigurable IPs, and the design of an *Operating System for Reconfigurable NoC* (OS4RNoC).

We focus on solving the OS4RNoC issue in this chapter, with special emphasis on how communication between hardware and software can be made efficient. To implement OS4RNoC, we extend an existing operating system, namely PetaLinux [2009] which was derived from uClinux [2008]. In an OS4RNoC, the communication can be classified into four different types, namely software module to software module (sw to sw communication), hardware module to hardware module (hw to hw communication), software module to hardware module (sw to hw communication), and hardware module to software module (hw to sw communication). In an OS4RNoC, sw to sw communication can be implemented using traditional IPC mechanisms. Hw to hw communication can be supported by an NoC. However, the communication between sw and hw can cause congestion in the NoC, which will degrade the overall system performance. There are some disadvantages with using traditional

Hw/Sw communication mechanism for RNoC. We will give an example to illustrate.

In a traditional embedded system, hardware accelerators (IPs) are used to improve the overall system performance. Each IP connects to the system bus through specific wrappers which is composed of control signals and FIFOs for communicating with the system and OS [Ces´ario et al., 2002; Hommais et al., 2001]. In our target system architecture, a simple FIFO-based architecture can be used for hardware-software communication. An example is shown in Figure 4, in which the overall communication architecture consists of a 4×4 mesh NoC with reconfigurable parts and one of routers is connected to an OS through FIFOs. It is shown in Figure 4 that to improve application performance, the OS4RNoC configures a CRC-32 hardware accelerator to a reconfigurable part. OS4RNoC sends data to the CRC-32 through an input FIFO. After CRC-32 finishes its computation, it sends the results through the output FIFO to the OS4RNoC. Note that the input and output FIFOs are connected to the NoC through a designated router. A specific *Network Interface Component* (NIC) is used to connect the router and the FIFOs. The FIFOs play the role of buffers for the hardware accelerators.

Since DPRS allows more than one hardware accelerator to be running simultaneously, we show two applications running concurrently in Figure 5. OS4RNoC configures a JPEG encoder application which consists of three reconfigurable modules, namely DCT, Quantization, and Encoder. Suppose the OS4RNoC injects data into the input FIFO using a round-robin scheduling algorithm, that is, CRC-32 and JPEG data are processed turn-by-turn. An example of an interleaved sequence of data in the output FIFO is shown in Figure 1.5. Note that the first piece of data in the output FIFO is generated by the CRC-32 module. However, if the currently reading process needs the data for the JPEG application, there will be a read miss. The OS4RNoC will then switch to a process of the CRC-32 application. The read miss will not only cause some delay, but might even cause communication deadlock. When more and more applications run on the NoC, the read miss rate will increase rapidly and the output FIFO will be soon filled with data. As a result, a data congestion will occur in the NoC, which propagates from the output FIFO router to the source reconfigurable modules. Eventually, the whole NoC will stop working due to severe congestion, thus affecting the overall system performance.

To solve the above problem, increasing the number of output FIFOs is a possible solution. When the number of output FIFOs is equal to the number of reconfigurable modules which com-

Figure 4. CRC-32 application runs on reconfigurable NoC

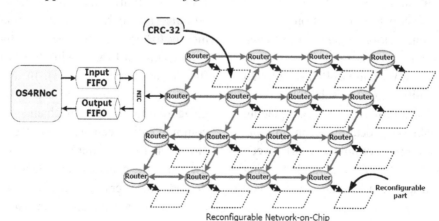

Figure 5. CRC-32 application and JPEG application run on reconfigurable NoC

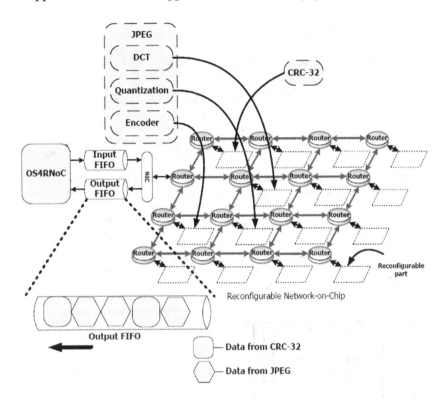

municate with OS4RNoC, the read miss rate will be very low because there will be no interleaving of data in the output FIFOs. However, resource utilization becomes quite poor in such a multiple FIFO mechanism and it is also difficult to estimate the number of output FIFOs required when the number of applications varies dynamically.

To solve the NoC congestion caused by the FIFO-based Hw/Sw communication and to utilize hardware efficiently, we propose a more efficient shared memory-based Hw/Sw communication mechanism for reconfigurable NoC (RNoC). Being randomly accessible, the shared-memory-based architecture avoids the problem of data that occurs in the FIFO-based mechanism. As shown in Figure 6, by classifying output data from NoC and storing the data of different applications to dedicated memory spaces allocated by the OS4RNoC. In this Thesis, we will focus on the communication mechanisms between software

and hardware tasks and propose a communication architecture with shared memory for RNoC.

BACKGROUND

To solve the issues in ASIC-based systems, DPRS has been proposed to provide greater flexibility along with reduction in power consumption. However, DPRS has also created issues of its own, due to reconfigurability. To ensure correct system reconfiguration, many researches focus on extending an existing OS to manage the reconfigurable resources. Santambrogio et al. [2008] proposed online partial dynamic reconfiguration management which provides simple software calls to perform reconfiguration such as module request, module release, module removal, and modules list. User applications can utilize these function calls to request or remove a module from the recon-

Figure 6. Communication architecture with shared memory on reconfigurable NoC

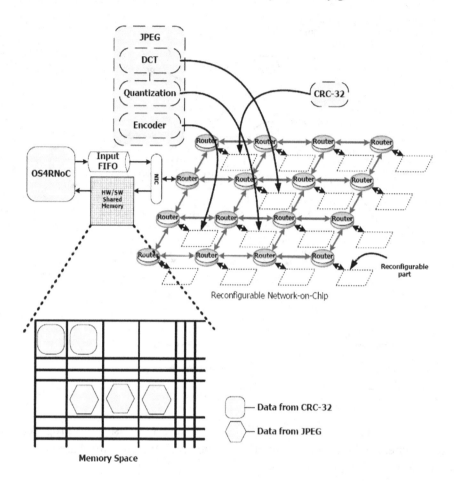

figurable system, or to get the list of the modules which are already configured in the DPRS. The authors also proposed a reconfiguration manager with two policies: cache policy and allocation policy. In the cache policy, a hard-removal means an unused module is deleted physically from the system by resetting the occupied resources, that is, configuring a blank bitstream. A softremoval means an unused module is deleted logically and the FPGA configuration of this module is not altered. In the allocation policy, the goal is to maximize the number of modules configured into a system at the same time. The modules are contrived and accessed by device drivers.

So and Brodersen [2008] developed an OS for reconfigurable system called Berkeley Operating system for ReProgrammable Hardware (BORPH), which models an executing instance of FPGA application as a hardware process. Traditional OS for reconfigurable system considers reconfigurable resources as a hardware device, and programmers use the reconfigurable resources through the device driver. BORPH provides a different way to access reconfigurable resources, by modeling them as hardware processes. Similar to the processes in a UNIX system, a hardware process can communicate with other processes through interprocess communication mechanisms. In addition, a BORPH Object File (BOF) format was proposed for the executable file of a hardware process. In a traditional UNIX system, a process is created by invoking two system calls: fork and exec. When a software process creates a hardware process, BORPH utilizes the same fork-exec sequence. To support reconfigurability, BORPH extended a standard Linux kernel to support the BOF file format on the BEE2 hardware platform [2008] which is composed of five FPGA chip devices.

Another OS support for reconfigurable system was presented by Williams and Bergmann [2004]. They mainly developed an ICAP device driver in the Linux OS. The ICAP module, developed by Xilinx, has an On-chip Peripheral Bus (OPB) interface and can be used for system self-reconfiguration. They use the standard driver architecture to integrate ICAP device with the Linux kernel. They developed a simple character-based device driver, which implements the read(), write(), and ioctl() system calls. The read function initiates a read by the ICAP into a user memory buffer. The write function specifies a number of bytes to be written by the ICAP from a user memory buffer. The ioctl function is an interface to perform device-specific control operations. An OS can perform self-reconfiguration with this ICAP device driver.

Bus-based reconfigurable systems are also proposed by some authors [G¨otz & Dittmann, 2006; Zhou et al., 2005]. G¨otz and Dittmann [2006] proposed a slot-based architecture for reconfigurable system. Slots represent reconfigurable regions which can be configured with hardware circuits at run time. In a reconfigurable system, busmacros can be used to ensure compatibility of signals between reconfigurable regions and static areas [Claus et al., 2007]. Thus, in a slot-based architecture, the slots are connected by a local bus and composed of busmacros [Claus et al., 2007]. In their approach, CPU executes most of the application tasks and only the critical application tasks run on the FPGA. To assign tasks to the CPU or the FPGA, the FPGA area overhead and CPU processor workload are taken into account. Tasks can migrate from software to hardware or vice-versa in the system. Every task has an estimated cost which represents the percentage of resources used by the task. The cost means both the circuit

area used by the task on FPGA and the processor workload utilized by the task on CPU. When there is an unbalanced workload between the FPGA and the CPU, a system reconfiguration takes place to balance the workload. When reconfiguring, some constraints have to be obeyed, first, the sum of the FPGA areas occupied by all tasks running on the FPGA should be no more than the total FPGA area, and second, the sum of the CPU workload of all tasks running on the CPU should be no more than the total CPU workload. They devoted to find out the reconfiguration order of tasks which must satisfy these constraints. A heuristic algorithm was proposed to support a feasible reconfiguration process. The basic strategy in the algorithm is to configure a task into FPGA which has the highest software cost and with the smallest hardware cost first such that the goal to reduce the CPU utilization is achieved.

Zhou et al. [2005] proposed managing reconfigurable resources by dividing reconfigurable resources into four states: used state, preconfiguration state, blank state, and configuring state. If least one running task is using the reconfigurable resource then this resource is at the used state. When the reconfigurable resource is at the preconfiguration state, all tasks using this resource are ready to run. Blank state means there is no task using the resource. When tasks are being configured, the reconfigurable resource is at the configuring state. An RTOS called SHUM-uCOS (Software-Tasks Hardware-Tasks Uniform Management uCOS) was developed by extending the MicroC OS II [Labrosse, 2002] for reconfigurable systems using a uniform multitask model. A set of API functions is provided for a designer to interact with the OS, which maintains a hardware task preconfiguration table to reduce the configuration cost at runtime. A scheduler is responsible for managing the states of hardware tasks and software tasks. A resource manager traces and records the dynamic creation and deletion of hardware tasks, providing information for scheduler to configure hardware tasks. A communication controller handles the low-level communication details. Also, a hardware task configuration database contains all the hardware task configuration data. A hardware-task configuration controller retrieves corresponding configuration data from a database, and configures the FPGAs after receiving the configuration command from the scheduler. Finally, the hardware task interface supplies the communication controller with the standard signals and protocols.

Network-on-chip has also been proposed as a promising communication architecture [Ogras et al., 2007] for future SoC systems. To increase the flexibility of ASIC-based systems and the low scalability of bus-based systems, many researches have focused on implementing NoC on an FPGA platform with reconfigurability, namely reconfigurable NoC. Because an NoC infrastructure is composed of several components, different researches have focused on different aspects of reconfigurable NoC.

The dynamic placement of modules was presented by Bobda et al. [2005], for a Dynamic Network-on-Chip (DyNoC) with a mesh topology. In the initial state, routers perform normal functionalities and each router is connected to one processing element. In addition, a processing element has direct connections to the neighbor processing elements. In general, a module could be as large as several processing elements, thus a new module design could extend over some routers in a DyNoC. One module needs just one route for communication. Thus, all routers covered by a module will be deactivated, except for the router at the upper right corner of the module. When a module finishes its work, the deactivated routers will be activated and reset to their initial states. A routing algorithm was also proposed for the DyNoC, which needs to handle. The obstacles encountered by a packet traveling from source to destination. The packets are routed around the obstacles which were created by large modules.

Pionteck, Koch, and Albrecht [2006] proposed a configurable network-on-chip, called CoNoChi that can support dynamic replacement of hardware

modules and dynamic insertion and removal of switches. Rana et al. [2008] proposed a dynamic topology by reconfiguring connection of circuits between switches. A lightweight circuit-switched architecture called programmable NoC (PNoC) was described in [Hilton & Nelson, 2005], which allows a system to reconfigure hardware modules and utilize parameterized routers to establish circuit switching connections.

OS-controlled networks-on-chip were proposed by some authors [Marescaux et al., 2003; Nollet et al., 2004]. Marescaux et al. [2003] used three different types of network, namely application data network, control network and reconfiguration network, to construct an NoC. Each network has its Network Interface Component (NIC). The OS can send Operation and Management (OAM) message on the control network to change the destination look-up table (DLT). Data NIC can support dynamic task relocation with changing destination look-up table. Data NIC is also responsible for monitoring communication resources per tile, such as the number of messages coming in and out of a specific tile. Data NIC gathers communication information in real time and sends it to the OS. Therefore, OS can manage communication resources per tile and guarantee the quality of service on a network. In addition, OS can control the maximum amount of messages an IP is allowed to send on the network per unit of time by an injection rate controller on Data NIC. However, the control network was implemented as a *shared bus* to limit resource usage and minimize latency. OS sends a control OAM message including payload data such as the content of a DLT and a command. For example, the Control NIC may read a command opcode such as UPDATE DLT and process it. The Control NIC also processes the statistics from Data NIC and communicates it to the OS. Control NIC is controlled by OS and can change the injection-rate windows on the Data NIC. Additionally, the control network suggests feasible partial reconfiguration processes. Control NIC implements a reset signal and bit masks to disable IP communication before partial reconfiguration to ensure the deletion of IP is not harmful for the system. After reconfiguration, the Control NIC clocks the IP. The reconfiguration network was implemented on a platform such as the Xilinx Virtex-II series with a native reconfiguration bus. The OS can drive the ICAP to request reconfiguration.

Nollet et al. [2004] connected an FPGA to a Compaq iPAQ PDA with a StrongARM processor by means of the iPAQ expansion port. The FPGA includes the slave processors, the NoC and the master processor interface component. They implemented a packet-switch 3×3 bidirectional mesh NoC as data NoC which is responsible for delivering data packets to PEs as data NoC. In addition, the control NoC is used for OS-control messages. The functionalities of control NIC and data NIC are described clearly by Marescaux et al. [2003]. Nevertheless, the master ISP runs the OS kernel. Not only monitoring the execution of the global system, the master OS also assigns tasks to the other slaves. Each slave OS has its own functionality. Besides, when the receiving PE is unable to process its input fast enough, the PE produces a blocked message. The master OS monitors the blocked message by polling the control NICs. With this information, the OS can decide the load of each PE and avoid link congestion. In addition, OS can change routing table to support adaptive routing. For a 3×3 network with 9 entries, changing a routing table requires on average 61 μs.

For reconfigurable NoC allows OS to reconfigure various hardware modules at runtime. Because of unknown running hardware modules, the traditional static FIFO-based Hw/Sw communication mechanism [Ces'ario et al., 2002; Hommais et al., 2001] is insufficient. Based on the above observations, few researches discussed Hw/Sw communication based on the reconfigurable NoC architecture. In this work, we focus on the Hw/Sw communication for reconfigurable NoC and propose an efficient communication mecha-

nism with shared memory to improve the overall system performance.

PRELIMINARIES

NoC is a promising communication architecture for future scalable SoCs. Though an ASIC-based NoC has higher performance, but it also suffers from a higher area overhead, lacks flexibility, has a higher production cost and a longer time-to-market. An FPGA device is more flexible and less costly than an ASIC. An FPGA device can now also support dynamic partial reconfiguration (DPR) to reduce reconfiguration time. To utilize the benefit of DPR in a scalable communication architecture, we will focus on the design of reconfigurable Network-on-Chip (RNoC). Most hardware IPs are designed to be attached to the buses in an existing communication architecture such as AMBA or CoreConnect. Thus for rapid system prototyping, our target system architecture will also include CoreConnect and access peripheral I/O devices. Although RNoC supports the communication among hardware modules attached to it, it lacks an efficient communication mechanism among software and hardware modules. Before introducing our proposed communication mechanisms for RNoC, we will describe the partial reconfiguration technology, the overview of NoC and the assumptions in the following.

Dynamic Partial Reconfiguration

Nowadays, FPGA devices such as the Xilinx Virtex family support dynamic partial reconfiguration. Based on the reconfigurability of FPGAs, a DPRS can be constructed. A DPRS allows changing applications at run time. Hardware circuits designed using the *early access partial reconfiguration* (EAPR) design flow, proposed by Xilinx [2006], are reconfigurable at run-time. As required by EAPR, a DPRS is partitioned into a static part and a reconfigurable part. The static part includes some elements such as a processor and some peripherals. The reconfigurable part is composed of reconfigurable elements such as application-specific processing elements. We call the reconfigurable elements *partial reconfigurable modules* (PRMs) that must be synthesized into partial bitstreams [2006] at compile time. FPGA resources that are dynamically configured by a *partial reconfigurable region* (PRR) constitutes a partial bitstream. Floor planning for PRRs can be made at design time using PlanAhead tool [2008]. Bitstreams for different reconfigurable elements can be loaded into a PRR at different times dynamically. The ICAP device realizes the run-time reconfiguration operation. All signals of PRMs, expect the clock signal, are connected to the static part through Busmacros for compatibility. The Busmacros are unidirectional, pre-placed, and pre-routed macros.

Overview of Network-on-Chip

Generally, *Network-on-Chip* (NoC) is composed of routers, links, and *processing elements* (PEs). In a circuit switching NoC, dedicated connection between PEs has to be established before a transmission. Though a circuit switching supports a high quality of service, it must wait for establishment time. In a packet switching NoC, data is sent as packets which include a header and payloads. The header records routing information. The payload contains messages from PEs. Each router connects to one PE through *Network Interface Component* (NIC), which is responsible for *packetization* and *depacketization* of messages from its corresponding PE. Router sends packets to correct destination through header information. The packet switching technologies can be classified *store-and-forward switching*, *virtual cut-through switching*, and *wormhole switching*. In store-and-forward switching, each router stores the whole packet and then sends the packet to the next router. Virtual cut-through switching allows routers to send packet to next router without stor-

ing the whole packet. The sizes of input FIFOs on routers in store-and-forward switching and virtual cut-throughput switching equal to the sizes of packets. In wormhole switching, packets are divided into small fragments called *flits*. The flits are always composed of one head flit and tail flits. Each router stores the whole flit and then sends the flit to the next router in wormhole switching. When a router serves a head flit, the router has to decide the direction of the whole packet. When the router serves tail flits, the router allows tail flits to follow the direction of the head flit. We assume that the size of the input FIFOs in a router with wormhole switching is equal to the size of a flit. This work is based on the infrastructure of the Hermes NoC [Moraes et al., 2004] as illustrated in Figure 7, which uses wormhole switching and has a mesh topology. The router in a Hermes NoC has four directions, called EAST, SOUTH, WEST, NORTH, and LOCAL. In this Thesis, the LOCAL direction connects a PRR. The router is composed of input FIFOs, a control multiplexer, and an arbiter. Every direction has one input FIFO to buffer input packets when the router is serving a head flit. The static xy routing algorithm which routes a packet in X direction first then Y direction is implemented in the control multiplexer. The arbiter decides which input packet is to be served. The size of a flit is eight bits. The packet size is flexible and the maximum size of one packet is five words, where one word consists of 16 bits. Thus, the maximum size of a packet is 80-bits. In our design, the first two flits record the destination address and the thread ID respectively. The thread ID is used to identify the software process, which is the owner of the packet. Figure 8 illustrates the packet format.

Hybrid Architecture for RNoC

The IBM CoreConnect bus architecture [1999] is adopted as the basic communication infrastructure in the Xilinx FPGA series. We built our target system with a Xilinx MicroBlaze processor,

Figure 7. Hermes NoC with reconfigurable Network-on-Chip

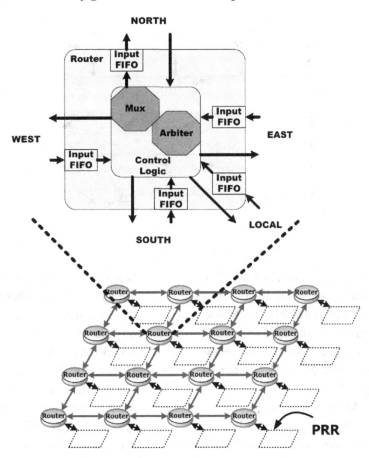

Figure 8. Packet format

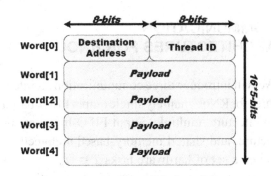

DDR2-SDRAM, and RS232-UAR connected by an OPB bus using the EDK toolkit [2008]. As shown in Figure 9, the *intellectual property interface* (IPIF) standard connection port was used to connect the RNoC to the same OPB bus

via a specific router. Each of the other routers is connected to a PRR. The OS4RNoC runs on the MicroBlaze processor and the filesystem is loaded into DDR2-SDRAM. We use the RS232-UART device as the standard input and standard output such that the debug messages can show on the screen. The OS4RNoC can access the RNoC by drivers which are implemented using the IPIF.

Basic Assumptions

To evaluate the performance and the cost of different communication architectures for RNoC, we need to make the following assumptions.

- **Hardware Application:** We suppose that a reconfigurable hardware application is

Figure 9. Connect RNoC to CoreConnect architecture

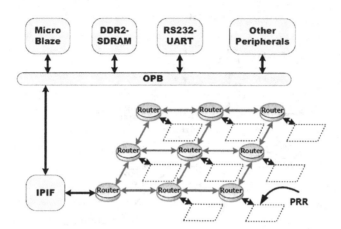

implemented by one or several PRMs. The intercommunication between PRMs in a hardware application is achieved by RNoC (hw to hw communication). It is assumed that the number of PRMs in a hardware application is less than the total number of PRRs in a target RNoC.

- **Hardware Task:** We model the hardware applications, to be configured on a RNoC, as a set of hardware tasks, where each task t has a set of attributes including start time $S(t)$, execution time $E(t)$, execution count $C(t)$, finished time $F(t)$, where $F(t) = S(t)+E(t) *C(t)$, and maximal required communication memory space $M(t)$ during $S(t)$ to $F(t)$. The total number of PRMs of concurrent hardware tasks is less than the total number of PRRs in a RNoC.
- **Lifetime of Hardware Tasks:** The lifetime of a hardware task (t) is the time between its start time ($S(t)$) and its finished time ($F(t)$).
- **Hardware Thread:** A hardware thread is responsible for communication between a software process and a hardware task. A hardware thread can be partitioned into two parts, namely the hardware writing thread and the hardware reading thread. The hardware writing thread sends data to RNoC through the RNoC driver, and the

hardware reading thread receives data from the RNoC driver. There is only one RNoC driver in a system. We schedule hardware threads to access the RNoC driver using round-robin scheduling.

- **Hardware Task Placement:** We assume placement decisions of hardware tasks are made in advance either by the user or the system. After placing a hardware task, the address on RNoC of the PRM corresponding to this hardware task is available to the OS4RNoC.

COMMUNICATION ARCHITECTURES FOR RNOC

We will introduce three communication architectures for RNoC, namely single output FIFO-based architecture, multiple output FIFO-based architecture, and shared memory-based architecture. Given a set of hardware tasks $T = \{t_1, t_2, ..., t_n\}$ with attributes (see section "Basic Assumptions"), we design three communication architectures for buffering the data from RNoC to OS4RNoC. We also analyze the performance and the area overhead for each of the three architectures.

Single Output FIFO-Based Architecture

The initial architecture we designed for RNoC was a single output FIFO-based architecture. The single output FIFO-based architecture is illustrated in Figure 10. In this architecture, only one output FIFO can be used by hardware tasks, thus there is no need for the OS4RNoC to consider communication memory allocation and management. After the OS4RNoC initialized a hardware task control block (HwTCB), configured and placed corresponding PRMs on RNoC, a hardware task starts to run. A HwTCB has two fields, namely the thread ID and the destination address as shown in Figure 11. The thread ID is the thread ID of a hardware thread which accesses this hardware task. The destination address records where the data of this hardware task to be sent. As described in Section 3.4, a hardware thread is responsible for communication between a software process and a hardware task. A hardware thread can be partitioned into two parts, namely a hardware writing thread and a hardware reading thread. Figure 11 illustrates that the input data of the hardware thread which thread ID is $0x00000003$ has to be sent to position (2,0) on RNoC. When a software process writes data to the hardware task, the hardware writing thread of this hardware task will invoke the RNoC driver with input data. The RNoC driver is responsible for writing data to the RNoC and receiving data from the RNoC. When the RNoC driver is invoked by a hardware writing thread, the RNoC driver writes the thread ID and the destination address in the HwTCB of that thread to the second and the first flits of a packet, respectively. After the RNoC driver writes the input data from the hardware writing thread to the payloads of the packet, the RNoC driver sends this packet to the RNoC. When a software process reads data from the hardware task, the hardware reading thread of this hardware task will also invoke the RNoC driver. In a single output FIFO-based architecture, the RNoC driver check the packet in the output FIFO. If the thread ID in the HwTCB is equal to the thread ID in the first packet header in the output FIFO, the RNoC driver will pop this packet and return the data to the hardware reading thread. If the thread ID is not the same, the hardware reading thread has to yield its right to read, and the RNoC driver does not pop any packet from output FIFO and waits for the next hardware reading thread to invoke.

Given a set of hardware tasks $T = \{t_1, t_2, ..., t_n\}$, the size of the output FIFO can be estimated in advance. The goal of designing the output FIFO is that the size of the output FIFO is large

Figure 10. FIFO-based hardware/software communication architecture for RNoC

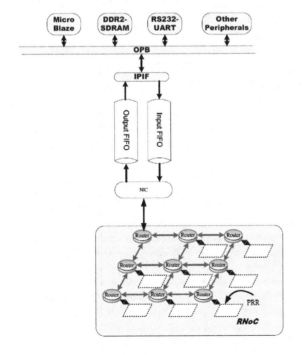

Figure 11. Hardware task control block

Thread ID	0x00000003
Destination Address	0x00000020

enough to hold the output data from all concurrently running hardware tasks at any execution time point. Supposed lifetime interval $I(t_i)$ is a time interval $[S(t_i), F(t_i)] = \{ x \in N \mid S(t_i) \leq x \leq F(t_i)\}$. We define a lifetime intervals graph as $LG=(V, E)$, where every vertex $t_i \in V$ represents a hardware task t_i and there exists an edge $e_{i,j} \in E$ if and only if $I(t_i) \cap I(t_j) \neq \varnothing$. A complete subgraph $LG' = (V', E')$ of LG, where $V' \subseteq V$ and $E' \subseteq E$ represents a subset of hardware tasks that are all executing concurrently and the lifetime interval $\cap I(t_k) \neq \varnothing$, where $t_k \in V'$. Thus, we can estimate the peak amount of memory required by the set of hardware tasks. A simple algorithm to calculate the maximum amount of buffer memory required for T is shown in Figure 12, which will be used as the size of the output FIFO. Though the single output FIFO-based architecture is easy to implement, it suffers from poor overall system throughput because of the out-of-order data sequence in the output FIFO when it is accessed in a round-robin fashion. To avoid this condition, an alternate communication mechanism, namely multiple output FIFO architecture can be adopted.

Multiple Output FIFO-Based Architecture

To decrease the influence of the output data sequence on system performance, we increase the number of output FIFOs by creating a multiple output FIFO-based architecture. Figure 13 illustrates such an architecture which consists of a multi-FIFO controller and an input buffer connected to OPB via IPIF and to RNoC via a NIC. A multi-FIFO controller is composed of four parts, namely a Hw/Sw communication table, a read controller, a write controller, and several output FIFOs. Given a set of hardware tasks $T = \{t_1, t_2, ..., t_n\}$, we increase the number of output FIFOs to the number of hardware tasks. As a result, each hardware task has a dedicated output FIFO for communication. There is no interleaving of data and thus the throughput is independent of data sequence. In a multiple output FIFO-based architecture, as shown in Figure 14, a HwTCB records the thread ID, the destination address, and the FIFO number. The hardware thread which thread ID is $0x00000003$ utilizes the FIFO number 1 to communicate. The OS4RNoC is responsible for

Figure 12. Algorithm to estimate the size of single output FIFO

```
input  : a set of Hw tasks: (T = {t₁,t₂,...,tₙ})
output : the maximum required memory: max_mem
1.1  begin
1.2  |   max_mem ← 0;
1.3  |   current ← 0;
1.4  |   start ← min{S(tᵢ), 1 ≤ i ≤ n};
1.5  |   end ← max{F(tᵢ), 1 ≤ i ≤ n};
1.6  |   for current = start to end do
1.7  |   |   tmp_mem ← 0;
1.8  |   |   foreach HwTask t do
1.9  |   |   |   if current ∈ I(t) then
1.10 |   |   |   |   tmp_mem ← tmp_mem + M(t);
1.11 |   |   |   end
1.12 |   |   end
1.13 |   |   if tmp_mem > max_mem then
1.14 |   |   |   max_mem ← tmp_mem
1.15 |   |   end
1.16 |   end
1.17 end
```

Figure 13. The multiple output FIFO-based architecture

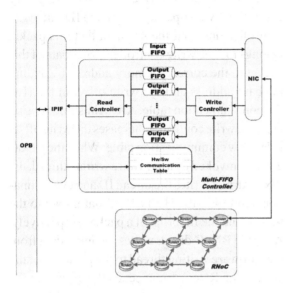

deciding a dedicated output FIFO for the hardware thread to communicate and updating the thread ID and the FIFO number in the HwTCB and Hw/Sw communication table, respectively. The Hw/Sw communication table consists of a mapping between thread IDs and FIFO numbers as illustrated in Figure 15. When a packet arrives at the multi-FIFO controller, the write controller sends this packet to its dedicated FIFO by comparing the thread ID in the packet header with the thread IDs in the Hw/Sw communication table. When a match is found, the corresponding FIFO number indicates the thread dedicated FIFO. When the RNoC driver is invoked by a hardware writing

Figure 15. The Hw/Sw communication table in the multiple output FIFO-based architecture

Thread ID	FIFO NO.
0x00000003	0x00000001
0x00000004	0x00000005
⋮	⋮

Figure 14. The Hw task control block in a multiple output FIFO-based architecture

Thread ID	0x00000003
Destination Address	0x00000020
FIFO Number	0x00000001

thread, the RNoC driver writes the thread ID and the destination address in the HwTCB of that thread to the second and the first flits of a packet, respectively. After the RNoC driver writes the input data from the hardware writing thread to the payloads of the packet, the RNoC driver sends this packet to the RNoC. When a software process reads data from the hardware task, the hardware reading thread of this hardware task will also invoke the RNoC driver. In a multiple output FIFO-based architecture, the RNoC driver sends the FIFO number in the HwTCB to the read controller, and then the read controller retrieves the data from the dedicated FIFO and sends them to the RNoC driver.

Because every hardware task in a multiple output FIFO-based architecture has a dedicated output FIFO, the size of the total memory required is equal to the sum of the communication memory required by each hardware task ($\sum_{i=1}^{n} M(t_i)$). The main drawback of multiple FIFO-based architecture is the high memory overhead due to each hardware task having a dedicated output FIFO. To overcome this drawback, we design the third communication architecture for RNoC, namely shared memory-based architecture.

Shared Memory-Based Architecture

In the shared memory-based architecture, we replace the output FIFOs with a shared memory as shown in Figure 16. The shared memory controller is composed of a read controller, a write controller, a Hw/Sw communication table, and a shared memory. The shared memory is divided into several continuous memory spaces called segments. The Hw/Sw communication table records the thread ID, the base address of a segment, the size of a segment, the read offset and the write offset as shown in Figure 17. The read offset represents the memory address offset in a segment where the read controller should read data from. The write offset represents the address offset in a segment where the write controller should write data into. Figure 18 illustrates that the HwTCB in a shared memory-based architecture records the thread ID, the destination address, and a base address of a segment. When a hardware task starts to execute, the OS4RNoC allocates a segment to this Hw task for communication by updating the thread ID, the base address, and the memory size and clearing the offset fields in the Hw/Sw communication table. The OS4RNoC also initializes a HwTCB

Figure 16. The shared memory-based architecture

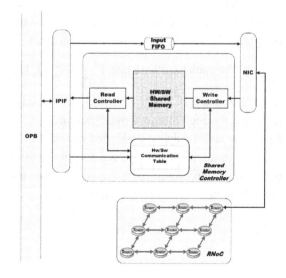

and writes the base address of the segment into the block. When a packet arrives at the shared memory controller, by comparing the thread ID, the write controller finds out the segment that this packet belongs to. The write controller writes data of this packet to the correct memory address by adding the base address and the write offset in the Hw/Sw communication table. After finishing writing data, the write controller increases the write offset in Hw/Sw communication table. When the RNoC driver is invoked by a hardware writing thread, the RNoC driver writes the thread ID and the destination address in the HwTCB of that thread to the second and the first flits of a packet, respectively. After the RNoC driver writes the input data from the hardware writing thread to the payloads of the packet, the RNoC driver sends this packet to the RNoC. When a software process reads data from the hardware task, the hardware reading thread of this hardware task will also invoke the RNoC driver. In a shared memory-based architecture, the RNoC driver sends the thread ID in the HwTCB to the read controller, and then the read controller will look up the Hw/Sw communication table, find the current address, and return data to the RNoC driver. After finishing reading data, the read controller increases the read offset in Hw/Sw communication table. When the write offset or the read offset is greater than the size of the segment, the write offset and the read offset will be cleared by the write controller and the read controller, respectively. The data flow in a segment is like a circular queue. When a hardware tasks terminates, the OS4RNoC will reallocate this segment.

Given a set of hardware tasks $T = \{t_1, t_2, ..., t_n\}$, we have to evaluate how many segments we should partition the shared memory into and decide the size of each segment. Suppose we have a list of segments $SL = \{S_1, S_2, ..., S_m\}$, where each segment s has a set of attributes including segment size $SSize(s_i)$ and a set of hardware tasks $HT(s_i)$ which utilize the s_i segment for communication. Hardware tasks with overlapping time intervals

Figure 17. The Hw/Sw communication table in a shared memory-based architecture

Thread ID	BaseAddr.	MemSize	Read Offset	Write Offset
0x00000003	0x0000000A	0x00000020	0x00000000	0x00000003
0x00000004	0x0000002B	0x00000020	0x00000002	0x00000006
⋮	⋮	⋮	⋮	⋮

can use the same segment. To reduce the memory overhead, we have to find a list *SL* of segments in shared memory such that it has the minimal number of segments and the minimum required communication memory space and can allow all hardware tasks to store output data. A clique partitioning approach [Cormen et al., 2001] can be used to determine the set of hardware tasks that could be allocated into the same segment. Recall that lifetime interval $I(t_i)$ is a time interval $[S(t_i), F(t_i)] = \{ x \in N \mid S(t_i) \leq x \leq F(t_i)\}$. Given a graph $G = (V, E)$, where every vertex $t_i \in V$ represents a hardware task t_i and there exists an edge $e_{i,j} \in E$ if and only if $I(t_i) \cap I(t_j) = \varnothing$. A complete subgraph $G' = (V', E')$ or clique of G, where $V' \subseteq V$ and $E' \subseteq E$ represents a subset of hardware tasks that are not executing concurrently and the lifetime interval $\cap\ I(t_k) = \varnothing$, where $t_k \in V'$. We have to find the minimal number of cliques, and to evaluate the minimum amount of required communication memory space. A heuristic algorithm, left-edge algorithm [Hashimoto & Stevens, 1988], can solve the clique partitioning problem. But the left-edge algorithm lacks the solution to the minimum communication memory space. We enhance the left-edge algorithm to find the minimal number of cliques and the minimum required communication memory space. The detailed algorithm is shown in Figure 19. Given a list of hardware tasks, the algorithm first sorts the hardware tasks in ascending order of the start times ($S(t)$) (line (2.4)). For each sublist of hardware tasks which have the same start times, the algorithm sorts the hardware tasks in descending order of the maximal required communication memory spaces ($M(t)$)

(line (2.6)). The algorithm makes several iterations. In each iteration, the algorithm creates a new segment (line (2.10)) and examines the lifetime intervals of hardware tasks from the first element to the last element in the sorted list (line (2.15)). The algorithm then assigns a hardware task to the segment if its lifetime interval does not overlap with the lifetime intervals of the already assigned hardware tasks to that segment (line (2.22~2.27)). If the required memory space of the newly assigned hardware task is greater than the segment size, the algorithm increases the segment size to the required memory space of the new assigned hardware task (line (2.33)). These assigned hardware tasks are then deleted from the sorted list of hardware tasks. The process is repeated until all hardware tasks have been assigned to segments and the required memory spaces is the sum of the segment sizes ($\sum_{i=1}^{m} SSize(s_i)$). According to the list of segments *SL*, we can implement a static memory management on the OS4RNoC.

Figure 18. Hw task control block in the shared memory-based architecture

Thread ID	0x00000003
Destination Address	0x00000020
Base Address	0x0000000A

Figure 19. The left-edge algorithm

```
     input  : a list of Hw tasks: (T = {t₁,t₂,...,tₙ})
     output: a segment list: SL, and total required memory size: memSize
2.1  begin
2.2  |  SL ← nil;
2.3  |  memSize ← 0;
2.4  |  Sort(T) in ascending order of the start times;
2.5  |  foreach sublist L ∈ T, where Hw tasks {tᵢ,...,tⱼ|1 ≤ i ≤ j ≤ n} ∈ L have the
     |  same start times do
2.6  |  |  Sort(L) in descending order of the communication memory spaces;
2.7  |  end
2.8  |  while T is not empty do
2.9  |  |  isInsert ← True;
2.10 |  |  New segment s;
2.11 |  |  SSize(s) ← M(t₁);
2.12 |  |  Insert t₁ into s;
2.13 |  |  Delete t₁ from T;
2.14 |  |  if T is not empty then
2.15 |  |  |  foreach Hw tasks t ∈ T do
2.16 |  |  |  |  foreach Hw tasks st ∈ HT(s) do
2.17 |  |  |  |  |  if I(t) overlaps I(st) then
2.18 |  |  |  |  |  |  isInsert ← False;
2.19 |  |  |  |  |  |  Break;
2.20 |  |  |  |  |  end
2.21 |  |  |  |  end
2.22 |  |  |  |  if isInsert = True then
2.23 |  |  |  |  |  if M(t) > SSize(s) then
2.24 |  |  |  |  |  |  SSize(s) ← M(t);
2.25 |  |  |  |  |  end
2.26 |  |  |  |  |  Insert t into s;
2.27 |  |  |  |  |  Delete t from T;
2.28 |  |  |  |  else
2.29 |  |  |  |  |  isInsert ← True;
2.30 |  |  |  |  end
2.31 |  |  |  end
2.32 |  |  end
2.33 |  |  memSize ← memSize + SSize(s);
2.34 |  |  Insert s into SL;
2.35 |  end
2.36 end
```

ARCHITECTURE EVALUATION AND EXPERIMENTS

We implemented the algorithms (see section "Communication Architectures for RNoC") to evaluate the communication overheads of the three communication architectures for RNoC. To evaluate the performance of the three different communication architectures for RNoC, we implemented parameterized processing elements to simulate real applications with different traffic patterns.

Evaluation of Communication Memory Overhead

To evaluate the communication memory overhead, suppose there are eight hardware tasks attached to a reconfigurable NoC. The total system evaluation time was set to 1000 cycles. With different fixed lifetime intervals, we generated chosen communication memory requirements from 1 to 128 bytes randomly and start time with uniform distribution from 0 to (1000 – lifetime interval) for the eight hardware tasks one hundred times. Figure 20 illustrates the evaluation results of the communication memory overheads for the architecture with normalized lifetime intervals. The horizontal axis represents the length of normalized lifetime intervals. The vertical axis represents the average communication memory overhead. With increasing length of lifetime interval, the average communication memory overhead of single output FIFO-based architecture and of shared memory-based architecture both increase. The communication memory overhead of multiple output FIFO-based architecture is the same at any length of lifetime intervals. When the length of lifetime interval is longer than 0.5, the lifetime intervals will always overlap, so the communication memory overheads of the three architectures are the same. Based on the requirements for lifetime intervals from 0.05 to 0.5 as illustrated in Table 1, the average communication memory overheads for the single output FIFO-based architecture, the multiple output FIFO-based architecture, and the shared memory-based architecture are 348.677 bytes, 514.969 bytes, and 379.033 bytes, respectively. From this experiment, we can observe that the average communication memory overhead for the multiple output FIFO-based architecture is 1.476 times the average communication memory overhead for the single output FIFO-based architecture. The average communication memory overhead for the shared memory-based architecture is 1.087 times the average communication memory overhead for the single output FIFO-based architecture.

In summary, the communication memory overhead of shared memory-based architecture is close to that of single output FIFO-based architecture. The multiple output FIFO based architecture needs high communication memory at any given length of lifetime intervals.

Performance Analysis

We performed performance experiment results on the Xilinx ML410 development board. ML410 has a Virtex-4 XC4VFX60 FPGA chip with 26624

Figure 20. Average communication memory overhead with fixed different lengths of lifetime intervals

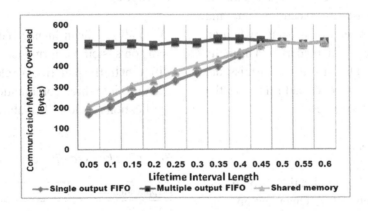

Table 1. Average communication memory overheads

Architectures	Memory Overhead (average in 0.05 ~0.5)	Normalized to Single FIFO- based
Single output FIFO-based	348.677	1
Multiple output FIFO-based	514.969	1.476
Shared memory-based	379.033	1.087

slices, 53248 flip flops, 53248 4-input LUTs. We configured a softcore processor, namely Xilinx MicroBlaze, into the FPGA chip to run the OS4RNoC. The operating frequency of MicroBlaze is 100MHz. We attached the RNoC to CoreConnect bus Architecture, ran OS4RNoC on the MicroBlaze, and use the IPIF to implement the RNoC driver to achieve hardware/software communication.

We implemented three different communication architectures for the RNoC. Because of the constraints of development tools, we used the parameterized processing elements (PEs) to simulate the behavior of reconfigurable PEs on the RNoC. We used the real applications attributes to perform our experiments. We also used the ModelSim simulator to get the execution cycles for each application. We compared the performance of different combinations of applications and different traffic patterns on the RNoC under the three different communication architectures. We suppose the router at position (0,0) is attached to the OS4RNoC for all traffic patterns. The examples include JPEG, AES-128, DES-64, 3-DES, and RSA-64. We evaluate the performance by measuring execution time and the read miss rate. A read miss occurs when an application software tries to read its data from the NoC output buffer but fails to do so. The miss rate is the ratio of the number of read misses to the total number of read accesses.

JPEG Application

In this example, the JPEG application is composed of three PEs, namely DCT, Quantization, and Encoder. The DCT takes 16 cycles to execute, 16 cycles to receive input data, and 32 cycles to transfer output data; the quantization takes 16 cycles to execute, 32 cycles to receive input data, and 16 cycles to transfer output data; the encoder takes 16 cycles to execute, 16 cycles to receive input data, and 4 cycles to transfer output data.

Encryption and Decryption Applications

The Advanced Encryption Standard-128 (AES-128) cipher has a key size of 128 bits. In this implementation, AES-128 takes an 8-byte data and an 8-byte key to calculate per iteration. A key expander is responsible to expand an 8-byte key into a 16-byte key. AES-128 takes 13 cycles to execute. Data Encryption Standard-64 (DES-64) takes an 8-byte data and an 8-byte key to calculate per iteration. DES-64 takes 48 cycles to execute. Triple DES applies the DES cipher algorithm three times to each data block. Triple DES takes an 8-byte data and three 8-byte keys to calculate per iteration. Triple DES takes 155 cycles to execute. RSA was proposed in 1978 by Ron Rivest, Adi Shamir, and Leonard Adleman at MIT. RSA is an algorithm for public-key cryptography. RSA takes an 8-byte data, an 8-byte exponent, and an 8-byte modulus to calculate per iteration. The execution cycle is different according to different exponent and modulus. In this example, we suppose RSA takes 790 cycles to execute.

Analysis

We analyzed the influence of different traffic patterns on system performance. We instanced nine different traffic patterns. We classified Pattern1 to Pattern6 into three combinations. Table 2 shows applications as described above in ascending or-

Table 2. The execution cycle of each application

Application Name	Execution Cycles
AES-128	13
DES-64	48
Triple DES	155
DCT	64
Quantization	64
Encoder	36
JPEG (DCT, Quantization, Encoder)	164
RSA	790

Figure 21. The combinations and patterns

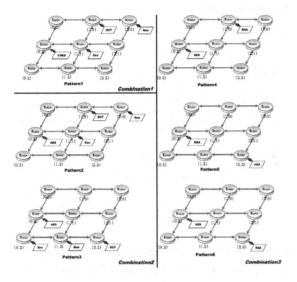

Figure 22. The execution time of Pattern1 to Pattern6

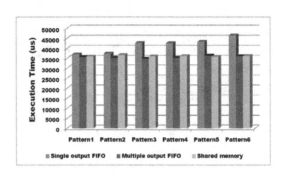

der of execution cycle. Combination 1 includes a JPEG application and a Triple DES application. Combination 2 includes a JPEG application and an AES-128 application. Combination 3 includes an AES-128 application and an RSA application. The variance in execution cycle of applications is the smallest in Combination 1 and is the largest in Combination 3. The placement pattern of each combination is shown in Figure 21. The input data size of each application is 4096 bytes. Figure 22 shows the execution time of Pattern1 to Pattern6. Along with the variance in execution cycles of concurrent applications increasing, single output FIFO-based architecture suffers from a poor performance because of the out-of-order output data. In Pattern6, the execution time on the multiple output FIFO-based architecture is lesser by 22.6% than on the single output FIFO-based architecture, and the execution time on the shared memory-based architecture is lesser 22.4% than on the single output FIFO-based architecture.

In the second experiment, we increased the number of concurrent applications. Pattern7 includes three applications, Pattern8 includes four applications, and Pattern9 includes five applications. The traffic patterns for these three

patterns are shown in Figure 23. Figure 24 shows the execution time of Pattern7 to Pattern9. We observe the execution time is the longest on the three architectures in Pattern9. In Pattern8, the multiple output FIFO-based architecture demonstrates a performance speedup of 1.789 times compared to that of the single output FIFO-based architecture. The shared memory-based architecture demonstrates a performance speedup of 1.748 times compared to that of the performance of single output FIFO-based architecture.

Figure 25 shows the miss rate of each pattern on the single output FIFO-based architecture. The miss rates of Pattern1 and Pattern2 are both zero. The miss rate of Pattern9 is 50%. A higher miss rate indicates that a large portion of the read operations are unsuccessful. Thus, the performance of Pattern 9 is the worst because of the highest miss rate.

CONCLUSION

We implemented three communication architectures for reconfigurable NoC, and compared their communication memory requirements and their performance. Experiment results show that the multiple output FIFO-based architecture has a performance speedup of 1.789 times compared to the single output FIFO-based architecture, but it suffers from a high communication memory

Figure 23. The patterns of Pattern7, Pattern8, Pattern9

Figure 24. The execution time of Pattern7, Pattern8, and Pattern9

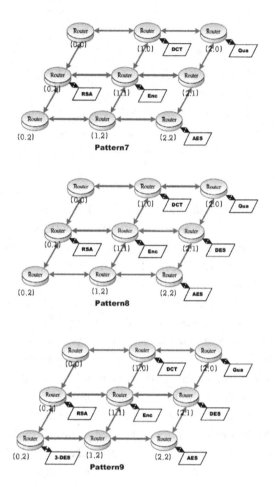

Pattern7

Pattern8

Pattern9

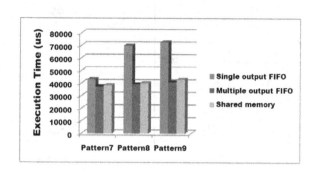

overhead. The shared memory-based architecture has a performance speedup of 1.748 times, while the average communication memory overhead is only 1.087 times the average communication memory overhead of the single output FIFO-based architecture. Thus, shared memory-based architecture exhibits performance as high as the multiple FIFO-based architecture and require memory as small as that of the single FIFO-based architecture. In this research, we supposed that only one specific router at position (0,0) serves the communication between operating system and RNoC. In the future, we can change the position of the specific router on a RNoC and increase the number of specific routers. For different positions of specific routers and the number of specific

Figure 25. The missrate of patterns on the single output FIFO-based architecture

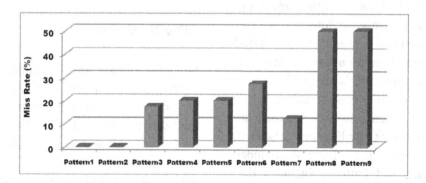

routers, the hardware tasks placement algorithm has to be decided in different strategies to meet the quality of services on RNoC.

REFERENCES

BEE2. (2008). *BEE2- Berkeley Emulation Engine 2*. Retrieved from http://bee2.eecs.berkeley.edu/

Benini, L., & Micheli, G. D. (2002, January). Networks on Chips: a new SoC paradigm. *IEEE Computer, 35*(1), 70–78.

Bobda, C., Ahmadinia, A., Majer, M., Teich, J., Fekete, S. P., & Veen, J. (2005, August). DyNoC: A dynamic infrastructure for communication in dynamically reconfigurable devices. In *Proceedings of the 15th IEEE International Conference on Field Programmable Logic and Applications*, (pp. 153–158).

Ces'ario, W., Baghdadi, A., Gauthier, L., Lyonnard, D., Nicolescu, G., Paviot, Y., et al. (2002, June). Component-based design approach for multicore SoCs. In *Proceedings of the 39th Design Automation Conference*, pages 789–794. ACM Press.

Claus, C., Zhang, B., Huebner, M., Schmutzler, C., & Becker, J. (2007, May). An XDL-based busmacro generator for customizable communication interfaces for dynamically and partially reconfigurable systems. In *Proceedings of the Workshop on Reconfigurable Computing Education at ISVLSI 2007*.

CoreConnect. (1999). *CoreConnect Bus Architecture*. Retrieved from http://www-01.ibm.com/chips/techlib/techlib.nsf/literature/CoreConnect_Bus_Architecture

Cormen, T. H., Leiserson, C. E., Rivest, R. L., & Stein, C. (2001). *Introduction to Algorithms* (2nd ed.). Cambridge, MA: The MIT Press.

Duato, J., Yalamanchili, S., & Ni, L. M. (1997, January). *Interconnection Networks: an engineering approach*. San Francisco, Morgan Kaufman.

Embedded linux/microcontroller project (2008). *Embedded linux/microcontroller project*. Retrieved from http://www.uclinux.org/

G¨otz, M., & Dittmann, F. (2006, September). Reconfigurable microkernel-based RTOS: mechanisms and methods for run-time reconfiguration. In *Proceedings of the IEEE International Conference on Reconfigurable Computing and FPGAs*, (pp. 1–8).

Hashimoto, A., & Stevens, J. (1988). Wire routing by optimizing channel assignments within large apertures. In Proceedings of 25 years of DAC: Papers on Twenty-five years of electronic design automation, (pp. 35–49).

Hilton, C., & Nelson, B. (2005, Auguest). A flexible circuit-switched NOC for FPGA-based systems. In *Proceedings of the 15th IEEE International Conference on Field Programmable Logic and Applications*, (pp. 191–196).

Hommais, D., P´etrot, F., & Aug´e, I. (2001, June). A tool box to map system level communications on Hardware/Software architectures. In *Proceedings of the 12th International Workshop on Rapid System Prototyping*, (pp. 77–83).

ITRS. (2008). ITRS. In *Proceedings of the Winter Public Conference Presentations*. Retrieved from http://www.itrs.net/Links/2008Winter/2008WinterPresentationsITWG/Presentations.html

Labrosse, J. J. (2002). *MicroC OS II: The Real Time Kernel*. San Francisco: CMP Books.

Marescaux, T., Mignolet, J.-Y., Bartic, A., Moffat, W., Verkest, D., Vernalde, S., & Lauwereins, R. (2003, September). Networks on chip as hardware components of an OS for reconfigurable systems. In *Proceedings of the 13th International Conference on Field Programmable Logic and Applications* [LNCS]. *Lecture Notes in Computer Science, 2778*, 595–605.

Moraes, F. G., Calazans, N., Mello, A., Möller, L., & Ost, L. (2004). HERMES: an infrastructure for low area overhead packet-switching networks on chip. *VLSI Integration, 38*(1), 69–93. doi:10.1016/j.vlsi.2004.03.003

Nollet, V., Marescaux, T., Verkest, D., Mignolet, J.-Y., & Vernalde, S. (2004, June). Operating system controlled network-on-chip. In *Proceedings of the IEEE/ACM Design Automation Conference*, (pp. 256–259). ACM Press.

Ogras, Ü. Y., Marculescu, R., Lee, H. G., Choudhary, P., Marculescu, D., Kaufman, M., & Nelson, P. (2007, September). Challenges and promising results in NoC prototyping using FPGAs. *IEEE Micro, 27*(5), 86–95. doi:10.1109/MM.2007.4378786

PetaLogix Developer Portal. (2009). *UserGuide - PetaLogix Developer Portal.* Retrieved from http://developer.petalogix.com/wiki/UserGuide

Pionteck, T., Koch, R., & Albrecht, C. (2006, August). Applying partial reconfiguration to networks-on-chips. In *Proceedings of the 16th IEEE International Conference on Field Programmable Logic and Applications*, (pp. 1–6).

Platform studio and the EDK (2008). *Platform studio and the EDK.* Retrieved from http://www.xilinx.com/ise/embedded design prod/platform studio.htm

Rana, V., Atienza, D., Santambrogio, M. D., Sciuto, D., & De Micheli, G. (2008, October). A reconfigurable network-on-chip architecture for optimal multi-processor SoC communication. In *Proceedings of the 16th IFIP/IEEE International Conference on Very Large Scale Integration*, (pp. 321–326). New York: Springer Press.

Santambrogio, M. D., Rana, V., & Sciuto, D. (2008, September). Operating system support for online partial dynamic reconfiguration management. In *Proceedings of the 18th International Conference on Field Programmable Logic and Applications*, (pp. 455–458).

So, H. K.-H., & Brodersen, R. W. (2008, February). A unified hardware/software runtime environment for FPGA-based reconfigurable computers using BORPH. *ACM Transactions on Embedded Computing Systems, 7*(2), 14. doi:10.1145/1331331.1331338

Williams, J. A., & Bergmann, N. W. (2004, June). Embedded Linux as a platform for dynamically self-reconfiguring Systems-on-Chip. In *Proceedings of International Conference on Engineering of Reconfigurable Systems and Algorithms*, (pp. 163–169). CSREA Press.

Xilinx. (2006). Early access partial reconfiguration user guide. *ISE 8.1.01i.*

Xilinx (2008). *Xilinx: PlanAhead.* Retrieved from http://www.xilinx.com/ise/optional prod/planahead.htm

Zhou, B., Qiu, W., & Peng, C. (2005, September). An operating system framework for reconfigurable systems. In *Proceedings of the 15th International Conference on Computer and Information Technology*, (pp. 781–787). IEEE Computer Society Press.

Section 2
Design Methods for
Reconfigurable NoC Design

Chapter 5
Design Methodologies and Mapping Algorithms for Reconfigurable NoC–Based Systems

Vincenzo Rana
Politecnico di Milano, Italy

Marco D. Santambrogio
Politecnico di Milano, Italy

Alessandro Meroni
Politecnico di Milano, Italy

ABSTRACT

This chapter describes in details the different approaches and design methodologies that can be employed in order to create reconfigurable Network-on-Chip-based systems. The target architecture can be mainly defined either as a homogeneous or as a non-homogeneous grid of tiles. Furthermore, in addition to these architectures, it is also possible to identify a regular non-homogeneous solution, which is a sort of mix of the previous two. A second distinction can be done based on the reconfiguration capabilities that the target system can support. In particular, by using one of the previously introduced architectures, it is possible to develop a reconfigurable system, based on the NoC paradigm, in which the communication infrastructure, the mapping of the computational cores or both can be dynamically configured at run-time.

INTRODUCTION

Nowadays, most of the reconfigurable embedded systems are designed for Field Programmable Gate Arrays (FPGAs) devices. An FPGA is an integrated circuit that can be programmed after it has been manufactured. FPGAs are similar in principle to, but have a vastly wider potential application than, programmable read-only memory (PROM) chips. FPGAs are used by engineers in the design of specialized ICs that can later be produced hard-wired in large quantities for distribution to computer manufacturers and end users. In addition to this, FPGAs can also be used to create systems which hardware

DOI: 10.4018/978-1-61520-807-4.ch005

configuration can be dynamically changed while they are up and running.

The first step that is necessary to perform, in order to create an FPGA-based reconfigurable embedded system, is the partitioning of the physical device in a set of disjoint regions. The shape and the distribution of these regions over the physical device represent the underlying reconfigurable architecture on which the system can be developed. The first part of this chapter will be devoted to the analysis of the different choices that can be made in order to create an optimal reconfigurable architecture for NoC-based embedded systems.

Once the underlying architecture has been defined, it is necessary to further split, as described in the second part of this chapter, the set of regions into two subsets: static regions and reconfigurable regions. In this way it is possible to define the components of the final system that will be considered as static and the ones that will be considered as reconfigurable. At run-time, in fact, it will be possible to modify the content of each reconfigurable region, while the static ones will remain fixed for the whole life of the system. This choice will deeply influence both the flexibility and the performance of the final system, in addition to the complexity of the design. A first design solution consists in configuring at run-time only the elements of the network (the switches and their interconnections), while maintaining fixed the location of the computational cores. A second solution is the choice to dynamically change at run-time the mapping of the computational cores on the communication infrastructure, which is considered as fixed. These two opposite solution can also be combined together in order to obtain a very high level of flexibility and adaptability, even if it considerably increases the time required to finalize the design of the system, its complexity and the run-time reconfigurations management task.

BACKGROUND

FPGAs Reconfiguration

FPGAs configuration capabilities allow a great flexibility in hardware design and, as a consequence, they make it possible to create a vast number of different reconfigurable systems. These can vary from systems composed of custom boards with FPGAs, often connected to a standard PC or workstation, to standalone systems including configurable logic block (CLB) and general purpose processors (GPP), to System-on-Chip's, completely implemented within a single FPGA mounted on a board, with only few physical components for I/O interfacing.

The easiest way in which an FPGA can be reconfigured is called *complete*. In this case the configuration bitstream, containing the FPGA configuration data, provides information regarding the complete chip and it configures the entire FPGA, that is why this technique is called complete. With this approach there are no particular constraints that have to be taken into account during the reconfiguration action. The main disadvantages of an approach based on the complete reconfiguration technique are the overhead introduced into the computation by the reconfiguration and the fact that is necessary to stop the system while a complete reconfiguration process is being performed. In order to cope with this situation a *partial* reconfiguration approach has been proposed. Partial reconfiguration is useful for applications that require the load of different designs into the same area of the device or the flexibility to change portions of a design without having to either reset or completely reconfigure the entire device.

There are different models of reconfiguration that can be classified according to the following scheme (an extension of the one proposed in (John Williams, Neil Bergmann (2004))):

- *Who* controls the reconfiguration;
- *Where* the reconfiguration unit (called *reconfigurator*, the element that is in charge of reconfiguring the FPGA) is located;
- *When* the configurations are generated;
- *Which* is the granularity of the reconfiguration;
- In *what* dimension the reconfiguration operates.

The first division (*who* and *where*) is between external and internal reconfiguration. In the first scenario, the reconfiguration is managed by an external entity, usually a PC, or by a dedicated processor. Internal reconfiguration, instead is performed completely within the FPGA boundaries; for this to be possible, the device must have a physical component dedicated to reconfiguration, such as the ICAP component in Xilinx FPGAs. It is important to stress that a scenario can be truly termed internal only when both the controller and the element in charge of physically realizing the reconfiguration, are placed inside the FPGA.

The generation of the configurations (*when*) can be done in a completely static way (at design time) by determining all the possible configurations of the system. Each module must be synthesized and all possible connections between modules and the rest of the system must be considered. Other possibilities are run-time placement of pre-synthesized modules, which requires dynamic routing of interconnection signal, or completely dynamic modules generation. This last option is currently impracticable, since it would require run-time synthesis of modules from VHDL (or other hw description language) code – a process requiring prohibitive times in an online environment.

Reconfiguration can take place at very different granularity levels (*which*), depending on the size of the reconfigured area. Two typical approaches are smallbits and module based: the first one consists in modifying a single portion of the design, such as single Configurable Logic Blocks (CLB) or I/O blocks parameters (as described in (Vince Ech, Punit Kalra, Rick LeBlanc, Jim McManus (2003))), while the second one involves the modification of a larger FPGA area by creating hardware components (modules) that can be added and removed from the system: each time a reconfiguration is applied, one or more modules are linked or unlinked from the system.

The last property is the dimension (*what*). One can distinguish between two different possibilities: mono-dimensional (1D) and bi-dimensional (2D) reconfiguration (Xilinx Inc. (2006)). For current FPGA devices, data is loaded on a column-basis, with the smallest load unit being a configuration bitstream frame, which varies in size based on the target device. Active partial reconfiguration of Xilinx Virtex devices, or simply partial reconfiguration, is accomplished in either slave SelectMAP mode or Boundary Scan, JTAG mode. Instead of resetting the device and performing a complete reconfiguration, new data is loaded to reconfigure a specific area of the device, while the rest of the device is still in operation. In a truly 2D reconfiguration it is possible to reconfigure an arbitrary portion of the FPGA without affecting the execution of the rest of the implementation. Most FPGAs, instead, require that in order to reconfigure a portion of a column of reconfigurable cells the whole column must stop its operations – the reconfiguration can act on a specific 2D portion of the column, but it still affects the execution of the whole column.

Target Architecture

Considering the partial dynamic reconfiguration paradigm applied with the EAPR flow (Xilinx Inc. (2006)), a suitable target architecture has to be defined as composed by two different regions representing a *static area* and a *reconfigurable area*. The communication infrastructure will be placed in the reconfigurable area and busmacros will ensure reliable communication channels between the two areas. In addition to this, the *reconfigurable area* of target FPGA-based architecture needs to be further divided. In particular, it has to be partitioned into smaller rectangular regions called *tiles*.

On the one hand, the static area of the architecture will be filled with both all the I/O interfaces[1] and all the components of the system that have to be always present in the final design. On the other hand, the reconfigurable regions will be filled up with either computational cores (master or slave components with their network interfaces) or communication elements (such as NoC switches). In this way the *communication layer* and the *computational layer* can be completely decoupled.

Furthermore, all the communication channels between the *static part* and the *reconfigurable part* and among the *reconfigurable regions* can be guaranteed to be reliable, even during reconfiguration processes, by means of *busmacros*, that are placed along each edge of each *reconfigurable region*, as described in (Xilinx Inc. (2004)) and (Xilinx Inc. (2006)).

Obviously, the reconfigurable regions can be configured at run-time with modules that share the same interface, which has to be defined at design time accordingly to the characteristics of the target system.

1D and 2D Placement and Reconfiguration Constraints

A *placement constraint* can be considered as a constraint used to identify the area where a functionality has to be mapped. In such a context it is possible to refer to 1D or 2D placements according to the fact that the height of each computational core, used to specify the desired system, is equal to the height of the reconfigurable device. If computational cores do not require the entire height of the reconfigurable device (column), we can refer to a 2D placement scenario.

Considering reconfiguration, it is possible to refer to a 1D reconfiguration scenario if the basic reconfigurable unit has a height equal to the height

of the reconfigurable device, on the other case we can speak of a 2D scenario. Therefore it is possible to define as a reconfiguration constraint the kind of reconfiguration, 1D or 2D, that we are allowed to use within a specific context.

Considering Xilinx devices such as Virtex (Xilinx Inc., 2003), Virtex II and Virtex II Pro (Xilinx Inc., 2007, *Virtex-II Pro and Virtex-II ProX Virtex-II Pro and Virtex-II Pro X FPGA User Guide*), it is not possible to define two reconfigurable regions that share two different parts of the same column. It is compulsory to assign a column to a single reconfigurable region: this observation explains why, even using a 2D placement, it is not possible to perform a 2D reconfiguration. In the newest Xilinx devices such as Virtex 4 (Xilinx Inc., 2007, *Virtex-4 user guide*) and Virtex 5 (Xilinx Inc., 2007, *Virtex-5 user guide*; Xilinx Inc., 2007, *Virtex-5 configuration user guide*) FPGAs, it is possible to configure regions characterized by both a 1D or a 2D placement, using a 1D or 2D reconfiguration respectively, providing the designers with a greater flexibility in choosing the best floorplanning for their applications.

Relocation Technique

Relocation is a powerful technique that can reduce the amount of memory required to store the partial reconfiguration bitstreams (Ferrandi, F.; Morandi M.; Novati M.; Santambrogio M. D.; Sciuto D. (2006); Morandi M.; Novati M., Santambrogio M. D.; Sciuto D. (2008)). A bitstream is a file able to configure a specific computational core in a specific location of the programmable device. It is important, at design time, to assign to each module a placement on the physical device. Unfortunately, due to runtime conditions, it might happen that the area assigned to a computational core can be already occupied by a different component. In such a scenario, without the relocation, it is not possible to serve the request for the desired computational core until the already mapped one ends its execution. This stall will worsen the system's performance. In such a context, introducing *relocation* as a solution can provide interesting results. *Relocation*, in fact, makes it possible to dynamically change a run-time a configuration bitstream in order to configure the same computational core in a different location. It is important to note that this modification of the bitstream does not require a re-synthesis, since what changes is only the position of the core and not its physical implementation.

Homogeneity

In order to use the relocation technique in a reconfigurable architecture it is first necessary to introduce the *homogeneity* concept. We can face two different kinds of *homogeneity*: *platform homogeneity*, and *placement constraints homogeneity*. The former refers to the possibility of identifying homogeneous regions inside a reconfigurable device according to its physical implementation. Xilinx FPGAs are composed of different elements i.e. it is possible to find CLBs, I/O Blocks, BRAM Blocks, Multipliers in a generic Virtex II pro device. This situation leads to the identification of homogeneous regions i.e., portions of the device repeated inside the device itself e.g., BRAM Blocks columns. *Placement constraints homogeneity* refers to the situation where the designer identifies the homogeneous regions. Two CLBs are homogeneous by definition, by nature, while the decision to group sets of CLBs to define an area, that can be identified several times, with the same composition, inside the device, is a volunteer placement solution that can be used to force a sort of *placement constraints induced homogeneity*.

2D Reconfigurable Homogeneous and Non-Homogeneous Architectures

When designing a reconfigurable NoC-based system, the first distinction can be found in the kind of architecture utilized to support dynamic reconfiguration capabilities. In particular, it is possible to exploit either a homogeneous grid of tiles or a non-homogeneous one, as shown in Figure 1 for a 2D reconfiguration scenario.

In particular, each one of these tiles can be defined as the minimum amount of adjacent configurable resources required to implement a given computational core layout. Usually, these tiles are defined as rectangular portions of the device, even if it is also possible to use any shape that can be obtained as a sum of rectangular areas.

2D Reconfigurable Homogeneous Approach

The first approach that will be analyzed is the homogeneous one, shown in Figure 1 (a), which

Figure 1. Homogeneous (a) and non-homogeneous (b) approaches

a)

b)

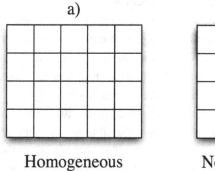
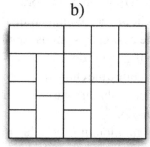

Homogeneous   Nonhomogeneous

is characterized by a very high level of flexibility, since each component of the system can be placed in any reconfigurable region (that is a single tile of the previously introduced grid). In addition to this, the homogeneous approach makes it possible to fully exploit the benefits of the *relocation technique*, thus increasing the overall timing performance and decreasing the amount of memory required to store the partial bitstreams. As shown in Figure 2, in fact, the computational cores (A, B, C, D, E, F, G and H) shown in (b) can be easily relocated to (d), since the subset of tiles used in (a) are completely compatible with others sets of tiles with the same shape (another matrix with the same number of columns and rows), such as the ones used in (d).

The main drawback of this approach arises when the difference between the area usage of the largest computational core and the area usage of the smallest one is not negligible, as shown in Figure 2 (a); in this case, each reconfigurable region has to be defined in order to hold the biggest computational core, so, when a smaller one (e.g., a single switch) is configured, part of the selected reconfigurable region is wasted. In Figure 2 (b) and (d) it is possible to see that the area usage, with the previously described approach, is around 46% of the whole device (11 tiles used over the 24 available tiles). It is possible to overcome to this limitation by defining a reconfigurable region as a portion of the device able to hold an island of computational cores, rather than a single computational core, as shown in Figure 2 (c) and (e). In this way, it is possible to compact all the computational cores in a certain number of islands characterized by a similar size, thus considerably

Figure 2. A set of computational cores and switches (a) deployed on a homogeneous grid of tiles with (c) and without (b) the definition of islands of computational cores. The same set of computational cores and switches can be relocated on another portion of the device (d) (e), since the target architecture is completely homogeneous

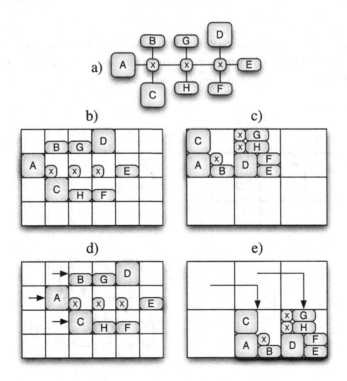

decreasing the area wasting. In the example shown in Figure 2, since the area usage of (c) and (e) is around 33% of the whole device (2 tiles used over the 6 available tiles), a reduction of around 13% of the area usage is obtained thanks to the definition of islands of computational cores. Obviously, if a single core of a particular island has to be exchanged at run-time with another core, the whole reconfigurable region corresponding to the island has to be reconfigured, since it is not possible to configure only a portion of a single reconfigurable region.

2D Reconfigurable Non-Homogeneous Approach

A second approach is the non-homogeneous one, which makes it possible to specifically define different sets of reconfigurable regions to hold a particular subset of computational cores. Thus, this approach makes it possible to tune the size of the each reconfigurable region in order to reduce the wasting of reconfigurable resources, even if this choice considerably limits the possibilities

to exploit the *relocation technique*. Figure 3 (a) shows how the same set of computational cores and switched presented in Figure 2 (a) can be deployed on a non-homogeneous grid of tiles, reducing the area usage to around the 19% of the whole device, but making it impossible to exploit the relocation technique, since there is not, in another portion of the same device, another set of tiles completely compatible with the ones used to hold the set of computational cores and switches in Figure 3 (a).

A very interesting scenario is the regular non-homogeneous approach in which just two different sets of reconfigurable regions are defined: one for the computational cores (or for the islands of computational cores) and one for the network switches. This approach is shown in Figure 3 (b) and is characterized by an area usage of 44% with respect to the whole device. In this approach it is possible to exploit the *relocation technique*, as shown in Figure 3 (c), area wasting can be kept low due to the optional use of islands of computational cores and the level of flexibility is high, since only two kinds of reconfigurable regions have been defined.

Figure 3. A set of computational cores and switches deployed on a non-homogeneous (a) and on a regular non-homogeneous (b) grid of tiles. The same set of computational cores and switches can be relocated on another portion of the device (c), in the regular non-homogeneous approach

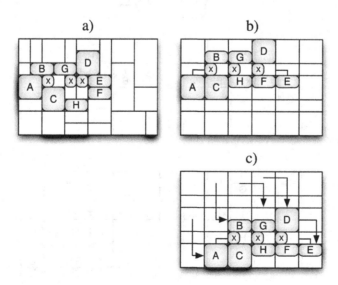

Comparison between the Homogeneous and the Non-Homogeneous Approach

In order to choose which is the best solution, between the homogeneous and the non-homogeneous approach, that has to be followed to implement a target NoC-based embedded system, it is necessary to evaluate a wide set of characteristics, such as the area usage of the whole system, the amount of available memory to store the partial bitstreams, the level of re-use of the architecture that the designer wants to achieve and the particular internal structure of the system itself.

Area Usage of the System

In order to reduce the amount of configurable resources needed to implement the target system, it is possible to reduce the area wasting due to the non-optimal utilization of the tiles. In order to reduce this area overhead, the best solution would be to employ a non-homogeneous approach, in which each tile is shaped on the computational core that it has to hold. As shown in the previous section, this is the best approach from the area overhead point of view, even if good results can be obtained also with the homogeneous or regular non-homogeneous approach that exploit the concept of islands of computational cores. Finally, the regular non-homogeneous approach that does not use islands of computational core makes it only possible to avoid area wasting due to the different size of computational cores and network switches, while the homogeneous approach, without islands, is by far the worst solution from this point of view.

Available Memory

One of the main characteristics that it is necessary to evaluate in order to select the most appropriate approach is the amount of available memory on the target device. This memory will be used to store all the partial bitstreams, each one able to configure a particular computational core of the system in a determined location on the physical device. If the amount of available memory is very limited, it is necessary to employ the relocation technique, in order to make it possible to store a single bitstream for each computational core of the system, independently from the location where it will be configured at run-time; the bitstream, in fact, will correspond to a single location on the physical device, but, as previously described, the relocation technique will make it possible to derive all the other bitstreams (all related to the same computational core, but each one able to configure it on a different specific location) at run-time.

For this reason, if the amount of memory of the target device is not sufficient in order to hold all the partial bitsreams, the homogenous approach has to be preferred to the non-homogeneous approach. It is very important to note that the regular non-homogeneous approach behaves exactly like the homogeneous approach from the point of view of the memory usage, since all the computational cores are configured at run-time on the same subset of tiles that are characterized by the same shape. The only drawback of the regular non-homogeneous approach is that the number of portions of the device where it is possible to relocate a computational core is smaller with respect to the homogeneous approach.

Table 1 presents the number of partial bitstreams necessary to configure each reconfigurable core on each reconfigurable region. This number is directly proportional to the memory usage, since all these partial bitstreams have to be stored in order to be used at run-time when required.

Level of Re-Use and Adaptability

On the one hand, the level of re-use represents the aptitude of a developed system to be used as the starting point for a new system with different requirements. On the other hand, the level of

Table 1. Number of partial bitstreams needed when applying the homogeneous and the non-homogeneous approaches with a variable number of reconfigurable cores and reconfigurable regions

Number of cores	Number of reconfigurable regions	Homogeneous approach	Non-homogeneous approach
1	1	1 partial bitstreams	1 partial bitstreams
10	1	10 partial bitstreams	10 partial bitstreams
1	10	1 partial bitstreams	10 partial bitstreams
100	10	100 partial bitstreams	1000 partial bitstreams
10	100	10 partial bitstreams	1000 partial bitstreams
100	100	100 partial bitstreams	10000 partial bitstreams

adaptability represents the aptitude of a developed system to be modified at run-time in order to cope with new upcoming requirements, for instance due to the changing working scenario.

If the designer wants to achieve a high level of re-use and adaptability, a homogeneous approach has to be preferred, since it provides the final system with a very high level of flexibility, which makes it possible to easily modify the location of the computational cores and of the network switches (since all the tiles are equivalent). The same consideration is valid also for the regular non-homogeneous approach, since each computational core can be easily moved to another location within the subset of tiles that share the same shape. On the other side, a non-homogeneous approach makes it possible to optimize the underlying grid of tiles to a specific configuration of the system. This optimization drastically reduces the degree of freedom in which the system can be modified at run-time, thus considerably decreasing the level of re-use and adaptability of the developed system.

Internal Structure of the System

By analyzing the internal structure of the system it is possible to obtain useful information to drive the selection of the approach to follow to create the grid of tiles. If the system, for instance, is composed by clusters of computational cores connected to a single switch (the switches can be connected among them to form any topology, such as ring, star, mesh, etc...), either a homogeneous or a non-homogeneous approach based on islands of computational cores should be the best solutions, since they exploit the intrinsic structure of the system under development.

On the opposite, if the system consists of a mesh of switches, each one connected to a single computational core, packing the computational cores into islands could reduce the performance of the system if it is not possible to group together the computational cores belonging to the same portion of the mesh; in particular, if it is necessary, in order to create islands of the same size, to pack together computational cores belonging to very distant location of the mesh, it will not be possible (or, at least, it will be very hard) to connect the islands among them respecting all the timing constraints of the original mesh.

DESIGN METHODOLOGIES

In addition to the kind of architecture used to support dynamic reconfiguration capabilities, a second distinction can be performed on the components of the system that can be dynamically reconfigured. To better understand how this feature deeply influences the design of the whole system, this section will be focused on the description of the different approaches that can be followed, with particular attention to their advantages and their drawbacks.

The first approach that will be analyzed is the design of a system in which all the computational cores always maintain the same location, while the Network-on-Chip components can be configured at run-time in order to guarantee an optimal communication infrastructure for each working scenario. In this approach, only the switches of the network and their interconnections are placed in the reconfigurable regions, while both computational cores and network interfaces are deployed in the static regions.

A second possibility is to design a system in which the underlying network is considered as static and the mapping of the computational cores on it changes at run-time accordingly to the requirements of the application that has to be executed on the device. The main advantage of this approach is that, since all the switches are deployed on static regions, it is much easier to handle issues such as flow control and deadlocks avoidance.

Finally, in the last approach that will be analyzed, both the location of the computational cores and the topology of the network can be dynamically reconfigured. This approach determines both a very high degree of freedom in the design of the reconfigurable system and a very high level of flexibility in the reconfiguration processes, even if it introduces a considerable complexity overhead

in the management of the whole reconfigurable architecture.

Network-on-Chip Reconfiguration

A typical scenario in which the reconfiguration of the communication infrastructure can take place is shown in Figure 4. In this example, it is possible to identify three applications (*application A*, *application B* and *application C*) that are characterized by different communication needs. In particular, a specific NoC can be developed and optimized in order to meet all the constraints of a single application: as it is possible to see in Figure 4, *NoC1* has been optimized for *application B*, *NoC2* for *application B* and *NoC3* for *application C*.

Another possibility consists in the definition of a NoC that is able to support all the applications, basing its development on a set of trade-off that make it possible to achieve a suboptimal communication infrastructure for each application, even if it is not possible to perform deep optimizations that concern only one of the input applications. This static (since it will not be changed at run-time) communication infrastructure is named *Static NoC* in Figure 4.

It is possible to find in literature several works (Evain S., Diguet J.-P., Houzet D. (2004); Ciordas C., Hansson A., Goossens K., Basten T. (2006); Srinivasan K., Chatha K. S. (2006); Srinivasan K., Chatha K. S. (2006); Murali S., Benini L., De

Figure 4. Minimum time interval between two applications (that have to be executed on the system) that makes it possible to hide the configuration latency necessary to change the underlying Network-on-Chip

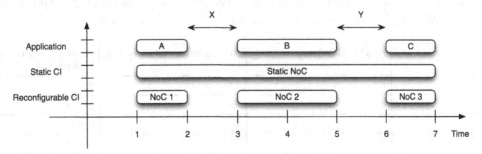

Micheli G. (2005); Jingcao Hu, Marculescu R. (2003); Ye T. T., De Micheli G. (2003); Benini L. (2006); Srinivasan K., Chatha K. S., Konjevod G. (2007)) that address the problem of the automatic generation of efficient NoCs (such as monitoring-aware or layout-aware NoCs) that are specifically developed for a particular application (or set of applications). Thus, they essentially focus on the definition of optimized NoCs, even if these communication infrastructures are always considered as static, since NoC reconfiguration is not taken into account at all.

As shown in (Rana V., Atienza D., Santambrogio M. D., Sciuto D., De Micheli G. (2008)), the employment of a set of specific NoCs instead of a single static NoC makes it possible to obtain considerable improvements in the average latency and in the area usage of the communication infrastructure, while reducing its power consumption. Referring to Figure 4, we consider a video processing application of 32 cores (*application A*), a Video Object Plane Decoder of 34 cores (*application B*) and an image processing application of 23 cores (*application C*). We refer the readers to (Bertozzi D., Jalabert A., Murali S., Tamhankar R., Stergiou S., Benini L., De Micheli G. (2005)) for the communication characteristics of these benchmarks. For the previously introduced applications, the *static NoC* consists of 6 switches (1 switch of 8x8, 2 switches of 9x9, 2 switches 10x10 and 1 switch of 11x11), while both *NoC 1* (for *application A*) and *NoC 3* (for *application C*) consists of 4 switches (3 switches of 10x10 and 1

switch of 11x11) and *NoC 2* (for *application B*) consists of 4 switches (1 switch of 10x9, 2 switches of 10x10 and 1 switch of 10x11). As shown in Table 2, the *static NoC* option is characterized by a higher area usage, a higher average power consumption (evaluated as proposed by Angiolini F., Meloni P., Carta S., Benini L., Raffo L. (2006)) and a higher average latency, with respect to the three ad-hoc NoCs specifically designed for each application. Using the specific NoCs, it can be reported reductions of 34% in latency and 24% in power consumption. Finally, the overall latency for the reconfiguration of the NoC to be used at run-time is very limited (in the order of tens or at maximum of hundreds of milliseconds), making it applicable in real-life scenarios, context in which applications can be dynamically switched among them by users.

In order to make it possible to implement and to use at run-time a NoC specifically optimized for the application that is currently running on the system, it is necessary to dynamically change the communication infrastructure itself, for instance by means of reconfiguration processes. This is possible only if the latencies, indicated with X and Y in Figure 4, between two subsequent applications are larger (or at least equal) than the reconfiguration latency necessary to configure on the device the new NoC.

In Figure 5 it is possible to find a schema of an embedded system based on a reconfigurable NoC. The system consists of two static regions and a single reconfigurable region. The two static

Table 2. Experimental results related to the size, the average latency and the average power consumption of the Static NoC and of the three specific NoCs (NoC 1, NoC 2 and NoC 3)

NoC	Number of switches	Average latency (clock cycles)	Average power consumption (mW)
Static NoC	6	5.96	278.021
NoC 1	4	3.9	211.789
NoC 2	4	4	204.308
NoC 3	4	4.07	216.519

Figure 5. Basic schema of a NoC-based embedded system in which the underlying communication infrastructure has been deployed on a reconfigurable region in order to be dynamically reconfigured at run-time

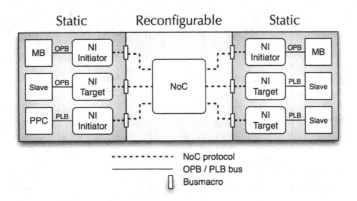

regions are filled with several computational cores, such as processors (that can be either soft-core, Microblaze, or hard-cores, PowerPC) and generic slave components, with their network interfaces (NI initiators for the master cores and NI target for the slave cores) and with buses (On-Chip Peripheral Bus, OPB, and Processor Local Bus, PLB).

The system presented in Figure 5 can be easily extended in order to create a more complex one. Figure 6 shows a complex system that consists of four Microblaze, one PowerPC and five slave components, interconnected with a 3x3 reconfigurable mesh. This system has been deployed on an architecture based on a regular non-homogeneous grid of tiles that can be dynamically configured by columns (1D partial reconfiguration). As it is possible to see from Figure 6, the NoC topology (in particular, one of the three sub-network deployed in the three reconfigurable regions) can be dynamically changed at run-time, in order to connect the computational cores in a different way (for instance, by decreasing the latency or by increasing the throughput of a path between two computational cores).

In conclusion, this approach is particularly suitable for a scenario in which the designer has to design a system that has to execute a set of applications that do not significantly differ for the number and the kind of computational cores (master and slave components), but rather for the communication needs (such as the applications introduced in the previous example). In this case, it is possible to dynamically rearrange the underlying communication infrastructure, leaving unaltered the location of the computational cores, in order to optimally serve the application that is currently running on the system.

Computational Cores Reconfiguration

This approach is strictly related to the mapping of the cores onto the underlying NoC. In the previous approach, in fact, the mapping of the computational cores is done just once; after that, each time a new application has to be executed on the system, it is possible to change only the underlying communication infrastructure. On the opposite, if the reconfigurable part of the system it is composed by the computational cores, it is necessary to perform an optimized mapping of the cores for each application to be run on the system.

Mapping of cores onto NoC tiles has been explored in many works, such as in (Murali S., De Micheli G. (2004)). In (Murali S., Coenen M., Radulescu A., Goossens K., De Micheli G.

Figure 6. Example of an embedded system based on a 3x3 reconfigurable mesh, deployed on a 1D regular non-homogeneous grid of tiles

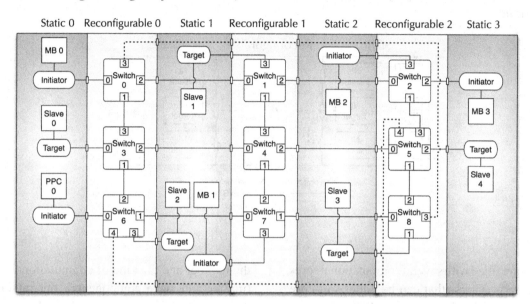

(2006)), the authors present mapping of multiple applications onto NoCs in order to minimize the communication energy dissipated across the NoC. There have been several works on reconfiguration of modules on FPGAs. In (Fekete S. P., van der Veen J.C., Ahmadinia A., Gohringer D., Majer M., Teich J. (2008)), the authors present methods to defragment space on the FPGA. An adaptive software/hardware reconfigurable system is presented in (Becker J., Hubner M., Hettich G., Constapel R., Eisenmann J., Luka J. (2007)). A hardware/software co-synthesis approach for FPGAs is presented in (Li Shang, Robert P. Dick, Niraj K. Jha (2007)). Adapting the NoC at runtime to suit the application requirements using FPGAs is presented in (Hubner M., Braun L., Gohringer D., Becker J. (2008)). Several works have addressed the placement and floorplan issues of reconfigurable FPGAs. In (Chen W., Wang Y., Wang X., Peng C. (2008)), a placement method to reduce reconfiguration data is presented and in (Yi Lu, Marconi T., Gaydadjiev G., Bertels K., Meeuws R. (2008)), the authors present a task placement algorithm. A multi-layer floorplanning

of reconfigurable regions is presented in (Singhal L., Bozorgzadeh E. (2006)).

Mapping Problem

Given an empty (not assigned) topology for a given system, the mapping problem consists of finding the best mapping between an element and an empty communication infrastructure slot. The best mapping is given considering constrains, such as latency and throughput. Generally speaking, the formulation of the mapping problem is characterized by a *Communication Graph*. Such a graph is used to represent the communication among the elements of a system. Furthermore, for each communication between two elements there can be constraints that will be evaluated by a mapping algorithm. A graph G is generally defined by a set of nodes N and a set of arcs A. An arc is defined using a pair of nodes, called the endpoints of the arc.

Definition 1. A communication graph (CG) is an undirected graph G=<N, A> where the set of nodes N = $\{e_0, e_1, ..., e_n\}$ represents the set of processing elements, e_i, in one application and the set of arcs A = {{e_0, e_1}, {e_1, e_2},..., {e_i, e_j}} represents the communication between two nodes of the graph.

Definition 2. A communication constraint between the processing element i and j (using the index notation, to identify the corresponding processing element. Therefore e_i is i and the e_j is j), is associated to each communication of the set A and it is defined as follows: $c_{i,j}$.

Definition 3. An empty (given) topology is defined as a set S = {s_0, s_1,..., s_m} of m free slots that can be assigned to n processing elements with m $\geq$ n.

Given a processing element e_i, a free slot s_p and a binary variable $x_{i,p}$ = {0,1} that indicates weather a processing element is mapped or not on a free slot, the following constraints must be guaranteed:

- all the processing elements must be mapped:

$$\sum\nolimits_{i=0, p=0}^{n, m} x_{i,p} = n$$

- one processing element cannot be mapped more than once:

$$\sum\nolimits_{p=0}^{m} x_{i,p} = 1 \ \forall \ i \in N$$

- no more than one processing element can be mapped on a single free slot

$$\sum\nolimits_{i=0}^{n} x_{i,p} = 1 \ \forall \ p \in S$$

- the constraints for each communication in *A* must be respected.

The problem to map a set of computational cores onto a NoC topology can be seen as the problem to assign to each computational core a single switch of the NoC in order to minimize both the average latency of the communication paths over the network and the maximum throughput on a single link between two different switches of the network, respecting the communication constraints of the application itself. In Figure 7 it is possible to see two different basic NoC topologies: a 2x3 mesh (a) and a ring (b). The switches are represented with a x (*x1, x2, x3, x4, x5, x6*), while the empty slots (that can be filled with either a master or a slave component) are indicated with a s (*s1, s2, s3, s4, s5, s6*). Considering an application that consists of 6 computational cores (elements *e0, e1, e2, e3, e4, e5* and *e6*), it is possible to create a *communication graph* that indicates the maximum number of hops that can be present on the path between two different elements, as shown in Figure 4 (c). Not all the possible mappings of the computational cores on the previously introduced NoC topology will make it possible to meet the latency constraints specified by the *communication graph*. In Figure 7 (d) it is possible to see a feasible mapping solution of the application on the 2x3 mesh, while in Figure 7 (e) it is possible to see a feasible mapping of the computational cores of the application on the ring topology.

Related Works

This section presents a several mapping algorithms that have been proposed in literature. In particular, Hu, J. and Marculescu, R. (2003 and 2005) proposes a mapping algorithm for energy aware placement of the cores considering the communication volume between cores and needed bandwidth. The presented branch-and-bound (BnB) algorithm can automatically map the computational cores onto a generic regular NoC architecture such that the total communication energy is minimized. At the same time, the performance of the mapped system is guaranteed to satisfy the specified constraints through bandwidth reservation. The energy aware mapping problem

Figure 7. Two different NoC topologies (a mesh (a) and a ring (b)) on which the computational cores of an application, which communication patterns (maximum number of hops between two different elements) are shown in (c), are mapped ((d) and (e))

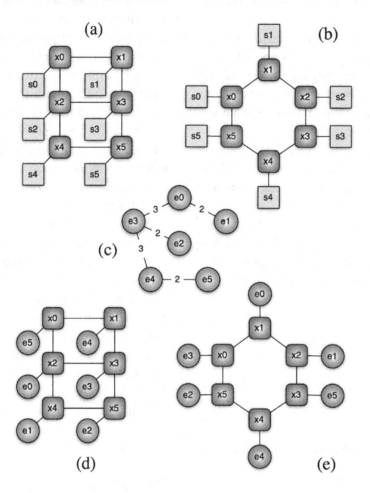

has been defined with the creation of an Application Characterization Graph (APCG) representing the constrained communication graph among all the computational cores, and the Architecture Characterization Graph (ARCG) that represents the routing paths between two different tiles of the device. With these graphs and constraints the authors propose an objective function that minimizes the communication energy consumption under performance constraints. The paper focuses on the tile-based architecture interconnected by a 2D mesh network with X-Y routing, but the proposed algorithm can be adapted to other regular architectures with different network topologies and different static routing schemes.

Wenbiao, Z., Zhang, Y., and Mao, Z. (2007) propose a mapping and routing technique for the mesh based NoC design problem with the goal of minimizing the energy consumption and normalized worst link-load. The proposed heuristic algorithm, PLBMR, is a *particle swarm optimization* (PSO^2) based two phases process, the mapping of the cores onto the NoC to minimize the NoC communication energy consumption, and the allocation of the routing paths to keep the link-load balance. A comparison with other algorithms, such GA and BnB, has been performed and the experimental results indicate that the proposed PLBMR is a viable alternative for solving the mapping and routing path's allocation for NoC.

High-Level Specification of Communication Constraints

In order to specify the communication requirements of each application to be executed on the target system, it has to be possible for the designer to define a communication graph for each application. The set of applications that will be executed at run-time can be in characterized, in general, by very different communication needs.

For instance, it has to be possible to define, for each application, a set of master cores and a set of slave cores. These computational cores represent the components of the target system that are required in order to ensure the correct execution of the considered application (since we are defining a reconfigurable system, the number and the kind of master and slave cores can dynamically vary at run-time accordingly to the application that is currently running on the system). Once masters and slaves have been defined, it has to be possible to create a set of connections among them. Each connection between a master core and a slave core specifies that these two cores can communicate at run-time. Each connection can be characterized by several communication constraints. For instance, it has to be possible to specify the following communication requirements:

- A *latency value*, that specifies the maximum latency that the communication infrastructure can introduce in the communication between the two computational cores. It can be specified in *s*, *ms* or *number of hops*. For instance if two computational cores communicate and have a latency constraint of 3 hops, it means that the maximum number of switches that can

be placed between the two computational cores is 3;

- A *throughput* (or *bandwidth*) *value*, that defines the throughput that the communication infrastructure has to support between the two connected computational cores. It can be specified in *bit/s*, *Kbit/s*, *Mbit/s* or with a *percentage* with respect to the total workload of the network. This constraint can be very useful in the definition of the *routing* of the NoC, since it makes it possible to find the path from the source computational core to the destination computational core that is able to support the communication throughput between them.

Using the information previously described, it is possible to define a *communication graph* for each target application, that summarizes all the communication requirements of the given application and that can be used in order to exploit the knowledge on the behavior of the target applications during the task placement, the mapping and the routing phases.

Another metric that can be useful to specify in order to tune the design of the target system is the *switch throughput upper bound*; this value (that can be specified in *bit/s*, *Kbit/s* or *Mbit/s*) represents the maximum throughput that each switch can handle. This value is obviously strictly dependent on the technology that the designer is planning to employ and it will be used in order to discriminate feasible solutions to infeasible ones when the routing among the computational cores is being defined.

It is important to note that it is also possible to deploy on the device more than one application (such as the three applications X, Y and Z of Figure 8) at the same time (depending on the size of the reconfigurable device), by simply creating an unconnected communication graph (e.g., using the communication graphs of applications X, Y and Z in order to create an unconnected communication graph, consisting of three distinct blocks, for a fictitious application V, as shown in Figure 8) and using it every time all the applications have to be deployed together on the target device.

Mapping Algorithms

As previously hinted, the mapping task consists in finding the best association among the computational cores (both slave and master cores) and the switches ports of the underlying communication infrastructures, taking into account all the latency and bandwidth constraints assigned to all the communication link between the computational cores of the system.

The first step that has to be performed in order to proceed with the mapping of the computational cores is the creation of a matrix in which it is possible to store the values of the maximum latency that is acceptable for each connection between two computational cores of the system. This *NxN* data structure, where *N* is the sum of the number of master cores and the number of slave cores of the given application, has to be generated for each application by analyzing its *communication graph* and will be very useful in order to speedup the execution of the algorithms that actually perform the mapping of the cores. As previously introduced, it is possible to perform the mapping task by employing either *exhaustive* or *heuristic* (e.g., *evolutionary*) algorithms (Cozzi D., Farè C., Meroni A., Rana V., Santambrogio M. D., Sciuto D. (2009)). In particular, the exhaustive algorithms

try to find the best feasible mapping solution for the given application, while the heuristic ones aim at defining good sub-optimal mapping solution, and they are utilized when it is not possible to perform (usually due to a timing constraint) an exhaustive search.

Exhaustive Algorithms

Exhaustive algorithms can be utilized only when dealing with very small applications (usually not more than 12 computational cores), since they evaluate all the feasible mapping solutions for a given application, computing the average latency on all the connections between its computational cores and storing this value for each solution that has been found. Thus, not only a large amount of memory it is required by these algorithms, but also a very long execution time has to be allocated in order to make it possible to complete the evaluation of all the mapping solutions.

In order to overcome these limits, the exhaustive algorithms can be provided with a set of rules and policies that make it possible to exclude all the unfeasible solutions in an early stage of the process, thus considerably reducing the computation time required for the completion of the mapping task. In particular, once a single computational core has been mapped on a single switch port of the communication infrastructure, the rules that can been adopted for the selection of the subsequent computational core to be mapped are the following ones:

Figure 8. Communication graphs of applications X (a), Y (b), Z (c) and of the fictitious application V that consists of the three previous applications (d)

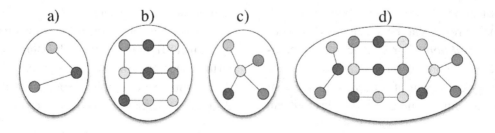

- *First rule*: if some computational cores have already been mapped on the communication infrastructure, then the subsequent computational core to be mapped has to be selected among the computational cores that are directly connected to (at least) one of the already mapped computational cores. This rule makes it possible to try to complete all the dependencies among the mapped and the unmapped computational cores before starting with the mapping of cores that are unconnected to the computational cores that have been already mapped on the NoC;

- *Second rule*: if there are more than one computational core that are directly connected to (at least) one of the mapped cores (as specified by the *first rule*), then the computational core with the highest number of connections with the already mapped computational cores has to be selected as the subsequent computational core that has to be mapped on the NoC;

- *Third rule*: if no computational cores have already been mapped or if there are no cores that are directly connected to (at least) one of the mapped computational cores, then the subsequent computational core to be mapped has to be selected among the computational cores that have the highest number of connections with the lowest latency constraint (in other words, that present the highest number of connections with a very strict constraint).

Heuristic Algorithms

Among all the heuristic algorithm that can be employed in order to perform the mapping task, evolutionary algorithms (such as genetic algorithms) seem to be the most suitable ones, since they have been proved to be effective algorithms that are able to generate very good solutions in a very short time even when the size of the given application is considerably large (even more than

64 computational cores). These algorithms can be developed either by modifying and adapting existing algorithms (such as NSGA2 (Deb K., Pratap A., Agarwal S., Meyarivan T. (2002); Khoa Duc Tran (2005); Ghomsheh V. S., Khanehsar M. A., Teshnehlab M. (2007))) or by creating ad-hoc genetic algorithms designed to solve specific mapping problems.

NSGA2

This section aims at describing a genetic algorithm, called NSGA2ver, that has been developed in order to find optimized mapping solutions. The *NSGA2ver* algorithm is a multi-objective genetic algorithm and it is suitable for a wide set of applications, even for very large ones (more than 64 computational cores). It is essentially based on an evolutionary multi-objective optimization approach (adopting the *Non-dominated Sorting Genetic Algorithm version 2, NSGA2* (Deb K., Pratap A., Agarwal S., Meyarivan T. (2002); Khoa Duc Tran (2005); Ghomsheh V. S., Khanehsar M. A., Teshnehlab M. (2007))).

In *NSGA2ver*, each individual represents a possible mapping solution among the computational cores and the switches ports. The *crossover* operation can be performed, between two individuals, in the following way: all the computational cores that are mapped in the same position in both the individuals will be mapped in the same position also in the offspring, while all the other computational cores will be randomly mapped in the other available locations. The *mutation* operation can be performed on a single individual and it has been defined as the random swap between the locations of two different computational cores. *NSGA2* has been developed to minimize the following objective functions, which define the *fitness* function:

- *First objective function*: the number of connections between to computational cores that violate the latency constraint defined for that connection;

- *Second objective function*: the average latency on all the connections between two computational cores that will communicate at run-time when the given application will be executed on the target reconfigurable system.

The minimization of these two objective functions provides very good sub-optimal solutions to the mapping problem. These solutions are feasible only when there are no constraints violations, thus it is very important to reduce the *first objective function* to zero, while the *second objective function* has to be minimized in order to improve the quality (in terms of overall timing performance) of the mapping solution.

Custom Genetic Algorithms

This section aims at presenting two examples of custom genetic algorithms (GA1ver and GA2ver) that have been specifically designed in order to speed up the previously presented mapping task for a certain set of applications.

The *GA1ver* algorithm is a custom single-objective genetic algorithm. Individuals, crossover operation and mutation operation of *GA1ver* have been developed similarly to the ones described for *NSGA2ver*. The main difference between *GA1ver* and *NSGA2ver* is that the first one has been developed in order to minimize a single objective function, that is the average latency on all the connections between the couple of computational cores that have to communicate in the target reconfigurable system (this metric is used to define the fitness of each individual). It is important to note that, in *GA1ver*, all the individuals represent feasible solutions (mapping solutions in which there are no constraints violations). Obviously, both crossover and mutation have to be redefined accordingly to the new definition of individual, thus both of them are applied as many times as required in order to produce a feasible solution (since they are intrinsically random, each time they are applied it is possible to obtain a different

output). This algorithm is suitable for small or medium applications (12 - 20 cores).

In order to further speedup the mapping task, it has been developed another version of this algorithm that does not check, during the iterations, if a solution is feasible or not. Removing this check it has been reached a good improvement with respect to the total execution time. Once the execution of the algorithm is completed, thus the set of mapping solutions have been successfully obtained, a *filter function* discriminates weather a solution if feasible or not considering the communication graph constraints. The modified version of *GA1ver* produces the *GA2ver* algorithm, another custom single-objective genetic algorithm with a slightly different coding of individuals. Individuals, crossover operation and mutation operation of *GA2ver* have been developed similarly to the ones described for *NSGA2ver*. As for *GA1ver*, the main difference between *GA2ver* and *NSGA2ver* is that the first one has been developed in order to minimize a single objective function, that is the average latency on all the connections between the couple of computational cores that have to communicate in the target reconfigurable system (this metric is used to define the fitness of each individual). It is important to state that, in *GA2ver*, latency constraints are not taken into account, since it can be proved by experimental results that reducing the average latency on all the connections can easily bring to the satisfaction of a high number of latency constraints (usually all of them). This algorithm can be effectively applied on applications that present a medium size (usually it provides very good solutions in a reasonable time when applied on applications that consist of 16 - 32 computational cores).

Network-on-Chip and Computational Cores Reconfiguration

As shown in Figure 9, if both the NoC and the computational cores can be reconfigured at run-time, it is strictly necessary to impose a strong

Figure 9. Synchronization between the reconfiguration of the computational cores and the reconfiguration of the underlying Network-on-Chip

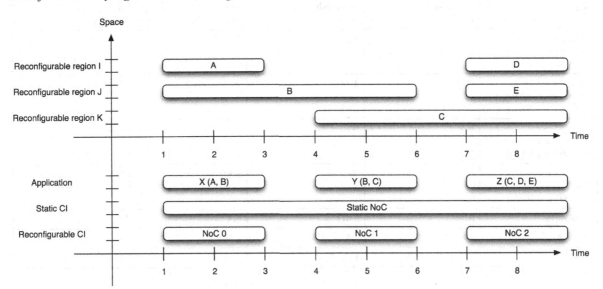

synchronization between the two reconfiguration processes in order to simultaneously obtain a system with both all the computational cores needed by the application that has to be run on the system and the optimal communication infrastructure to interconnect them.

In the example shown in Figure 9, three different applications have to be deployed on the target system. In particular, application X needs the computational cores A and B, application Y needs the computational cores B and C, while application Z needs the computational cores C, D and E. We assume, for the sake of simplicity, that the target reconfigurable architecture it is composed by only three reconfigurable regions and that each core can be deployed on any reconfigurable region. In Figure 9 it is possible to see that, if application X, Y and Z have to be executed in this order, it is necessary to initially configure the device with both computational cores A and B (for instance in the reconfigurable regions I and J) and with the *NoC 0* communication infrastructure. Thus, it is necessary to configure, for instance in the reconfigurable region K, the computational core C and to substitute the *NoC 0* communication

infrastructure with *NoC 1*. These configuration processes have to be obviously performed before the application Y is deployed on the system, since it needs both core C and *NoC 1* in order to be correctly executed. Once application X finishes its execution, it is possible to unload core A and to configure, in the same reconfigurable region (reconfigurable region I), the computational core D, that will be required by the subsequent application. Finally, in order to start the execution of application Z, it is necessary to configure the core E in the reconfigurable region freed by the unloading of core B, that is necessary for application Y but not for application Z. The last configuration process concerns the configuration of the *NoC 2* communication infrastructure instead of *NoC 1*.

In conclusion, the approach presented in this section combines the approaches presented in the previous sections, thus presenting a very high level of complexity. In order to manage all the reconfiguration processes needed by this approach, it is necessary to develop a specific NoC for each application to be run on the system and to define, on this specific NoC, an optimized mapping of the computational components needed by the ap-

Table 3. Timing comparison among different mapping algorithms

Algorithm	2x2 mesh (sec)	3x3 mesh (sec)	5x5 mesh (sec)	6x6 mesh (sec)
PLBMR [Wenbiao, Z., Zhang, Y., and Mao, Z. (2007)]	0.3	1.5	16.4	**44.9**
BnB [Wenbiao, Z., Zhang, Y., and Mao, Z. (2007)]	**0.234**	**1.345**	18.564	55.678
SA [Hu, J. and Marculescu, R. (2003 and 2005)]	NA	NA	181.67	Hours
BnB [Hu, J. and Marculescu, R. (2003 and 2005)]	NA	NA	**6.52**	NA
Exhaustive	**0.0002**	**0.011**	NA	NA
GA2ver	NA	NA	**1.584**	**3.312**
Gain	1170	122.27	4.11	13.55

plication itself. Thus, this approach is well suited for a scenario in which the applications that have to be deployed on the system need disjointed sets of computational cores (or, at least, very different sets of cores) that cannot be mapped on the same communication infrastructure, but that need optimized NoC, since their communication needs considerably vary from application to application.

Experimental Results

Different state-of-the-art algorithms have been analyzed in order to compare the timing performance of the proposed mapping algorithms with the ones of the algorithms that can be found in literature.

Table 3 presents the execution time (For the exhaustive and the genetic algorithms, the execution times have been taken on a MacBook Pro 2.16 GHz with 2 GB of RAM) of different mapping algorithms, varying the number of core elements. The listed algorithms, even if considering different constraints, perform the cores mapping on a 2D mesh communication infrastructure, the most common NoC topology that can be found in literature. The timing results of the proposed

algorithms have been evaluated by running them on the same input applications used by the state-of-the-art approaches (when these applications were explicitly shown) and considering the average value over 100 executions.

CONCLUSION

In order to design a reconfigurable NoC-based system, it is firstly necessary to define the reconfigurable architecture on which the system will be deployed. This choice can be performed with respect both to the characteristics of the target platform (such as the amount of available memory to store partial bitstreams) and to the goals that the designer wants to achieve (such as the total area usage of the system and its level of run-time adaptability).

Once the underlying architecture has been defined, it is necessary to select the components of the system that will be possible to reconfigure at run-time. This second choice strictly depends on the intrinsic characteristics of the applications that have to be executed on the system. For instance, if they share most of their computational cores, differing only for the communication needs among these cores, probably the most suitable solution consists in configuring at run-time only the communication infrastructure, while keeping fixed the locations of the computational cores. On the opposite, if the applications need disjoint sets of computational cores that can be mapped on a shared communication infrastructure (usually it is possible to exploit a mesh topology), it is possible to fix the underlying NoC, while mapping and configuring all the computational cores needed by an application every time it has to be executed on the system.

REFERENCES

Angiolini, F., Meloni, P., Carta, S., Benini, L., & Raffo, L. (2006, March 6-10). Contrasting a NoC and a Traditional Interconnect Fabric with Layout Awareness. In *Proceedings of Design, Automation and Test in Europe, 2006.* (DATE '06), (vol.1, pp.1-6).

Becker, J., Hubner, M., Hettich, G., Constapel, R., Eisenmann, J., & Luka, J. (2007)... *Dynamic and Partial FPGA Exploitation. IEEE, 95*(2), 438–452.

Benini, L. (2006, March 6-10). Application Specific NoC Design. In *Proceedings of Design, Automation and Test in Europe, 2006.* (DATE '06), (vol.1, pp.1-5).

Bertozzi, D., Jalabert, A., Murali, S., Tamhankar, R., Stergiou, S., Benini, L., & De Micheli, G. (2005, February). NoC synthesis flow for customized domain specific multiprocessor systems-on-chip. In Proceedings of IEEE Transactions on Parallel and Distributed Systems (vol.16, no.2, pp. 113-129).

Chen, W., Wang, Y., Wang, X., & Peng, C. (2008, July 29-31). A New Placement Approach to Minimizing FPGA Reconfiguration Data. *In Proceedings of the International Conference on Embedded Software and Systems, 2008* (ICESS '08), (pp.169-174).

Ciordas, C., Hansson, A., Goossens, K., & Basten, T. (2006). A Monitoring-Aware Network-on-Chip Design Flow. In *Proceedings of the 9th EURO-MICRO Conference on Digital System Design: Architectures, Methods and Tools* (DSD 2006), (pp.97-106).

Cozzi, D., Farè, C., Meroni, A., Rana, V., Santambrogio, M. D., & Sciuto, D. (2009, May). Reconfigurable NoC design flow for multiple applications run-time mapping on FPGA devices. In *Proceedings of the 19th ACM/IEEE Great Lakes Symposium on VLSI*, (pp. 421-424).

Deb, K., Pratap, A., Agarwal, S., & Meyarivan, T. (2002, April). A fast and elitist multi-objective genetic algorithm. *IEEE Transactions on NSGA-II. Evolutionary Computation, 6*(2), 182–197. doi:10.1109/4235.996017

Evain, S., Diguet, J.-P., & Houzet, D. (2004, November 18-19). A generic CAD tool for efficient NoC design,", 2004. In *Proceedings of 2004 International Symposium on Intelligent Signal Processing and Communication Systems* (ISPACS 2004), (728-733).

Fekete, S. P., van der Veen, J. C., Ahmadinia, A., Gohringer, D., Majer, M., & Teich, J. (2008). Offline and Online Aspects of Defragmenting the Module Layout of a Partially Reconfigurable Device. [VLSI]. *IEEE Transactions on Very Large Scale Integration Systems, 16*(9), 1210–1219. doi:10.1109/TVLSI.2008.2000677

Ferrandi, F., Morandi, M., Novati, M., Santambrogio, M. D., & Sciuto, D. (2006, November). Dynamic Reconfiguration: Core Relocation via Partial Bitstreams Filtering with Minimal Overhead. In *Proceedings of the International Symposium on System-on-Chip* (SoC 06), (pp. 33-36).

Ghomsheh, V. S., Khanehsar, M. A., & Teshnehlab, M. (2007). Improving the non-dominate sorting genetic algorithm for multi-objective optimization. *In Proceedings of CISW 2007*, (pp. 89–92).

Hu, J., & Marculescu, R. (2003, January 21-24). Energy-aware mapping for tile-based NoC architectures under performance constraints. In *Proceedings of Design Automation Conference, 2003* (ASP-DAC 2003), (pp. 233-239).

Hu, J., & Marculescu, R. (2003). Energy-aware mapping for tile-based NoC architectures under performance constraints. In *Proceedings of the Design Automation Conference, 2003* (ASP-DAC 2003), (pp. 233–239).

Hu, J., & Marculescu, R. (2003). Exploiting the routing flexibility for energy/performance aware mapping of regular NoC architectures. In *Proceedings of the Design, Automation and Test in Europe Conference and Exhibition* (pp. 688–693).

Hu, J., & Marculescu, R. (2005, April 4). Energy- and performance-aware mapping for regular NoC architectures. In Proceedings of IEEE Transactions on Computer-Aided Design of Integrated Circuits and Systems (pg 24).

Hubner, M., Braun, L., Gohringer, D., & Becker, J. (2008, April 14-18). Run-time reconfigurable adaptive multilayer network-on-chip for FPGA-based systems. In *Proceedings of the IEEE International Symposium on Parallel and Distributed Processing, 2008* (IPDPS 2008), (pp.1-6).

Khoa Duc Tran. (2005). Elitist non-dominated sorting GA-II (NSGA-II) as a parameter-less multi-objective genetic algorithm. In Proceedings of SoutheastCon, (pp. 359–367).

Morandi, M., Novati, M., Santambrogio, M. D., & Sciuto, D. (2008, April). Core allocation and relocation management for a self dynamically reconfigurable architecture. *In Proceedings of the IEEE Computer Society Annual Symposium on VLSI* (ISVLSI 08), (pp. 286 – 291).

Murali, S., Benini, L., & de Micheli, G. (2005, January 18-21). Mapping and physical planning of networks-on-chip architectures with quality-of-service guarantees. In *Proceedings of the Design Automation Conference, 2005* (ASP-DAC 2005). (vol.1, pp. 27-32).

Murali, S., Coenen, M., Radulescu, A., Goossens, K., & De Micheli, G. (2006, March 6-10). A Methodology for Mapping Multiple Use-Cases onto Networks on Chips. In *Proceedings of Design, Automation and Test in Europe, 2006* (DATE '06), (vol.1, pp.1-6).

Murali, S., & De Micheli, G. (2004, February 16-20). Bandwidth-constrained mapping of cores onto NoC architectures. In *Proceedings of Design, Automation and Test in Europe Conference and Exhibition, 2004* (vol.2, pp. 896-901).

Rana, V., Atienza, D., Santambrogio, M. D., Sciuto, D., & De Micheli, G. (2008, October). A Reconfigurable Network-on-Chip Architecture for Optimal Multi-Processor SoC Communication. In *Proceedings of the 16th IFIP/IEEE International Conference on Very Large Scale Integration* (pp. 321-326).

Shang, L., Dick, R. P., & Jha, N. K. (2007, March). SLOPES: Hardware–Software Cosynthesis of Low-Power Real-Time Distributed Embedded Systems With Dynamically Reconfigurable FPGAs. In. *Proceedings of the IEEE Transactions on Computer-Aided Design of Integrated Circuits and Systems*, *26*(3), 508–526. doi:10.1109/TCAD.2006.883909

Singhal, L., & Bozorgzadeh, E. (2006, August 28-30). Multi-layer Floor planning on a Sequence of Reconfigurable Designs. In Proceedings of the International Conference on Field Programmable Logic and Applications, 2006 (FPL '06), (pp.1-8).

Srinivasan, K., & Chatha, K. S. (2006, March 27-29). A methodology for layout aware design and optimization of custom network-on-chip architectures. In *Proceedings of the 7th International Symposium on Quality Electronic Design* (ISQED '06), (pp.6-357).

Srinivasan, K., & Chatha, K. S. (2006, October 22-25). Layout aware design of mesh based NoC architectures. In Proceedings of the 4th international conference on Hardware/software codesign and system synthesis (CODES+ISSS '06), (pp.136-141).

Srinivasan, K., Chatha, K. S., & Konjevod, G. (2007). Application Specific Network-on-Chip Design with Guaranteed Quality Approximation Algorithms. In *Proceedings of Design Automation Conference, 2007* (ASP-DAC '07), (pp.184-190).

Wenbiao, Z., Zhang, Y., & Mao, Z. (2007). Link-load balance aware mapping and routing for NoC. *WSEAS Trans. Cir. and Sys.*, 6(11), 583–591.

Xilinx Inc. (2003). *Virtex Series Configuration Architecture User Guide.* (Technical Report XAPP151), (March 2003).

Xilinx Inc. (2004). XAPP290 - Two flows for partial reconfiguration: Module based or Difference based, September 2004.

Xilinx Inc. (2006), Early Access Partial Reconfiguration User Guide, March 2006.

Xilinx Inc. (2007). *Virtex-II Pro and Virtex-II ProX Virtex-II Pro and Virtex-II Pro X FPGA User Guide.* Xilinx Inc., March 2007.

Xilinx Inc. (2007). *Virtex-4 user guide.* Technical Report UG70, Xilinx Inc., March 2007.

Xilinx Inc. (2007). Virtex-5 user guide. Technical Report UG190, Xilinx Inc., February 2007.

Xilinx Inc. (2007). Virtex-5 configuration user guide. Technical Report UG191, Xilinx Inc., February 2007.

Ye, T. T., & De Micheli, G. (2003, June 24-26). Physical planning for on-chip multiprocessor networks and switch fabrics. In. *Proceedings of IEEE International Conference on Application-Specific Systems, Architectures, and Processors, 2003*, 97–107.

Yi, L., Marconi, T., Gaydadjiev, G., Bertels, K., & Meeuws, R. (2008, April 14-18). A self-adaptive on-line task placement algorithm for partially reconfigurable systems. In *Proceedings of the IEEE International Symposium on Parallel and Distributed Processing* (IPDPS 2008), (pp.1-8).

KEY TERMS AND DEFINITIONS

FPGA: Integrated circuit that can be programmed after it has been manufactured.

Mapping: The problems of finding the best association between the elements of a system and the slots of a communication infrastructure. Switch throughput upper bound: maximum throughput that each switch can handle.

NSGA2: Non-dominated Sorting Genetic Algorithm version 2, a multi-objective genetic algorithm.

Placement Constraint: Constraint used to identify the area where a functionality has to be mapped.

Platform Homogeneity: Possibility of identifying homogeneous regions inside a reconfigurable device according to its physical implementation.

Relocation: Technique that makes it possible to dynamically change a run-time a configuration bitstream in order to configure the same computational core in a different location.

ENDNOTES

[1]   The components that use the I/O pins to communicate with the external of the FPGA device.

[2]   PSO system combines local search methods and with global search methods, attempting to balance exploration and exploitation.

Chapter 6

From MARTE to Reconfigurable NoCs:
A Model Driven Design Methodology

Imran Rafiq Quadri
CNRS, France

Majdi Elhaji
CNRS, France

Samy Meftali
CNRS, France

Jean-Luc Dekeyser
CNRS, France

ABSTRACT

Due to the continuous exponential rise in SoC's design complexity, there is a critical need to find new seamless methodologies and tools to handle the SoC co-design aspects. We address this issue and propose a novel SoC co-design methodology based on Model Driven Engineering and the MARTE (Modeling and Analysis of Real-Time and Embedded Systems) standard proposed by Object Management Group, to raise the design abstraction levels. Extensions of this standard have enabled us to move from high level specifications to execution platforms such as reconfigurable FPGAs. In this chapter, we present a high level modeling approach that targets modern Network on Chips systems. The overall objective: to perform system modeling at a high abstraction level expressed in Unified Modeling Language (UML); and afterwards, transform these high level models into detailed enriched lower level models in order to automatically generate the necessary code for final FPGA synthesis.

INTRODUCTION

Since the early 2000s, System-on-Chip (SoC) has emerged as a new methodology for embedded systems design. In a SoC, the computing units (programmable processors, hardware functional units), memories, I/O devices, communication channels, etc.; are all integrated into a single chip. Moreover, multiple processors can be integrated into a SoC (Multiprocessor System-on-Chip, MPSoC) in which

DOI: 10.4018/978-1-61520-807-4.ch006

the communication can be achieved through Network on Chips (NoCs). These SoCs are generally dedicated to target application domains (such as multimedia video codecs, software-defined radio and radar/sonar detection systems) that require intensive computations. According to Moore's law, rapid evolution in hardware technology doubles the number of transistors in an Integrated Circuit (IC) nearly every two years. As the computational power increases, more functionalities are expected to be integrated into the system. As a result, more complex software applications and hardware architectures are integrated, leading to a *system complexity* issue which is one of the main hurdles facing SoC co-design. The fallout of this complexity is that the system design (particularly software design) does not evolve at the same pace as that of hardware due to issues such as development budget limitations, reduction of product life cycles and design time augmentation. This evolution of balance between production and design has become a critical issue and has finally led to the *productivity gap*. System reliability and verification are also the other issues related to SoC industry and are directly affected by the design complexity. An important challenge is to find efficient design methodologies that raise the design abstraction levels to reduce overall complexity, while effectively handling issues such as accurate expression of inherent system parallelism: such as application loops; and hierarchy.

Network on Chips is considered as an emerging paradigm for resolving the problems related to current highly integrated complex SoCs (Benini, L., and Micheli, G 2001). A SoC may have tens or hundreds of IP (Intellectual Property) cores with each running at different clock cycles resulting in *asynchronous clocking*. NoCs thus adopt a globally asynchronous, local synchronous (GALS) approach and help to improve the performance: such as throughput; and scalability as compared to other communication structures such as point to point signal wires and shared buses. They are an ideal choice for MPSoC architectures as they

allow separation of the *communication* and the *computation* concerns while allowing IP reuse by utilization of standard interfaces.

Currently High Level Synthesis (HLS) (or Electronic System Level) is an established approach in SoC industry. This approach raises the design abstraction level to some degrees as compared to traditional hand written HDL (Hardware Description Languages) implementations. The gap between the high abstraction levels and the low abstraction levels is often bridged using one or several *Internal Representations* (IRs) (Guo et al 2005). The behavioral (algorithmic) description of the system is written in a high level language such as SystemC (OSI 2007) or a similar language, and is then refined into a RTL (Register Transfer Level) implementation using HLS tools. An effective HLS flow and associated tools must be flexible to cope with the rapid hardware/software evolution; and *maintainable* by the tool designers. The underlying low level implementation details are hidden from users and their automatic generation reduces time to market and fabrication costs. However, usually the abstraction level of the HLS tools is not elevated enough to be totally independent from low level details. Normally, the set of concepts related to an IR are generally difficult to handle due to absence of formal definitions of key concepts and their relations. The text based nature of a system description also results in several disadvantages. Immediate recognition of system information such as related to hierarchy, data parallelism and dependencies is not possible; differentiation between different concepts is a daunting task in a textual description and makes modifications complex and time consuming.

Model Driven Engineering (Planet MDE 2007) (MDE) is an emerging domain and can be seen as a *High Level Design Flow* in order to resolve the issues related to SoC co-design. MDE enables system level (application/architecture) modeling at a high specification level allowing several abstraction stages (i.e. IRs). Thus a system can be viewed globally or from a specific point of view of the system, allowing to separate the system model into parts according to relations between system concepts defined at different abstraction stages. This *Separation of Views* (SoV) allows a designer to focus on a domain aspect related to an abstraction stage thus permitting a transition from *solution space* to *problem space*. Using a graphical modeling language i.e. UML (Unified Modeling Language) for system description increases the system comprehensibility. This allows designers to provide high-level descriptions of the system that easily illustrate the internal concepts (task/data parallelism, data dependencies and hierarchy). These specifications can be *reused, modified* or *extended* due to their graphical nature. Finally MDE's model transformations allow to generate executable models (or executable code) from high level models bridging the gap between these models and execution platforms.

FPGAs (Field Programmable Gate Arrays) are considered an ideal solution for SoC implementation due to their reconfigurable nature. Designers can initially implement, and afterwards, reconfigure a complete SoC on FPGA for the required customized solution. Thus FPGAs offer a migration path for final ASIC (Application Specific Integrated Circuit) implementation. For FPGAs, the on chip interconnection can be either bus or NoC based.

MARTE (OMG 2007a) (Modeling and Analysis of Real-Time and Embedded Systems) is an industry standard of Object Management Group (OMG), dedicated to model-driven development of embedded systems. MARTE extends UML, allowing to model the features of software and hardware parts of a real-time embedded system and their relations, along with added extensions (for e.g. performance and scheduling analysis). Although rich in concepts, MARTE lacks a design flow to move from high level modeling to execution platforms.

Gaspard (DaRT team 2009, Gamatié et al 2008b) is a MDE based MARTE compliant SoC co-design framework dedicated specifically

towards parallel hardware and software; and it allows to move from high level MARTE specifications to different execution platforms. It exploits the inherent *parallelism* included in repetitive constructions of hardware elements or regular constructions such as application loops. Gaspard also focuses on a limited application domain, that of *intensive signal processing* (ISP) applications.

The main contribution of this paper is to present a novel MDE based design methodology for implementing the aspects of Partial Dynamic Reconfiguration from an extended MARTE standard. This design flow successfully responds to the major issues, for both users and designers of a typical HLS flow. Applications are graphically specified at a high abstraction level with UML and factorized expressions of parallelism, multidimensional data arrays and powerful constructs of data dependencies are managed thanks to the use of the MARTE standard profile. The design flow allows to specify part of the reconfigurable system at a high abstraction level: notably the reconfigurable region and the reconfiguration controller. Afterwards, using model to model transformations, the gap between high level specifications and low implementation details can be bridged to automatically generate the code required for the creation of bitstream(s) for final FPGA implementation. Currently our approach focuses on a traditional SoC architecture; however it can be extended to include the aspects of NoCs also.

The rest of this chapter is organized as follows. Section 2 describes Network on Chip concepts while an overview of MDE is provided in section 3. Section 4 summarizes our MARTE compliant GASPARD framework and section 5 gives a detailed explanation of the deployment extension in MARTE. Section 6 illustrates our methodology related to modeling complex NoCs. Some case studies are present in section 7. Finally section 8 details the conclusion.

Figure 1. Disciplines for the design of SoCs

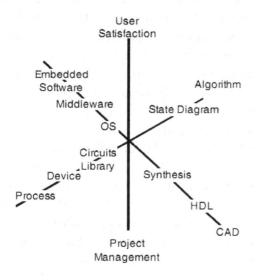

NETWORK ON CHIPS

Recently, the term SoC (System on Chip) has replaced VLSI (very-large-scale integration) as the key word in information technology (IT). The change of name is a reflection of the shift of focus from chip to system in the IT industry. These SoCs are widely used in portable and handheld systems such as smart phones and portable game devices. Therefore, SoC design discipline is extremely complicated and covers a variety of areas, such as marketing, software, computing system, and semiconductor IC design, as described in Figure 1. SoC development requires hexagonal expertise in not only technological areas such as IC technology, CAD, software, and algorithm but also in management techniques complicated team, project, and customer research.

The entire SoC design process of a system can be split into many stages; each stage having a design input and output. Each design activity requires inputs for simulation, analysis, or synthesis. In the top-down design, the design process starts from the system function requirements; these are the design inputs to the first design stage. Its design output is more detailed and more complex than the system requirement and one step closer

to the physical implementation. Each stage needs a model of computation (MoC) to map its design input to the output (see Figure 2). The model is composed of a set of simpler subsystems and a method, or the rules for integrating them, to implement the system functions. Each stage also has a different model with a different level of abstraction.

As the chip scale grows, current System on Chip (SoC) designs generally incorporate a number of processing cores to answer high performance requirements with reasonable power consumption [Vangal et al 20007, Lattard D et al 2007]. This design methodology has the advantage of achieving high performance with moderate design efforts because, once a processor is designed and verified, it can be replicated and reused. However, integrating a number of processors on a single chip does not necessarily mean a SoC design. Depending on the applications, SoC also requires the integration of numerous peripheral modules, such as on-chip memory, external memory controller, I/O interface, and so on. As a result, it is getting more and more important to provide efficient interconnections among numerous processing cores and peripheral modules of SoC. Traditional interconnection techniques, such as on-chip bus or point-to-point inter-connections are not suitable for current large-scale SoCs because of their poor scalability. In recent years, a design paradigm based on a Network on Chip (NoC) was proposed as a solution for interconnection techniques of large-scale SoCs [Dally et al 2001, Benini, L and Micheli, G.. 2002]. The modular structure of NoC makes chip architecture highly scalable, and well controlled electric parameters of the modular block improve the reliability and operation frequency of the on-chip interconnection network. After the proposal of the NoC design paradigm, research concerning NoC has fairly advanced.

NoCs bring the networking principles reserved for data transfer, normally used in large area networks, to the SoC domain. Developing NoC-based systems tailored to a particular application

Figure 2. Model of Computation

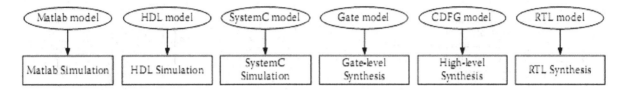

domain, satisfying the application performance constraints with minimum power-area overhead is a major challenge. With evolution in technology, as the geometries of on-chip devices reach the physical limits of operation, another important design challenge for NoCs will be to provide dynamic (run-time) support against permanent and intermittent faults and errors that can occur in the system. To meet the increasing communication demands, the bus based architectures normally used in MPSoCs, have evolved over time from a single shared bus to multiple bridged buses and to crossbar-based designs. Current state-of-the art bus architectures, such as the AMBA multi-layer, enable the instantiation of multiple buses operating in parallel, thereby providing crossbar architecture.

However, such architecture is inherently no scalable for large number of cores in the design. Networks on Chips (NoCs), has recently emerged as the design paradigm for designing such scalable micro networks for MPSoCs. A NoC consists of switches, links, and Network Interfaces (NIs) whichh connects a core to the network and coordinates the transmission and reception of packets from/to the core. A packet is usually segmented into multiple FLow control unITS (flits). The switches and links are used to connect the various cores and NIs together. The use of a NoC to replace bus-based wiring has several key advantages:

- Better scalability at the architectural and physical levels. NoCs can add bandwidth as needed and segment wires as required.
- Better performance under high loads. NoCs can operate at high frequencies, cope with large bandwidth demands, and parallelize traffic streams

- NoCs facilitate modularity by orthogonalizing the design of the communication architecture from the computation architecture, thereby leading to reduced design efforts.
- Quicker design closure. NoC are more predictable: They intrinsically provide wire segmentation, which helps ensuring that design will not be needed in the last phases of the design flow, when they are more costly.
- Higher energy-efficiency. To support the same traffic load, NoCs can operate at a lower frequency than bus-based systems and the data transfer can be finished faster. These can lead to a reduction in energy consumption of the system.

OSI to NoC Model

The NoC exploits a layered-stack approach to communicate between integrated processing elements (PE), which may be processors, dedicated functional blocks, or memories. The NoC decouples the communicational part from the computational part efficiently from an early design stage. This dissociation enables a parallel design of PEs and interconnection structures without any interference between them. In this subsection, we focus on the communicational part, the functions of which are provided by NoC. Figure 3(a) represents the OSI seven-layer model [Zimmerman, H. 1980] and its correspondence to the building blocks of NoC. The NoC involves the four bottom layers of

Figure 3. OSI Layer Model (top) and Layered Architecture of NoC (bottom)

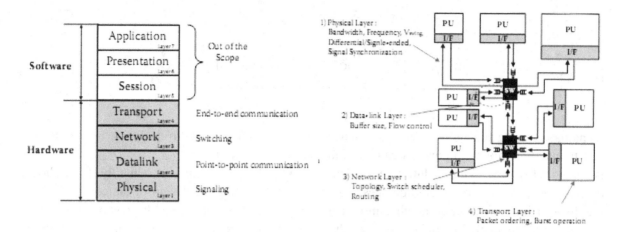

the OSI seven-layer model, realized as hardware on a fabricated chip. Figure 3(b) describes design issues to be considered for each of the OSI model layers. Because the physical layer defines electrical and physical specifications, its design issues include operational frequency of the wire link, signaling scheme, clock synchronization, and so on. In contrast to the traditional personal computer (PC) networks, NoC physical layer design has significant impact on power consumption and performance of SoC because a large volume of on-chip data transactions among PEs occur frequently and concurrently, compared to inter-PC communications.

Therefore, inefficient design of the physical layer easily results in a huge amount of wasted power, and poor performance. In addition, maximum clock frequency and wire width of the physical layer determine the theoretical bandwidth limit of the NoC design. After the first prototype chip fabrication of NoC [Lee, S et al 2003], there have been researches concerning the physical layers of NoC [Lee, K et al 2004, Panades and Greiner, A 2007].

The third layer of the OSI seven-layer model is the network layer. The major design issues of the network layer involve NoC topology selection, packet routing schemes, and quality-of-

service (QoS) guarantee. Topology of the NoC should be very carefully selected because of its significant impact on overall performance and power consumption. To reduce communication overheads, PEs with frequent data transactions need to be placed at a close distance, although sufficient bandwidth for every node should be supplied to avoid performance degradation. As for low-overhead routing schemes, source routing is generally adopted in NoC because only simple decoding logic is needed for packet routing at each router instead of large look-up tables. Finally, guaranteeing QoS is also important for efficient utilization of bandwidth. In the NoC, QoS guarantee is implemented by supporting priorities or constructing different classes of virtual channels. The Æthereal NoC [Rijpkema, E 2003] implements both Guaranteed Throughput (GT) and Best Effort (BE) router for worst case QoS guarantees. An arbitration look-ahead scheme, which aims at reducing packet switching latency, is also reported [Kwanho, K et al 2005].

The transport layer is the highest level of the OSI seven-layer model implemented by the NoC. In the NoC-based SoC design, each of the PEs and functional blocks should be designed according to NoC protocols to support interfaces to the on-chip network. NI modules, which perform packet generation and parsing, provide abstractions of end-to-end data transactions between PEs and other functional modules. In many NoC implementations, out-of-order packet transmission is not supported because of limited buffer resources on a chip.

Concepts Related to Network on Chip

The main objective of a network on chip to provide a system for exchanging data between different processing resources (or IP core). The network consists of nodes (also called switches or routers). The data passing between nodes through links (or channels) from point to point communication. The combination of nodes and links is the communication system. Each node includes a set of communications ports. These ports allow connection to the links connecting the nodes between them. Next architectures, resources are to nodes connected directly or through links. The nodes make routing decisions (where to send data) and arbitration (which transmit data).

Topology

A topology is the connection map between the constituent processing elements (PEs). There exist many network topologies, and the simplest one is a ring or a shared bus. A three-dimensional cube or torus is a more complicated topology, but a mesh topology is widely considered as a typical NoC topology, especially for the homogeneous SoC. The mesh topology, however, is neither a unique nor an optimal solution for the heterogeneous multiprocessor. You should choose an optimum topology for your system and application in terms of the system performance, power-consumption, performance/power, and area cost. Traditionally, interconnection networks can be categorized into two classes: direct and indirect network. The former provides a direct connection

between processing nodes. On the other hand, in an indirect network, the communication between any two nodes has to be carried through some switches. Most of the NoC topologies are indirect networks even for the mesh topology. However, as the distinction between these two classes of networks is blurring, the categorization becomes meaningless. Many network topologies have been proposed in terms of their graph-theoretical properties. Most of them were proposed for minimizing the network diameter for a given number of nodes and node degrees [Duato, J et al 2005]. However, very few of them have ever been analyzed and implemented in NoCs. Figure 4 shows a few of the most famous topologies and also an application-specific one as examples. There has been research on NoC topology exploration. Murali et al. have developed a tool for automatically selecting an application-speciic topology for minimizing average communication delay, area, and power dissipation [Murali, S. 2004]. Wang et al explore the technology-aware topology of various meshes/tori [Wang, H 2005]. Kreutz et al. present a topology evaluation engine based on a heuristic optimization algorithm [Kreutz et al 2005]. In these works, the candidate pool of topologies was limited to the typical, regular and homogeneous topologies such as a mesh, torus, hypercube, tree, or multistage network.

Designers of large-scale SoCs must be aware of the advantages and disadvantages of each architecture in order to select an appropriate candidate for their implementations. The metrics that are of interest can be broadly categorized as:

- Performance (latency, throughput, cross-section bandwidth).
- Energy consumption.
- Reliability (error detection and/or correction).
- Scalability.
- Implementation cost (area).

Figure 4. An example of network topologies

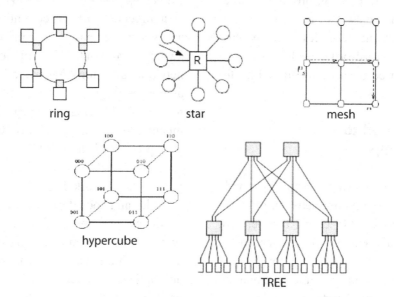

The topology of a network is described by a set of vertices, or nodes V connected by a set of edges, or channels. The set of edges E, where an edge

$$e_{x,y} = (x, y) \in E \mid x, y \in V$$

Connects a source node x to a destination node y. A network topology can therefore be represented by the graph:

$$G = (V, E)$$

Representing network topologies as graphs does not take into account physical implementation issue, but it does allow interesting properties to be studied such as:

- Node degree: The number of channels that connect a node to its neighbors.
- Diameter: The maximum distance between two nodes in the network.
- Regularity: A network is regular when all the nodes have the same node degree.
- Symmetry: A network is symmetric when it looks alike from every node.

Routing Algorithms

Routing algorithms define the path followed by each message or packet. The list of routing algorithms proposed in the literature is almost endless. The routing algorithm used influences many properties of interconnection networks. Hereinafter the most important ones are reported:

- *Connectivity.* It deals with the capability to route packets from any source node to any destination node.
- *Adaptivity.* Ability to find alternative paths for packets in the presence of contention.
- *Deadlock and livelock freedom.* Ability to guarantee that packets will not block or wander across a network forever.

Routing algorithms can be classier as:

- Deterministic routing always selects the same path between two nodes, even if there are multiple paths.
- Oblivious routing does not consider the state of the network when making decisions.

- Adaptive routing uses information about the state of the network to make routing decisions. These algorithms attempt to circumvent congestion points in the network in an effort to more evenly distribute traffic.

Packet-switched routers, based on the information in the packet header, determine an input → output configuration of the switch to forward the packet. Formally, a routing function is defined as:

$$R[p, i]: P \times I \rightarrow O^n$$

Given the routing information in the packet header p (P is the ensemble of possible packet headers) and the corresponding input port i from the ensemble of the input ports I of the router, it maps to an ensemble of output ports O $\{0,.....,$ n-1$\}$ of the same router. A deterministic routing function has n =1 and $\forall$ (p, i) $\in$ (P$\times$I), $\exists!$ o$\in$ O / R(p, i)= o. An adaptive routing function has 1 <n $\leq$ card (O), so for a given packet header p, multiple output ports O$\{0,...,$n-1$\}$ are possible. A selection function is then required to choose a single output port o from the routed ensemble O $\{0,...,$ n-1$\}$. The selection function typically does not depend on p, but on an internal state of the router, so that at different moments in time, a different output port can be selected for the same (p, i) $\in$ (P$\times$I) input pair.

Let us consider a 3 × 3 mesh and the XY routing algorithm. Function L_{xy} represents the routing logic of each node. It decides the next hop of a message depending on its destination. In the following definition, s_x or d_x denotes the coordinate along the X-axis, and s_y or d_y denotes the coordinate along the Y-axis.

```
function Lxy(s,d)
if s = d then return d
else if sx < dx then return (sx + 1,sy)
else if sx > dx then return (sx - 1,sy)
else if sx = dx and sy < dy then return
```

```
(sx,sy + 1)
else return (sx,sy - 1)
end if
end function
```

Problems on Routing

Deadlock, livelock are potential problems on routing algorithms. Routing is in deadlock when two packets are waiting each other to be routed forward. Both of the packets reserve some resources and both are waiting each other to release the resources. Routers do not release the resources before they get the new resources and so the routing is locked. Livelock occurs when a packet keeps spinning around its destination without ever reaching it. This problem exists in non-minimal routing algorithms. Livelock should be cut out to guarantee packet's throughput.

Switching Techniques

The switching technique determines how data flows through a router, from its input port to its output ports. There are three switching techniques, circuit, packet, and wormhole switching.

In *circuit switching*, a physical path consisting of a series of links and routers is reserved from the sending node to the destination node. The setup time refers to the time required to reserve the resources, and the tear-down time refers to the time required to release them. Circuit switching has a high initial latency due to the setup time, but it exhibits high throughput because the bandwidth is guaranteed due to the reserved resources. The disadvantage is that during the setup and tear-down times, when data is not being transmitted, the network resources are underutilized.

In *packet switching*, large messages are broken up into smaller pieces called packets. Each packets flows through the network independently, possibly along different routes, from sender to receiver. Each packet must be stored in its entirety before being forwarded to the next node on the network, called store-and-forward, which can result in large buffer requirements. Since no resources are explicitly reserved, there is the possibility that two or more packets may wish to use the same resources at the same time, called contention. When contention occurs, one packet is granted the resource, and all others must wait.

Network Flow Control

Network flow control, also called as routing mode, determines how packets are transmitted inside a network. The mode is not directly dependent to routing algorithm. Many algorithms are designed to use some given mode, but most of them do not define which mode should be used.

Store-and-Forward is the simplest routing mode. Packets move in one piece, and entire packet has to be stored in the router's memory before it can be forwarded to the next router. So the buffer memory has to be as large as the largest packet in the network. The latency is the combined time of receiving a packet and sending it ahead. Sending cannot be started before the whole packet is received and stored in the router's memory.

Virtual Cut-Through Routing is an improved version of store-and-forward mode. A router can begin to send packet to the next router as soon as the next router gives permission. Packet is stored in the router until the forwarding begins. Forwarding can be started before the whole packet is received and stored to router. The mode needs as much buffer memory as store-and-forward mode, but latencies are lower.

Wormhole Routing. In wormhole routing packets are divided to small and equal sized flits (flow control digit or flow control unit). A first flit of a packet is routed similarly as packets in the virtual cut-through routing. After first flit the route is reserved to route the remaining flits of the packet. This route is called wormhole. Wormhole mode requires less memory than the two other modes because only one flit has to be stored at once. Also the latency is smaller and a risk of

dead-lock is larger. The risk can be reduced by multiplexing several virtual ports to one physical port, so the possibility of traffic congestion and blocking decreases.

NoC Design Challenges

Designing an efficient NoC architecture, while satisfying the application performance constraints is a complex process. The design issues span several abstraction levels, ranging from high-level application modeling to physical layout level implementation. Some of the most important phases in designing the NoC include: modeling NoC topology for the application, mapping of cores onto the topology, finding paths and reserving resources, verifying performance of the system, developing simulation and synthesis models, and achieving reliable operation of the interconnect. In order to handle the design complexity and meet the tight time-to-market constraints, it is important to automate most of these NoC design phases. To achieve design closure, the different phases should also be integrated in a seamless manner. The NoC design challenge lies in the capability to design hardware-optimized, customizable platforms for each application domain.

Computer-aided synthesis of NoCs is particularly important in the case of application-specific systems on chip, which usually comprise computing and storage arrays of various dimensions as well as links with various capacity requirements. Moreover, designers may use NoC synthesis as a means for constructing solutions with various characteristics that can be compared effectively only when a detailed model is available. Thus, synthesis of NoCs can be used for comparing prototypes. Needless to say, synthesis may also be very efficient for designing NoCs with regular topologies. NoC architectures are pushing the evolution of traditional circuit design methodologies to deal effectively with functional diversity and complexity. At the application level, the key design challenge is to expose task-level parallelism and to formally capture concurrent communication in models of computation. Then high-level concurrent tasks have to be mapped to the underlying communication and computation resources. At this level, an abstract model of the hardware architecture is usually exposed to the mapping tool, so that area and power estimates can be given in the early design stage, and different objective functions (e.g., minimization of communication energy) can be considered to evaluate the feasibility of alternative mappings.

MODEL DRIVEN ENGINEERING

MDE revolves around three focal concepts: *Models*, *Metamodels* and *Model Transformations*. A model is an abstract representation of some reality and has two key elements: *concepts* and *relations*. Concepts represent "things" and relations are the "links" between these things in reality. A model can be observed from different abstract point of views (views in MDE). The abstraction mechanism avoids dealing with details and eases re-usability. A metamodel is a collection of concepts and relations for describing a model using a model description language; and defines syntax of a model. This relation is analogous to a text and its language grammar. Each model

is said to *conform* to its metamodel at a higher definition level. A metamodel can be viewed as an IR in an HLS flow. Finally, MDE permits to separate the concerns in different models, allowing reutilization of these models and to keep them human readable.

The MDE development process starts from a high abstraction level and finishes at a targeted level, by flowing through intermediate levels of abstraction via *Model Transformations* (MTs) (Sendall and Kozaczynski 2003); by which concrete results such as an executable model (or code) can be produced. MTs carry out refinements moving from high abstraction levels to low levels models and help to keep the different models synchronized. At each intermediate level, implementation details are added to the MTs. A MT as shown in Figure 5 is a compilation process that transforms a source model into a target model and allows to move from an abstract model to a more detailed model. Usually, the initial high level models contain only domain specific concepts, while technological concepts are introduced seamlessly in the intermediate levels. The source and target models each conform to their respective metamodels, thus respecting *exogenous* transformations (Mens, T., and Van Gorp, P 2006). A model transformation is based on a set of rules (either declarative or imperative) that

Figure 5. An overview of model transformations

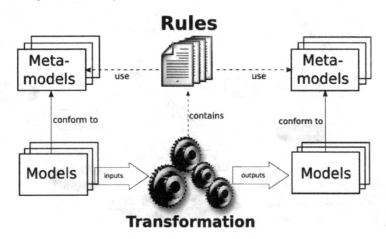

help to identify concepts in a *source* metamodel in order to create enriched concepts in the *target* metamodel. New rules extend the compilation process and each rule can be independently modified; this separation helps to maintain the compilation process. The advantage of this approach is that it allows to define several model transformations from the same abstraction level but targeted to different lower levels, offering opportunities to target different technology platforms. The model transformations can be either unidirectional (only source model can be modified; targeted model is re-generated automatically) or bidirectional (targeted model is also modifiable, requiring the source model to be modified in a synchronized manner) in nature. In the second case, this could lead to a model synchronization issue (Stevens, P 2007). For model transformations, OMG has proposed the Meta-Object Facility (MOF) standard for metamodel expression and Query/ View/Transformation (QVT) (OMG 2005) for transformation specifications.

GASPARD: MARTE COMPLIANT CO-DESIGN FRAMEWORK

Gaspard (DaRT team 2009, Gamatié et al 2008b) is a MDE oriented SoC co-design framework that utilizes a *subset* of the MARTE standard currently supported by SoC industry. In Gaspard as in MARTE, a clear *separation of concerns* exists between the hardware/software models, as shown in Figure 6.

Gaspard has also contributed in the initial MARTE conception. One of the key MARTE packages, the *Repetitive Structure Modeling* (RSM) package has bee inspired from Gaspard. Gaspard, and in turn RSM, is based on the Array-OL (Boulet, P 2007) model of computation

Figure 6. GASPARD framework with deployment added at the MARTE specification level

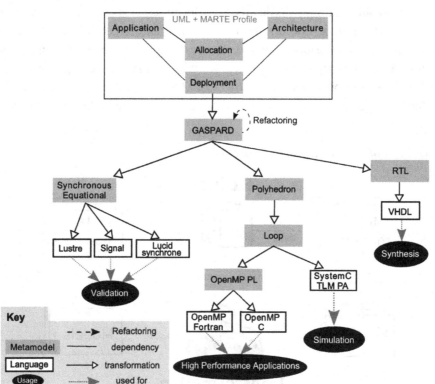

146

(MoC) that describes the *potential parallelism* in a system; and is dedicated to intensive multidimensional signal processing (ISP). Array-OL itself is a specification language and not an execution model. In Gaspard, data are manipulated in the form of multidimensional arrays. The absence of limited number of dimensions in data arrays allows to represent data in a manner typical of their manipulation in ISP applications. For example, video processing applications handle two spatial and one temporal dimension. Sonar chain is another kind of application, which handles spatial, temporal and frequency dimensions. RSM allows to models such applications.

RSM permits to describe the regularity of a system's structure (composed of repetitions of structural components interconnected in a regular connection pattern) and topology in a compact manner. Gaspard uses the RSM semantics to model large regular hardware architectures (such as multiprocessor architectures) and parallel applications. For application functionality, both data parallelism and task parallelism can be expressed easily via RSM. A *repetitive* component expresses the data-parallelism in an application (in the form of sets of input and output patterns consumed and produced by the repetitions of the interior part). A *hierarchical* component contains several *parts*. It allows to define complex functionalities in a modular way and provides a structural aspect of the application: specifically, task parallelism can be described using such a component.

The MARTE Hardware Resource Model (HRM) concepts are inspired heavily from the preexisting hardware concepts in Gaspard. Finally the Generic Component Modeling (GCM) concepts are used as the basis for component modeling. Gaspard currently targets a limited application domain, namely *control and dataflow oriented* ISP applications (such as multimedia video codes, high performance applications and anti-collision radar detection applications). The applications targeted in Gaspard are widely encountered in SoC domain and respect Array-OL semantics (Boulet, P 2007).

Gaspard also integrates the MARTE allocation mechanism (*Alloc* package) that permits to associate the applicative part of the system onto the available hardware resources (for e.g. mapping of a task or data onto a processor or a memory respectively). An example of an allocation is present in Figure 7. The figure clearly illustrates the utilization of the MARTE concepts presented before. The RSM package represents the hardware repetitions and the application loops concisely in a declarative way, while the *Alloc* package allows to map the application on to the hardware resources.

Although MARTE is suitable for modeling purposes, it lacks the means to move from high level modeling specifications to execution platforms. Gaspard bridges this gap and introduces additional concepts and semantics to fill this requirement for SoC co-design.

Gaspard also defines a notion of a *Deployment* specification level (Atitallah et al 2007) in order to generate compilable code from a SoC model. This level is related to the specification of *elementary* components (ECs): basic building blocks of all other components having atomic functions. Although the notion of deployment is present in UML, the SoC design has special needs, not fulfilled by this notion. Hence, Gaspard extends the MARTE profile to allow deploying of ECs. To transform the high abstraction level models to concrete code, detailed information must be provided. The deployment level associates every EC (of both the hardware and the application) to an implementation (code) hence facilitating Intellectual Property (IP) reuse. Each EC ideally can have several implementations: e.g. an application functionality can either be optimized for a processor (written in C/C++) or written in hardware (HDL) for implementation as an hardware accelerator. Hence this level is able to differentiate between the hardware and software functionalities; and allows to move from platform independent high level models to platform dependent models for eventual implementation. Deployment provides IP information to model transformations

Figure 7. An Allocation: mapping a part of a H.263 codec onto a QuadriPro architecture

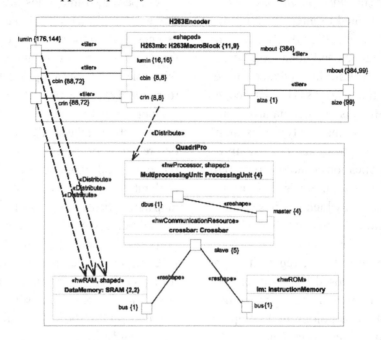

to form a compilation chain in order to transform the high abstraction level models (application, architecture and allocation) for different domains: formal verification, simulation, high performance computing or synthesis. Hence deployment can be seen a potential extension of the MARTE standard to allow a complete flow from model conception to automatic code generation. It should be noted that the different transformation chains: simulation, synthesis, verification etc., are currently unidirectional in nature.

Once Gaspard models are specified in a graphical environment, model transformations are carried out via a transformation tool. However, since the standardization of QVT, few of the investigated tools are powerful enough to execute large complex transformations such as present in the Gaspard framework. Also none of these engines is fully compliant with the QVT standard. An alternative solution to QVT is the Eclipse Modeling Framework or EMF (Eclipse a), that allows to create and modify models. In order to solve this dilemma, in 2006, an initial transformation tool called MOMOTE (MOdel to

MOdel Transformation Engine) was developed internally in the team that was based on EMFT QUERY (Eclipse b). MOMOTE is an enhanced Java framework that allows to perform model to model transformations. It is composed of an API and an engine. It takes source models as input and produces target models with each conforming to some metamodel. Another advantage of MOMOTE over the then existing transformation tools was that it supported external black box calls: e.g. native function calls, rule inheritance, recursive rule call and integration of imperative code. However, since that time, new tools such as QVTO and smartQVT have emerged that implement the QVT Operational language and are effective for handling the Gaspard model transformations. Currently, in order to standardize the model transformations and to render them compatible with the future versions of the MARTE standard; we have chosen QVTO as the future transformation tool for Gaspard. Current all the existing MOMOTE based transformation rules for each execution platform are being converted into QVTO based transformation rules.

MOCODE (MOdels to CODe Engine) is another internal Gaspard integrated tool that allows automatic code generation and is based on EMF JET (Java Emitter Templates) (Eclipse c). JET is a generic template engine for code generation purposes. The JET templates are specified by using a JSP (JavaServer Pages) like syntax and are used to generate Java implementation classes. Finally these classes can be invoked to generate user customized source code, such as Structured Query Language (SQL), eXtensible Markup Language (XML), Java source code or any other user specified syntax. MOCODE offers an API that reads input models, and also an engine that recursively takes elements from input models and executes a corresponding JET Java implementation class on them.

DEPLOYMENT LEVEL: A DETAILED OVERVIEW

In order to generate an entire system from a high level specification, all implementation details of every EC have to be determined. Low level details are much better described by using usual programming languages instead of graphical UML models. As explained before, the deployment level in Gaspard enables one to precise a specific implementation (IP) for each EC (of both application and architecture) among a set of possibilities.

The reason being that in a complex system on chip design, one functionality can be implemented in different ways. This is necessary for testing the system with different tools, or at different abstraction levels. For instance, different IPs can be provided for a given application component and may correspond to an optimized version for a specific processor or a version compliant with a given language. As compared to the earlier deployment concepts specified in (Atitallah et al 2007), the current deployment level has been modified to respect the semantics of traditional UML deployment diagrams.

The concept of *VirtualIP* has been introduced to express the behavior (functionality) of a given EC, independently from the compilation target. It links to all the possible implementations (IPs) for one EC. Finally, the concept of *CodeFile* is used to specify, for a given IP, the file corresponding to the source code and its required compilation options. The CodeFile thus identifies the physical path of the source code. It should be noted that the modeling of a CodeFile is not possible in the *UML composite structure diagram* but is carried out in the *UML Deployment diagram*. The desired IP is then selected by the SoC designer by linking it to the EC through the *implements* dependency.

Figures 8 and 9 show a clear description of the deployment level. The component *HuffmanCoding* is an elementary component of the Gaspard application (H.263 codec) present in Figure 6. At the deployment level, this elementary component has several possible implementation choices. These choices can be for the same execution platform (same abstraction level) in a given language, or can be for different ones. In the illustrated example, the component can be implemented for simulation in SystemC or can be implemented as hardware functionality: a hardware accelerator in an FPGA by synthesizable VHDL. The final *implements* dependency from the *Huffman-VHDL* component to the *HuffmanCoding* illustrate that this is the targeted implementation choice and the execution platform.

PROPOSED DESIGN FLOW

Currently in our design flow related to the synthesis and implementation at the RTL level, we are able to model the application part of the system, which is modeled via the MARTE profile. Afterwards the model transformations allow to generate that modeled application into the VHDL code, thus transforming the application into a hardware functionality that is afterwards implemented onto the targeted FPGA (while keeping the mul-

Figure 8. Deployment of the HuffmanCoding elementary component

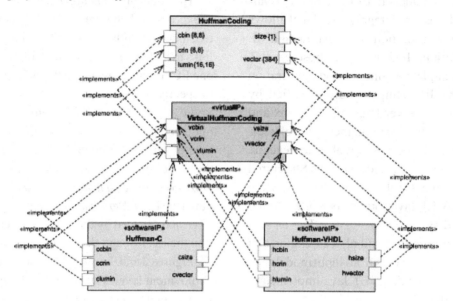

Figure 9. The CodeFile artifact determines the physical path for the code related to an IP

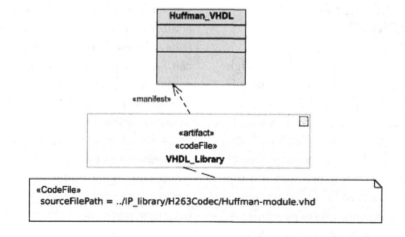

tidimensional arrays and repetitions specified at the modeling level). Figure 10 shows the model transformation view of our current design flow.

The limitation of the current design flow is that the hardware part of a SoC is not modeled at the high abstraction level; and the model transformations do not exist to automatically generate the code from the modeled examples.

In order to model complex NoC systems; and for their automatic code generation, we propose a design flow as shown in Figure 11. This flow will be able to model NoC algorithms (the application part) and the hardware structure of the NoC via the MARTE profile. Afterwards, the algorithms can be allocated to the hardware components (such as processors, network routers etc). Also the elementary components related to the modeled systems can be deployed onto user defined or third party IPs.

Our aim is not to replace the commercial FPGA tools but to aid them in the conception of a system. While tools like ISE and PlanAhead are

Figure 10. Current model transformations flow

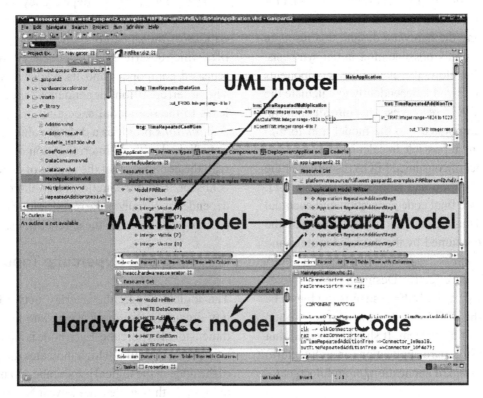

Figure 11. The proposed model driven design flow for NoCs

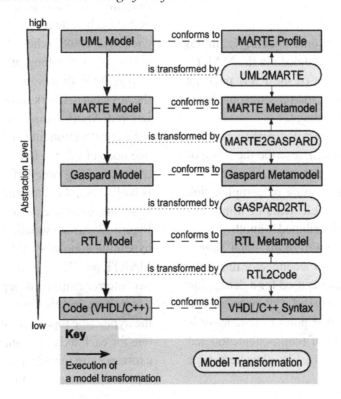

capable of estimating the configurable FPGA resources (CLBs and in turns the slices) required for implementing the hardware design, this resource estimation is only possible after initial synthesis. In our design flow, the elementary components can be synthesized independently to calculate the consumed FPGA resources. This information can be then incorporated into the model transformations, making it possible to calculate the approximate number of consumed FPGA resources of the overall application and architecture (at the RTL model) before final code generation and eventual synthesis. Thus the designer is able to compare the resources consumed by the modeled system and the total resources available on the targeted FPGA resulting in an effective Design Space Exploration (DSE) strategy. If the system is too big to be placed on the FPGA, the designer can carry out a *refactoring* of the system. It should be noted that a refactored Gaspard application (or architecture) remains a Gaspard application (or architecture).

CASE STUDY

In this chapter we present some modeling examples of NoCs which can be modeled via the MARTE profile, taking advantage of the RSM package specifically. For this we first present some basic concepts of the RSM package.

The shape of a pattern is described according to a *Tiler* connector which describe the tiling of produced and consumed arrays. The *Reshape* connector allows representing of complex link topologies in which the elements of a multidimensional array are redistributed in another array. The difference between a *Reshape* and a *Tiler* is that the former is used for a connector that links two parts while the latter is used for a delegation connector: between a port of a component and ports of its parts. Another point to remember is that the ports (interfaces) of a component modeled in Gaspard have the MARTE *FlowPort* stereotype by default.

The *interrepetition* dependency is used to specify an acyclic dependency among the repetitions of the same component, compared to a tiler, which describes the dependency between the repeated component and its owner component. The interrepetition dependency specification leads to the sequential execution of repetitions. A *defaultlink* provides a default value for repetitions linked with an interrepetition dependency, with the condition that the source of dependency is absent. The introduction of an interrepetition dependency serializes the repetitions and data can be conveyed between these repetitions.

Modeling of a Hypercube Topology

The logical hypercube overlay network topology organizes the applications into a logical n-dimensional hypercube. Each node is identified by a label (e.g., "010"), which indicates the position of the node in the logical hypercube. In an overlay network with N nodes, the lowest N positions of a hypercube are occupied (according to a Gray ordering). One advantage of using a hypercube is that each node has only [log N] neighbors [Liebeherr, J 1999], where N is the total number of nodes. Also, the longest route in the hypercube is [log N] A disadvantage of hypercube is that the physical network infrastructure is completely ignored. Another disadvantage is that the hypercube construction must be done sequentially, i.e. one node at a time. Therefore, for large groups it can take a long time before the overlay network is built. Also, the departure of a single node may require substantial changes to the overlay topology. Figure 12 shows modeling of a three dimensional hypercube topology as shown in Figure 4 via the MARTE profile. This modeling approach utilizes two interrepetition link dependencies. The *Router* component in this figure represents one PE in the hypercube topology and contains three ports, one for each axis. The shape value on the router expresses the repetition of the component that is repeated 8 times. The interrepetition link dependencies connect one instance of the component to another consecutive neighboring instance on the relative axis. The values on the interrepetition dependencies thus determine the exact connection in relation to the axis.

Modeling of a Mesh Topology

A mesh-shaped network consists of m columns and n rows. The routers are situated in the intersections of two wires and the computational resources are near routers. Addresses of routers and resources can be easily defined as x-y coordinates in mesh. Regular mesh network is also called as Manhattan Street network. It represents a topology for on-chip networks, which is often proposed in NoC related literature [Jantsch, A 2003]. For a mesh network, routers have at maximum four neighbours, the maximum distance in hops on an n × m mesh network is $(n - 1) + (m - 1)$. Figure 13 show modelling of a Mesh topology of a Mesh network as shown in Figure 4 via the MARTE profile.

As similar to the modeling approach in Figure 12, the RSM dependencies allow to express the mesh topology in a compact manner. The shape of 3,3 on the router component illustrates that this component is repeated 9 times.

Modeling of a Star Topology

In a star topology, the hop count is always 1 and every transaction goes through the central cross-bar switch. The star graph of order N, sometimes simply known as an "n-star" is a tree on n nodes with one node having vertex degree n-1 and the other n-1 having vertex degree 1. The central switch has a number of N I/O ports and the average distance between two PEs via the central switch is $(N)^{1/2}-1$. Figure 14 shows the modeling of a Star topology in MARTE.

The *VRouter* component contains a *Router* component that is itself repeated 8 times. Each repetition of the Router component has a port with a dimension of 1, while the VRouter component has a port with a dimension of 8. Thus a tiler connector is utilized to connect the different instances of the Router component to the VRouter component. Finally at the highest level of modeling, this VRouter component is connected to another component, the *Rout* component which also has a port with a dimension of 8. A simple connector is sufficient to connect the components VRouter and Rout to each other.

Figure 12. Modeling of a Hypercube topology

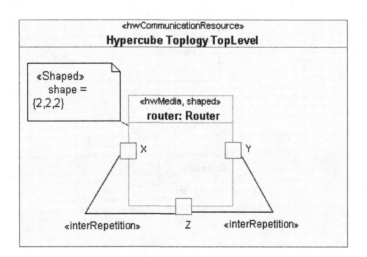

Figure 13. Modeling of a Mesh topology

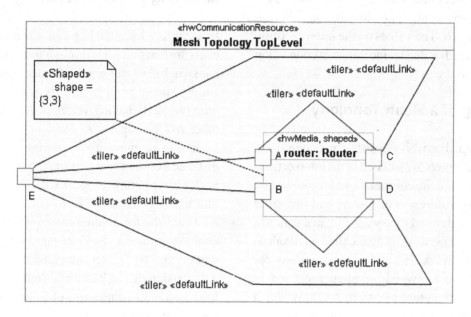

Figure 14. Modeling of a Star topology

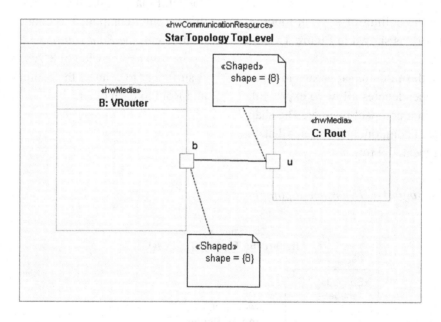

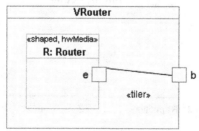

CONCLUSION

This paper presents a novel model driven methodology to move from high level MARTE specifications to complex Network on Chip architectures. Our methodology allows to specify complex NoC intensive signal processing applications such as a multimedia codecs and digital filters in a graphical language, which via model transformations, are implemented as hardware functionalities in a targeted FPGA. These functionalities retain the inherent task and data parallelism specified at the modeling level. Currently we are in the process of developing the model transformations which will be able to take the modeled systems as input and generate the code which corresponds to the user requirements and the high abstraction level models.

REFERENCES

Atat, Y., & Zergainoh, N. (2007). Simulink-based MPSoC Design: New Approach to Bridge the Gap between Algorithm and Architecture Design. In. *Proceedings of ISVLSI, 07, 9–14.*

Atitallah., et al. (2007). Multilevel MPSoC simulation using an MDE approach. In Proceedings of SoCC 2007.

Benini, L., & Micheli, G. (2001). *Network on Chips: A new SoC Paradigm.* IEEE Computer.

Berthelot, (2008). A Flexible system level design methodology targeting run-time reconfigurable FPGAs. *EURASIP Journal of Embedded Systems, 8*(3), 1–18. doi:10.1155/2008/793919

Boden., et al. (2008). GePARD - a High-Level Generation Flow for Partially Reconfigurable Designs. In *Proceedings of ISVLSI 2008.*

Boulet, P. (2007). *Array-OL Revisited, Multidimensional Intensive Signal Processing Specification.* (Tech. rep.)., INRIA. Retreived from http://hal.inria.fr/inria-00128840/en/

Carver, (2008). *Relocation and Automatic Floorplanning of FPGA Partial Configuration Bit-Streams. (Tech. rep.).* Redmond, WA: Microsoft Research.

Cesario., et al. (2002). Component-Based Design Approach for Multicore SoCs. In *Proceedings of Design Automatic Conference,* (DAC'2002), (pp. 789).

Dally., et al. (2001). Not Wires: On-Chip Interconnection Networks. In *Proceedings of the IEEE Design Automation Conf.,* (pp. 684–689).

Damasevicius, R., & Stuikys, V. (2004). Application of UML for hardware design based on design process model. In *Proceedings of ASP-DAC'04.*

DaRT team. (2009). *GASPARD SoC Framework.* Retrieved from http://www.lifl.fr/DaRT

Dorairaj, (2005). PlanAhead Software as a Platform for Partial Reconfiguration. *Xcell Journal, 55,* 68–71.

Duato, J. (2007). *Interconnection Networks.* San Francisco: Morgan Kaufmann.

Eclipse. (n.d.). *Eclipse Modeling Framework.* Retrieved from http://www.eclipse.org/emf

Eclipse. (n.d.). *Eclipse Modeling Framework Technology.* Retrieved from http://www.eclipse.org/emft

Eclipse. (n.d.). *EMFT JET.* Retrieved from http://www.eclipse.org/emft/projects/jet

Gailliard., et al. (2007). Transaction level modelling of SCA compliant software defined radio waveforms and platforms PIM/PSM. In Proceedings of Design, Automation & Test in Europe, DATE'07.

Gamatié., et al. (2008b). A model driven design framework for high performance embedded systems. *Research Report RR-6614*, INRIA. Retrieved from http://hal.inria.fr/inria-00311115/en

Guo., et al. (2005). Optimized generation of datapath from C codes for fpgas. In Proceedings of Design, Automation & Test in Europe, DATE'05 (112–117).

Jantsch, A., & Tenhunen, H. (2003). *Will Networks on Chip Close the Productivity Gap?* Boston: Kluwer Academic Publishers.

Kangmin., et al. (n.d.). A 51 mW 1.6 GHz on-chip network for low-power heterogeneous SoC platform. In *Proceedings of the IEEE Int. Solid-States Circuits Conference* (Digest of Technical papers), (pp. 152–518).

Kim, K., et al. (2005). An arbitration look-ahead scheme for reducing end-to-end latency in networks on chip. In *Proceedings of the IEEE Int. Symp. on Circuits and Systems*, (pp. 2357–2360).

Koch., et al. (2006). An adaptive system-on-chip for network applications. In *Proceedings of IP-DPS 2006*.

Koudri., et al. (2008). Using MARTE in the MOP-COM SoC/SoPC Co-Methodology. In *Proceedings of MARTE Workshop at DATE'08*.

Kreutz, M., et al. (2005). Energy and Latency Evaluation of NoC Topologies. In *Proceedings of the Int. Symp. on Circuits and Systems*, (pp. 5866–5869).

Lattard., et al. (2007). A Telecom Baseband Circuit based on an Asynchronous Network on Chip (Digest of Technical Papers). In *Proceedings of the IEEE Intl. Solid State Circuits Conf.*, (pp. 258–601).

Lee, S.-J., et al. (2003). An 800 MHz star-connected on-chip network for application to systems on a chip. In *Proceedings of the IEEE Int. Solid-States Circuits Conf.*, (Digest of Technical papers), (pp. 468–469).

Liebeherr, J., & Beam, T. K. (1999). HyperCast: A protocol for maintaining multicast group members in a logical hypercube topology. In *Proceedings First International Workshop on Networked Group Communication (NGC '99)*, (Lecture Notes in Computer Science, Volume 1736, pp. 72-89).

McUmber., et al. (1999). UML-based analysis of embedded systems using a mapping to VHDL. In *Proceedings of the IEEE International Symposium on High Assurance Software Engineering (HASE'99)*, (pp.56–63).

Mens, T., & Van Gorp, P. (2006). A taxonomy of model transformation. In *Proceedings of the International Workshop on Graph and Model Transformation, GraMoT 2005* (Vol. 152., pp. 125–142).

Mohanty, (2002). *Rapid design space exploration of heterogeneous embedded systems using symbolic search and multi-granular simulation*. LCTES/Scopes.

Murali, S. (2004). SUNMAP: A Tool for Automatic Topology Selection and Generation for NOCs. In. *Proceedings of the Design Automation Conf.*, *2004*, 914–919.

OMG. (2005). *MOF Query /Views/Transformations*. Retrieved from http://www.omg.org/cgi-bin/doc?ptc/2005-11-01

OMG. (2007a). *OMG MARTE Standard*. Retrieved from http://www.omgmarte.org

OMG. (2007b). *OMG Unified Modeling Language (OMG UML)*, Superstructure, V2.1.2. Retrieved from http://www.omg.org/spec/UML/2.1.2/Superstructure/PDF/

OSI. (2007)... *System, C*, Retrieved from http://www.systemc.org/.

Panades, I., & Greiner, A. (2007). Bi-synchronous FIFO for synchronous circuit communication well suited for Network-on-Chip in GALS architectures. In *Proceedings of the 1st IEEE/ACM Int. Symp. On Networks-on-Chip*, (pp. 83–92).

Planet, M. D. E. (2007). *Portal of the Model Driven Engineering Community*. Retrieved from http://www.planetmde.org

Quadri, (2009). (in press). A Model Driven design flow for FPGAs supporting Partial Reconfiguration. *International Journal of Reconfigurable Computing*. New York. *Hindawi Publishing*.

Rijpkema, E. (2003). Trade offs in the design a router with both guaranteed and best-effort services for networks on chip. In *Proceedings of Design, Automation and Test in Europe Conference and Exhibition*, (pp. 350–355).

Sendall, S., & Kozaczynski, W. (2003). Model transformation: The heart and soul of model driven software development. *IEEE Software*, *20*(5), 42–45. doi:10.1109/MS.2003.1231150

Stevens, P. (2007). A landscape of bidirectional model transformations. In *Generative and Transformational Techniques in Software Engineering 2007*, (GTTSE'07).

Vangal., et al. (2007). On An 80-Tile 1.28TFLOPS Network-on-Chip in 65 nm CMOS (Digest of Technical Papers) *In Proceedings of IEEE Intl. Solid State Circuits Conference*, (pp.98–589).

Wang, H., et al. (2005). A Technology-aware and Energy-oriented Topology Exploration for On-chip Networks, *In Proceedings of the Conf. on Design Automation and Test in Europe*, (pp. 1238–1243).

Zimmermann, H. (1980). OSI Reference Model—the ISO Model of Architecture for Open Systems Interconnection. *IEEE Transactions on Communications*, *28*(4), 425–432. doi:10.1109/TCOM.1980.1094702

Chapter 7

Dynamic Reconfigurable NoCs:
Characteristics and Performance Issues

Vincenzo Rana
Politecnico di Milano, Italy

Marco Domenico Santambrogio
Politecnico di Milano, Italy

Simone Corbetta
Politecnico di Milano, Italy

ABSTRACT

The aim of this chapter is the definition of the main issues that arise when dealing with the design of a NoC-based reconfigurable system. In particular, after the definition of the target architecture, several factors, requirements and constraints that have to be taken into account during the design of reconfigurable NoCs will be described and analyzed. The second part of this chapter will focus on the main issues in dynamic reconfigurable NoCs design, such as the definition of a layered approach, of a packet-switched communication infrastructure, of a proper routing mechanism and of a communication protocol support. Finally, the last part of this chapter will deal with the description of the most relevant implementation details, such as the placement of the bus-macros, the design of the network switches and the physical implementation of the routing mechanism.

INTRODUCTION

In traditional System-on-Chip design it is possible to know in advance the actual communication requirements, the application needs and all the components (modules) needed to realize the desired architecture. They can be understood a priori, at synthesis-time by the analysis of the application specification. For this reason, once all the components of the system have been defined, they will

remain unchanged throughout the life cycle of the system. In dynamically changing environments, on the contrary, the design factors are likely to change, in that the dynamic nature of the target system does not allow to be fully understood at design time. This chapter describes and analyzes several different issues that arise when designing a reconfigurable NoC. In this particular scenario it is not possible to utilize a generic NoC infrastructure on a generic reconfigurable system, since the interactions between the issues of the two dif-

DOI: 10.4018/978-1-61520-807-4.ch007

ferent scenarios creates a new set of difficulties that need to be explicitly faced by the designer.

The main relevant design factors in NoC design are: latency, parallelism, resource usage. These factors are, in general, subject to a design trade-off, since independent optimization does not guarantee system-wide performance. Latency corresponds to the amount of time required for an output to be generated, given the inputs to that particular functional unit performing the computation. This value relies upon the complexity of the circuitry of the component, and on physical characteristics of the interconnects. The degree of parallelism captures the simultaneous communications that are performed between pairs of modules. This can be used to express and (qualitatively) define the communication performance. The resource usage is a hard constraint, since the bounded physical resources within the device have not to be exceeded. Cost-driven designs take care of the silicon cost when realizing SoC, and employing as few resources as possible reduces the design costs. Reconfiguration capabilities, however, can be used to (partially) solve this problem: the same chip area can be used in time division fashion, by allocating different modules in different time instants on the same area. A flexible communication infrastructure ensures several modules to be connected in different ways, and it guarantees the required level of connectivity. In this way, fault-tolerance could be achieved, resulting in a higher availability of the system. Dynamic reconfiguration at the communication level increases the adaptability level to the communication infrastructure resulting in the system to be able to scale the communication as required. Last, but not least, by exploiting reusability it is possible to decrease the (mean) time required to design the desired target system. Within this new context the designers have to face new issues, ad previously hinted. Among this new set of issues, the most relevant ones concern the definition of a novel approach to implement a reliable routing mechanism and reliable communication channels,

both during and after a reconfiguration process. When the NoC has to be configured at run-time, it is necessary to extend the routing protocol with a support that makes it possible to dynamically change the routing of the packets accordingly to the new topology of the underlying communication infrastructure. This means that also the number of switches that a single packet has to cross is not known at design-time, and also this issue has to be solved within the communication protocol.

From the communication channels point of view, it is necessary to find a way in which it is possible to connect different modules that can be placed in different reconfigurable regions and that can be characterized by a different implementation. In order to solve this problem it is necessary to define a shared interface and to employ particular components, the bus-macros, which can be placed on the boundaries between two reconfigurable regions, creating a sort of bridge between them that will not be affected by the reconfiguration processes. This makes it possible to anchor the signals of a reconfigurable module on predefined positions, which will not be changed during the evolution of the system, since bus-macros are never involved in reconfiguration processes. The placement of these bus-macros affects the definition of the reconfigurable architecture and has to be faced by taking into account several factors that influence the overall performance of the whole system.

BACKGROUND

Target Reconfigurable Architecture

The target reconfigurable architecture is based on the model presented in (Ferrandi, F.; Santambrogio, M.D.; Sciuto, D. (2005)), consisting of a static part and a reconfigurable one, that can be further split into several *reconfigurable regions*. The chip (FPGA device) is divided into two (or more) regions, the first, the *static part*, containing all those physical resources needed throughout the entire system life cycle, and the remainder, the *reconfigurable region*s, representing the dynamically reconfigurable portions of the system. The adoption of such a model simplifies the task of resources allocation, and it also adheres to tools' requirements[1]. In addition, a suitable fixed communication channel has to be established between the two regions, and this has been exploited by means of bus-macro: they represent a fixed communication channel across the reconfigurable regions boundaries. The capabilities of modern FPGA devices reconfiguration, being glitch-less, avoid those signals spanning the boundaries to be lost during reconfiguration processes.

Bus-Macros

As described by Xilinx Inc. (2006), bus-macros provide a means of locking the routing between different reconfigurable regions (or between a reconfigurable region and the static part), making it possible to connect the pins of their interfaces. As a result, all connections between different regions have to pass through a bus-macro, with the exception of the clock signal (global signals, GND and VCC, are handled automatically by the tools in a way that is transparent to the user). It is important to note that all reset signals must pass through a bus-macro too.

Xilinx bus-macros can be classified by their direction, their width and if they are synchronous or not. In particular, bus-macros can be used to route signals from the left to the right and from the right to the left, or from to the top to the bottom and from the bottom to the top (only on devices, such as Xilinx Virtex 4 and Virtex 5, supporting 2D reconfiguration). The physical width of these bus-macros can be 2 *Configurable Logic Block* (*CLB*) if they are *narrow* or 4 *CLB* if they are *wide*. In addition to this, the signals passing through a bus-macro can be registered or not. In the first scenario the bus-macro is synchronous, while it is asynchronous in the latter.

All Xilinx bus-macros, regardless of direction or physical width, provide eight bits of data bandwidth and enable/disable control. They are provided by with the *Early Access Partial Reconfiguration* (*EAPR*) software tools (Xilinx Inc. (2006)) in the form of pre-placed, pre-routed hard macros for several kinds of Xilinx devices.

Packet-Switching Routing

The term packet-switching refers to a precise mechanism by which data is forwarded along the network. Packet-switching is opposed to classical circuit-switching information flow, in that the basic information units that are sent in this context are called *packets*. A packet contains application-specific information that has to be read by the destination. Two terms can be used in place of packet, denoting a more precise characterization of the mechanism: *phit* and *flit*. The first term refers to the physical information unit that is sent across the network, and it represents the minimum amount of information that can be sent. The term *flit*, instead, is used to denote a message, i.e. a collection of (possibly) more than one *phit*. In general, a message is composed of several *phits*, each having a precise role in the communication; the number of *phits* that a *flit* is made of relies both on the network design and on the adopted communication protocol.

To support the information flow, a routing mechanism is adopted. The term *routing* corresponds to the choice of the output branch toward which an incoming packet should be directed to reach the desired destination. Routing is implemented by means of routing tables, containing the necessary information for a packet to reach the destination.

Furthermore, from now on, two terms will be used to denote the end-points of the communication. An *initiator* is the actor, within the communication architecture, that starts a communication request by sending the appropriate packets toward the desired destination; in this context, the destination is called *target*. Notice that these two terms are network-related, in that they refer to the *Network Interface* (NI) of the corresponding modules. The NI is used to connect the Core to the network. A Core is any kind of processing module in the system, having connectivity requirements.

Factors, Requirements and Constraints of Reconfigurable NoCs Design

Several factors guide the design of a complex system, both at the computational and communication layer. The main design factors of the communication fabrics are its performance, latency, area and computational overhead. In addition, a suitable communication infrastructure for dynamically reconfigurable environments should address flexibility and reliability.

Nevertheless, in dynamically reconfigurable systems the application requirements cannot be easily understood *a priori*. New applications can be loaded on the device *after* the system has been deployed, or existing applications can be removed from the target system during execution. A precise support has to be ensured in order to achieve such a capability; for these reasons the design of a valid communication infrastructure for dynamically reconfigurable architectures is a complex task, in which several competitive factors, requirements and constraints are to be considered (Mak, T.S.T.; Sedcole, P.; Cheung, P.Y.K.; Luk, W. (2006)). As opposed to non-dynamically reconfigurable systems, the design factors may change due to the inherently different features of the target environment. For this reason, some definitions slightly differ from their common formulation.

Latency

Latency corresponds to the amount of time required for an output to be generated, given the inputs to the functional unit. In case of packet-switched network fabrics, this corresponds to the time required for a packet to be processed by the routing element. This value relies upon the complexity of the circuitry of the component, and also on physical characteristics of the interconnections. It can be defined as an integral positive number representing the number of clock cycles that a data transfer is delayed when passing through a generic element (Pionteck, T.; Albrecht, C.; Koch, R.; Maehle, E.; Hubner, M.; Becker, J. (2007)). In addition, *path latency* of length *p* can be defined as the number of clock cycles delay when passing through exactly *p* elements (refer to Figure 1 (a)); the path latency depends upon the single latencies through the *p* elements, as defined in Equation 1, taken from (Pionteck, T.; Albrecht, C.; Koch, R.; Maehle, E.; Hubner, M.; Becker, J. (2007)).

$$path\_latency_p = \sum_{j=1}^{p} latency_j \qquad (1)$$

However, in case a pipelined architecture is employed, the aforementioned definition does no longer hold. As a matter of fact, the term latency would refer to the time delay between two subsequent inputs, not capturing the inherent parallelism of the approach; in this case, it is better to think of the *execution time*, related to the entire computation. This scenario is reported in Figure 1 (b).

Whether dynamic reconfiguration is employed or not, latency has to be minimized in order to provide the end-user with an efficient communication scheme. Latency directly impacts on the user perspective of the executing system.

Bandwidth

Bandwidth is the amount of information *per unit of time* that a channel can support. It relies on physical characteristics of the medium, so that improvement can be obtained through low (physical) level optimization. Impacting features on bandwidth are the system clock frequency and the width (in bit) of the channel.

Figure 1. Latency in a non-pipelined case (a), and in a pipelined case (b)

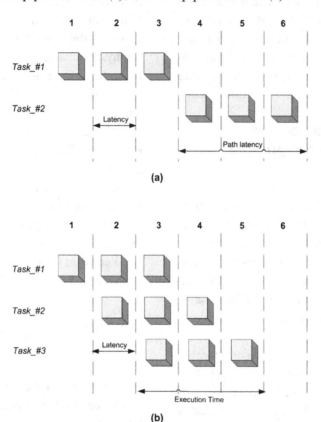

The *average* bandwidth can be defined as reported in Equation 2; T_0 is the period over which we are calculating the average, while *phi(t)* is the amount of data transferred at time t along that link j. In this case, without loss of generality, the summation is finite, representing a discrete time.

$$bandwidth_{j,T_0} = \frac{1}{T_0} \sum_{t=0}^{T_0} \phi(t) \qquad (2)$$

Bandwidth can be used to measure the efficiency (performance) of the links at local level; as a matter of fact, the measurement will keep trace of the only logical link that connects two modules. In order to get a better measurement of the performance of the infrastructure, throughput (explained next) should be analyzed.

Throughput

Throughput is used to characterize the *overall* performance of a communication architecture. Throughput is a system-wide concept, being the amount of data *per unit of time* that is delivered in the entire system. Equation 3 reports its expression, where k is the number of active elements in the network, T_0 is the period over which to calculate the throughput and bandwidth$_{j,T0}$ follows the definition of bandwidth as the amount of information sent across a local channel j, in a given period T_0.

$$throughput_{T_0} = \sum_{j=1}^{k} bandwidth_{j,T_0} \qquad (3)$$

Throughput represents the amount of parallelism that the network can achieve and sustain.

This parameter should be optimized, in that application may take advantage of highly efficient communication channels.

However, congestion situations - slowing down the application - may arise in the application, and they should be taken into consideration in the design of the network elements. Because of the true potential congestion of local subsystems, throughput can overestimate the system efficiency. For this reason, by definition, it captures the overall performance of a static system, since it is defined as a finite summation over the system channels. In the case of a dynamically reconfigurable system, it can be no longer used to estimate the network performance, since at different time instances a link could be present no more. In this context, a better evaluation is an "execution-wide" throughput, i.e. evaluating performance during the entire execution. The overcome would be the *curve of throughput* capturing the system throughput at any instant (or, at least, at any discrete instant, i.e. at snapshots level). The performance estimation with respect to a time period T_0 should then be defined as the actual amount of information that has been forwarded in the time period, as in Equation 4. $\omega(t)$ corresponds to the curve of throughput, and T_0 is the reference time interval. With this model, the execution of an application in a dynamically reconfigurable context can be seen as a sequence of *photos*, i.e. a sequence of snapshots of the system representing the parameter settings within the application.

$$throughput_{execution-wide} = \frac{1}{T_0} \int_0^{T_0} \omega(t)dt \qquad (4)$$

Each photo corresponds to a particular network setting and environment, having precise performance and Quality-of-Service features and requirements. The overall performance can be done via snapshots analysis, i.e. by performing quantitative analysis over the current photo of the system. In this way the dynamic behavior of the system is captured as a sequence of static phases, and the different methods for performance analysis seen for the static case can be here employed.

Resource Allocation

Programmable devices have bounded physical resources, so that they have bounded active areas. Cost-driven designs take care of the silicon usage when realizing on-chip systems. However, thanks to modern technologies, this parameter is slightly changing since the arrival of reconfigurable logic devices makes it possible to reuse the same chip (the same silicon area) for different purposes and at different times (Kao C. (2005)). In this way the same chip area can be reused, increasing the resource utilization efficiency.

The required amount of physical resource (area) relies on the design process quality, and on the complexity of the system. In static architectures, the area requirements can be optimized at design-time, and depending on the chip layout style that is employed the resource requirements can be minimized. For example full-custom technology can be adopted to address minimal resource usage, but meanwhile requiring high costs and design efforts (Mack, R.J. (1996)). On the other hand, in dynamically reconfigurable systems area constraints cannot be solved at design-time; during the normal execution of the application, both requirements and constraints may be subject to change.

In the context of communication infrastructure design, the resources utilization corresponds to the physical resources that are allocated to realize the desired functionality.

Scalability

Scalability is the degree of adaptability of the system performances as a function of the amount of required workload. Greater the scalability, greater is the chance for the system to behave with few requests as well as for instance with several interconnecting actors.

For a communication infrastructure to be scalable, it means that with the increasing number of modules requiring (concurrent) communication, the network scales its performances to support and sustain the required service. Considering the most widely used communication paradigm in SoC design, scalability is not achieved in point-to-point links and buses; as matter of fact, in the former case the resource usage would be too large, while in the second case the latency (waiting time for the connected cores) would be too high to be suitable for on-chip communication.

Reusability

Reusing pre-existing (possibly third-party) modules in a new design is one of the approaches by which cost reduction is achieved. Reusability is based on the concept of libraries. A library is a set of modules with specific or general purposes.

STATE OF THE ART

This section presents the related works in the context of Network-on-Chip design paradigm, with particular attention to dynamically reconfigurable architectures.

XPipes (Bertozzi, D.; Benini, L. (2004)) will be presented as the state-of-the-art solution for the general NoC approach, even though not supporting dynamic reconfiguration; on the other hand, CuNoC (Jovanovic, S.; Tanougast, C.; Weber, S.; Bobda C. (2007)), an extension to the DyNoC (Bobda C.; Ahmadinia A. (2005)) architecture, and CoNoChi (Pionteck, T.; Koch, R.; Albrecht C. (2006)) will be described as dynamically reconfigurable network--based architectures.

XPipes

One of the first implementations of a Network-on-Chip has been proposed by Bertozzi, D. and Benini, L. (2004). *XPipes* is a true packet-switched architecture, with a static routing mechanism. XPipes has been developed targeting high-performance and reliable communication for on-chip multi-processors environments, even if it is not able to support reconfiguration mechanisms. Also, the XPipes architecture is supported by the XPipes synthesis flow for automatic generation of domain-specific communication infrastructures.

XPipes is based on a layered design approach, in which the network is divided into several logical phases. The protocol is called *Smart Stack*, and it has three layers:

1. The *data-link* layer is based on the assumption that the underlying communication link has non-zero probability of error, and its purpose is to achieve reliability up to a predefined minimum. Error correction codes or automatic retransmission can be used.
2. The *network* layer is used to implement an end-to-end delivery control of the flowing information. At this level, atop the previous layer, routing and switching activity is performed. The term routing is referred to the choice of the route toward which the incoming packet should be sent, to reach the destination; switching, on the other hand, refers to the effective packet sending phase. Routing and switching are required in multi-channels communication networks, since the network has to decide which is the path to follow in order to deliver the packet.
3. Last, the *transport* layer is used to decompose messages into packets at the source and to compose packets into messages at the destination.

The communication protocol in XPipes is based on three main steps. At first, the sender core (called the *initiator*) starts sending the desired data toward the network. The initiator interface implementing the transport layer encapsulates the message into lower-level (network-specific) messages. The initiator, then, starts a session with the communication infrastructure. In the second stage, upon receiving a packet, the network element forwards the packet to an appropriate output port. At the end-point of the communication, the destination (called *target*) receives the packet and delivers the request to the upper-level layers, until the core-specific application space is reached.

In general, XPipes allows for three routing mechanisms to be used. With *store-and-forward* routing, a packet is forwarded to the output port as long as it is entirely stored in the switch. With *virtual cut-through* the latency requirements can be reduced by forwarding the packet as long as it is ensured that the next switch is able to accept the entire packet. Finally, a *wormhole* scheme can be used. In this case a header message contains the routing information and subsequent packets follow the same predefined routing path. In this last case, no buffering mechanism is required at any switch.

DyNoC

DyNoC (*Dynamic Network-on-Chip*) has been proposed by Bobda, C. et al (2005) and by Bobda C. and Ahmadinia A. (2005). Its purpose is to provide the end user with a valid framework to support flexible inter-cores communication. Also, DyNoC targets dynamic module placement within reconfigurable architectures (Bobda, C., Ahmadinia, A.; Majer, M.; Teich, J.; Fekete, S.; Van der Veen, J. (2005)).

The DyNoC architecture consists of a two-dimensional mesh. Each module is composed of a *site* and the required interface. The site is used to host user-defined functional units (also called *Processing Elements* or PEs), as required by the target application. These PEs get access to the communication infrastructure in terms of network elements: each of them is associated with a switch used to forward the packets across the network. Each routing element (aside from the boundaries) is connected to each of its 4 neighbors. This guarantees high connectivity in the system, but it

does limit the core placement. As a matter of fact, the placement of the functional core is required to span a 2-dimensional shape using the sites provided by the DyNoC architecture. Nevertheless, it is possible to define module placement spanning multiple slots.

The purpose of the DyNoC architecture is to make the cores reachable, at any instant of time and from any other module within the system. This is accomplished by the switching elements surrounding the cores. This is ensured by the so-called DyNoC Rule (Bobda, C., Ahmadinia, A.; Majer, M.; Teich, J.; Fekete, S.; Van der Veen, J. (2005) and Jovanovic, S.; Tanougast, C.; Weber, S.; Bobda C. (2007)) if the components are synthesized in such a way that they are surrounded only by communication primitives (the network elements), then the network is ensured to be strongly connected.

The adopted routing protocol is an improved version of the XY-Routing algorithm, the *SXY-Routing* (Surrounding XY-Routing). The XY-Routing algorithm is based on four basic directional moves: up or down, and right or left. These moves correspond to the directions that a packet could take during the data transfer, from the sender to the receiver. Upon receiving a packet, a switch decides where to forward it according to the relative position of the destination. This is achieved by using a centralized routing table. According to this position, one of the four directions is taken. The packet is forwarded toward that specific direction. In this (static) algorithm, the routing is performed either horizontally or vertically, and this is suitable only for those architectures with a fixed two-dimensional network.

The S-XY Routing algorithm is then able to dynamically bypass possible obstacles. The obstacles here represent placed modules in the architecture. The surrounding mechanism is achieved thanks to the presence of control information, used throughout the routing process. Thanks to this capability, the placement of a module does not impose any limitations on the routing, so that the placement is made flexible. The dynamic capabilities of DyNoC come from its support to dynamic module placement. In case a new module has to be inserted in the architecture, the routing components falling in the area defined by the new module are deactivated, so that the only available ones for routing are those placed in proximity to the module boundaries.

CuNoC

CuNoC (*Communication Unit Network-on-Chip*) (Jovanovic, S.; Tanougast, C.; Weber, S.; Bobda C. (2007)) has been proposed as an extension to DyNoC. CuNoC is based on the Communication Unit component, representing the input stage of a network switch. As opposed to general architectures, as for instance the one proposed by Bertozzi, D. and Benini, L. (2004), CuNoC does not use any kind of buffering mechanism at the input ports, and also it employs a priority-to-the-right rule for prioritizing the concurrent incoming requests. As it happens in DyNoC, in CuNoC it is possible to dynamically introduce a new module, mapping it to one ore more tiles of the resulting two-dimensional mesh, and such that it is always surrounded by network components, as required by the DyNoC Rule.

There are two different types of CU: the classic CU and the to-give-way CU (CUgw). Classical CUs receive incoming packets from the input ports, and decides where to route the information to reach the target core. A possible bottleneck situation may occur if all the output ports are not available, meaning that several concurrent communications are going on. To solve this problem, and to exit from the congestion situation the CUgw modules are used, the packets are forwarded from the CU to the adjacent CUgw, and these latter components will then forward the message to the adjacent CU according to the destination address.

CoNoChi

CoNoChi, *Configurable Network-on-Chip*, directly applies partial reconfiguration to the communication level of the target SoC. Aim of CoNoChi is to provide a flexible and dynamically reconfigurable infrastructure for dynamically reconfigurable SoCs (Pionteck, T.; Koch, R.; Albrecht C. (2006)). The rationale is that a flexible network leads to an easy placement of the hardware modules; also, they can be plugged in and unplugged from the system, at run-time and without affecting the remainder portion of the system, not subject to the reconfiguration task. The reconfiguration process of the network is controlled by a centralized (global) controller instance, required to correctly update the system state and the routing tables of the switching elements.

CoNoChi is based on a single network element, the switch, connecting pairs of modules to the network fabrics. Links can be used either to connect adjacent switches, or to connect a hardware core to a switch (Pionteck, T.; Koch, R.; Albrecht C. (2006)). This can be accomplished in two ways, i.e. through *direct connection* or through an *ad-hoc* interface. The use of an ad-hoc mechanism as interface, allows the network to exploit logical addressing mode. With this capability, a module is addressed by a unique identifier, being this identifier totally independent on the real physical placement of the core within the FPGA area. This feature is useful when dynamic core relocation is to be realized, so that physical replacement of the module can be performed in a transparent way.

In CoNoChi it is possible to dynamically insert (remove) modules and switches to (from) the network. Dynamic reconfiguration can be performed either by stalling the network or not. In the former case, there is no special requirements. Actually, it may be the case that some packets are in-transit. The requirement is that they have not to be lost. This is accomplished by either deleting or by assigning them to another existing functional unit.

ISSUES IN DYNAMIC RECONFIGURABLE NOCS DESIGN

In order to design a very high-performance recon-figurable NoC, it is absolutely necessary to keep into account all the key points presented in the previous section. For instance, in order to ensure flexibility and reliability, it is necessary to adopt a layered approach in the definition of the packet-switched communication infrastructure and to specifically define an optimized routing mecha-nism and a communication protocol support. Both the routing mechanism and the communication protocol have to be aware of the reconfiguration processes that affect the NoC (e.g., adapting its topology to the current working scenario), since they have to correctly handle the communication flow even during and after each reconfiguration process. Thus, the correct management of these issues is a crucial aspect that has to be taken into account when designing NoCs that have to support the reconfiguration mechanisms (e.g., allowing the run-time modification of the NoC topology).

A Layered Approach

System-on-Programmable-Chip (SoPC) design can be divided into two orthogonal tasks (Keutzer, K.; Newton, A.R.; Rabaey, J.M.; Sangiovanni-Vincentelli, A. (2000)): the design of the com-munication infrastructure and the design of the functional aspect (Processing Elements). The communication layer is composed of physical and logical resources to exchange information between pairs of modules. The computational layer, on the other hand, is composed of those logic elements implementing the desired functionality.

To manage the increasing complexity of the on-chip applications (Benini, L.; De Micheli, G. (2002)), a layered approach has been taken into consideration. The rationale is to decouple the communication layer from the computational one, in order to deal with the complexity issues in large-scale system design. In this way, the communication and the computational layers can be *independently* optimized to achieve high performance and Quality-of-Service. The inde-pendent optimization of two orthogonal aspects of the same system is a relevant feature, in that the overall complexity is reduced, and the efforts can be allocated in an optimized way.

Among all the approaches introduced in the previous section, only XPipes presents a clear definition of several layers, while DyNoC, Cu-NoC and CoNoChi do not focus on this aspect. In particular, as previously hinted, XPipes is based on three layers (data-link, network and transport) in order both to simplify the management and to improve the quality of the end-to-end com-munication.

Layers Definition

The first step of the design has been the defini-tion of the layers composing the communication protocol. As an example, let us consider (without any loss of generality) a communication protocol based on 4 layers, as shown in Figure 2. The stacked approach hides the communication details to the Cores; as a matter of fact, the network interface provides a transparent connection between the Processing Element and the underlying commu-nication infrastructure. This mechanism ensures flexibility and adaptability of the network. In this way, a change in the business logic could affect only the functional-level design, without (at a minimal regularity assumption) influencing the network-side logic. The same applies on the other direction, since if some details are changed within the communication mechanisms only the network interface implementation has to be changed accordingly.

The first layer is called *inter-switch layer*. This can be considered as the network layer for the ISO/OSI protocol case, for well-known data networks. The information units exchanged at this level are packets. Purpose of this layer is to provide a message exchange mechanism

Figure 2. The proposed layered approach

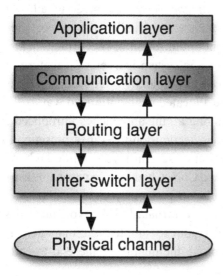

between pairs of network nodes, to route and forward information toward the desired destination module. This layer is implemented as the inter-nodes communication in the network components, such as switches and routers. The minimum amount of information unit that is exchanged at this level is the *phit*.

The *routing layer* allows the definition of the routing path for an incoming packet. The routing layer is (generally) built atop the network layer, and routing decision task is independent on the packet. Nevertheless, there exist an alternative possibility, i.e. embedding the routing policies layer at the interface level, i.e. at the application layer. This feature will be explained in detail in the following sections.

The *communication layer* is used to manage the end-to-end communication; purpose of this layer is to provide a suitable level of control over the entire communication between a Master and a Slave core. The communication is considered successful if packets sent from the initiator are correctly received from the target component, and a positive reply is sent back to the Master. This layer is implemented at the interfaces of the modules, with specific control mechanisms.

Being a packet-switched network, the requests coming from the Master core have to be translated into a sequence of messages. The *application layer* is employed for this goal. Requests coming from the application are translated into a sequence of network-specific packets. The formats of the packets are well defined in the network protocol. Once again, this task is ensured by the network interfaces by means of an encapsulating module.

A Packet-Switched Communication Infrastructure

To support a layered approach, the communication is based on packet-switching routing instead of circuit-switching. Packet-switching approach, as opposed to wire-based information flow (in which the information is sent along predefined physical channels) addresses flexibility, scalability, and reliability of the target communication (Dally, W.J.; Towles, B. (2001)). Nevertheless, these advantages do not come for free; as a matter of fact, a certain amount of computational overhead is required, both to interpret messages and to compose/decompose them into domain-specific information. For this reason, each module in the system consists of three parts: the business logic, implementing and realizing the desired functionality; the network interface, to compose (decompose) the requests (messages) into messages (requests); the routing scheduler, used to set the routing path that a packet will follow to reach the desired destination.

In packet-switched context, each request is encapsulated into a sequence of one or more packets, and each packet has a precise format and semantics, as defined by the communication protocol. Packets carry useful data and information to the destination; the destination is then responsible for generating a suitable reply. The definition of the communication protocol is then required, to support the entire communication and message exchange. It defines the rules that have to be followed to realize the desired communication.

All the previously presented NoCs support packet-switched routing. For instance, packets in XPipes are called *flits* (logical information units), each being a collection of (possibly) several *phits* (physical information units).

Routing Mechanisms

Routing in this context refers to the choice of the output branch toward which an incoming packet should be directed. The actual choice is then applied by means of forwarding mechanisms, i.e. enabling the desired output port of the switch. There exist different routing policies, for instance borrowed directly from geographic- (or local-) wide networks. Algorithms for this purpose can be either static or dynamic (Dally, W.J.; Towles, B. (2001)). In the former case the routing path is decided once for all *prior* to system initialization; predefined routing paths cannot be changed after the system execution starts. In the latter case, on the other hand, routing paths can be updated and changed dynamically, as system is up and keeps going on executing; in this case dynamic routing tables are employed.

Examples of on-chip dynamic routing tables can be found in modified versions of Distance Vector Routing or Link State Routing (Ali, M.; Welzl, M.; Hellebrand, S. (2005)). However, for performance-constrained systems these tech-niques may require too high computational overhead to be employed in on-chip environments.

Several different routing mechanisms have been employed in the NoCs that can be found in literature. For instance, DyNoC adopt the SXY-Routing, that is an improved version of the XY-Routing mechanism, while XPipes employ three different routing mechanisms (store-and-forward, virtual cut-through and wormhole) in order to achieve a higher level of flexibility and adaptability.

Problem Description

The problem of on-chip routing is more complex than geographical-like data networks routing, since real-time requirements pose a strict limitation to the solutions that can be adopted. On-chip routing (in the network sense) has to ensure flexibility and efficiency at the same time, as well as low resource usage and power consumption. Flexibility is achieved by means of dynamic routing mechanisms, while resource requirements are kept low using ad-hoc routing tables. Power consumption can be controlled by ensuring as low computational overhead as possible.

The routing problem is depicted in Figure 3. The network is composed of 4 switches, used to route the incoming packets. There are 3 independent systems interfaced with the network; these

Figure 3. Routing path problem definition

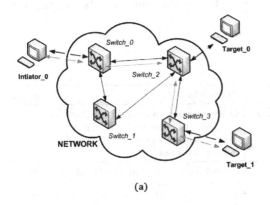

(a)

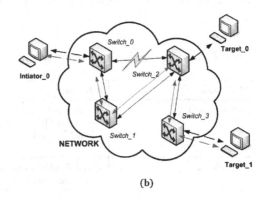

(b)

are depicted as common Computer Workstations in Figure 3. The interconnection between pairs of switches or between a network node and an end-point system are assumed to be full duplex.

The problem consists in the communication between the actor marked as *Initiator_0* and the destination end-point *Target_1*. The communication has to be set up as a sequence of physical links from the sender to the receiver. There exist different paths to reach *Target_1*. According to the performance requirements and to the network traffic, it should be possible to choose the most convenient path. For instance, an available path is the one marked with green lines in Figure 3 (a), in which the packets pass through *Switch_0*, *Switch_2* and *Switch_3*.

The routing path can be changed to the one reported in red in Figure 3 (b), by means of dynamic routing mechanism. In this case packets pass through *Switch_0*, *Switch_1* and *Switch_3*. This dynamic mechanism ensures flexibility. Suppose for instance that the link between *Switch_0* and *Switch_2* is broken, or it is busy; in this case the second routing path can be used, and the communication quality is maintained.

Target-Based vs. Initiator-Based Routing

The aforementioned routing mechanisms can be grouped into two classes: target-based, or initiator-based. In the former approach, the routing path is chosen according to the destination address, and independently on the Master core that initiated the communication request. In the latter case, instead, the routing path is chosen according to the pair initiator-target.

According to the chosen strategy, the implementation changes. In case target-based routing is employed, the routing tables are stored in the network nodes; in these tables there exists an entry for each destination core and for each of them the table stores the information of the routing path. The incoming packets are routed according to the value in the destination identifier field. For each incoming packet, and for each common target, the routing path is fixed (unless a dynamic update is performed) independently on the initiator. Opposed to this mechanism, in the initiator-based routing, the routing table is stored within each initiator. An entry exists for each destination, and the entire routing path is stored. The path is decided a priori with respect to the desired target, and this information is encapsulated in the packets. The routing path can be tailored to the initiator-target pair. The advantages and drawbacks of the aforementioned approaches are reported in Table 1.

A Hybrid Approach

When dealing with a reconfigurable NoC it is necessary to employ a hybrid approach that can be defined by mixing the previously presented routing mechanisms. This makes it also possible to take advantage from the features of both routing mechanisms presented so far. As a matter of fact, the routing tables are stored within each network node (switch); there exists an entry for

Table 1. Routing mechanisms comparison

Mechanism	Advantages	Drawbacks
Initiator-based	- Routing path can be tailored by the initiator - Switch overhead is low - Switch design has low complexity	- Each initiator is aware of the entire network - Update of the routing table is complex - Dynamic network management is complex, overhead increases
Target-based	- Great flexibility and transparency on the communication details	- Switch computational overhead is high, due to the reading of the routing tables

each destination core, and for each entry the output port of the switch to be followed in order to reach the destination is stored. In addition, each master network interface is capable of forcing the routing path toward the destination core. In this way, according to the current network parameters, one of the two mechanisms is used. In addition, the information about the routing path can be dynamically changed in the switch routing tables, and in the network interface. In this way, not only flexibility is achieved at the routing level of the communication, but it is also possible to solve all the problems that arise when the NoC topology is dynamically changed at run-time by means of reconfiguration processes.

To support both mechanisms, a special packet has to be used in order to force the routing path (this issue will be further analyzed in the following sections); the switch reading this packet will not use the local routing information, but the information is piggybacked to the incoming packet. Otherwise, a different packet is used, and the output port is selected according to the destination identifier, contained in the destination field of the incoming message. The Network Interfaces

(NIs) can be configured for one of the two routing mechanisms at communication start-up; as a matter of fact, a software-accessible register has to be used in order to store the routing type information, which will be used to enable the desired routing strategy. This mechanism allows the system to be configured as desired, so that flexibility can be easily achieved and NoC reconfiguration can be completely supported.

An example scenario of an initiator-based routing is reported in Figure 4. The initiator is aware of the only network reported within the dotted yellow box, and it has to communicate with the target module. The request is sent by the upper application layer of the stacked protocol, and the interface is responsible of generating the required packet (as it will be discussed in the next part of this chapter). The initiator, in addition, forces the routing path toward the target module, encapsulating the path within the packet. In this scenario, the data <S, E, E, E> is piggybacked to the packet.

Upon receiving this packet, a switch recognizes the message with force routing option and reads the output port toward which the *phit* should be forwarded. Notice that the switches along the

Figure 4. Force routing example

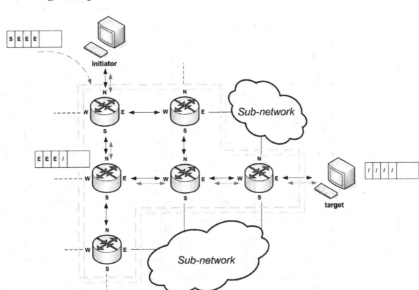

path can mask the presence of other network domains, reported as dotted clouds in Figure 4. This mechanism overrides (temporarily) the information contained in the routing tables, but it does not overwrite them. To do so, the *routing update* packet should be used, instead (as described in the following sections).

The adoption of the hybrid approach described in this section brings, in addition to all the previously described benefits, also some drawbacks in terms of complexity of the design and of area usage. Table 2 shows the impact of the packet size (the maximum number of bits contained in each phit), of the number of entries in the routing tables (the maximum number of target that can be stored in each routing table) and of the number of hops that can be encapsulated in a single packet (for routing force phits) when designing a NoC switch targeting a Virtex 4 device.

On the other hand, Table 3 shows the different contributions to the area usage of the sub-modules of a switch characterized by the following characteristics:

- 2 slot buffers;
- 4 input ports, each one with a single buffer;
- Packet size set to 27 bits;
- 64 entries in the routing tables (6 bits in the packet destination field);
- Number of hops per packet set to 4.

As it is possible to see from Table 2 and Table 3, the main drawback in the adoption of the hybrid approach is the increment of the area usage, even though the maximum size (with 32 bit per packet and 128 targets) is around only 500 slices, that is definitely reasonable for the implementation of a single switch.

Table 2. Resources requirements for Virtex 4 devices

				Slices		
Device Code	**Packet Size**	**Targets**	**Targets (bits)**	**Available**	**Used**	**Percentage**
XC4VFX12	27	32	5	5472	299	5.46%
		64	6	5472	364	6.65%
		128	7	5472	482	8.80%
XC4BSX25	27	32	5	10240	299	2.92%
		64	6	10240	364	5.92%
		128	7	10240	482	7.85%
XC4VLX15	27	32	5	6144	299	4.875%
		64	6	6144	364	5.92%
		128	7	6144	482	8.81%
XC4VFX12	32	32	5	5472	331	6.05%
		64	6	5472	395	7.22%
		128	7	5472	514	9.39%
XC4BSX25	32	32	5	10240	331	3.23%
		64	6	10240	395	3.86%
		128	7	10240	514	8.37%
XC4VLX15	32	32	5	6144	331	5.39%
		64	6	6144	395	6.43%
		128	7	6144	514	8.37%

A Topology Adaptive Network

In addition to the previous communication-related features, a reconfigurable NoC also has exports the ability to adapt its topology, making it possible to dynamically change its internal structure in order to meet new communication requirements (the ones required by the current application with respect to the current working scenario). Thus, all the benefits of topology-adaptive Network-on-Chip based solutions (Bartic, T.A.; Mignolet, J.-Y.; Nollet, V.; Marescaux, T.; Verkest, D.; Vernalde, S.; Lauwereins, R. (2005); Soares, R.; Silva, I.S.; Azevedo, A. (2004)) can be used to adapt the network to the (currently) on-going application. Reconfiguration can be used for instance to manage high-traffic situations during normal system execution. This feature is realized by means of dynamic switch insertion in the architecture, or by completely changing the entire topology at run-time.

As previously hinted, network nodes (switches) can be dynamically inserted into the system at run-time. Obviously, in order to guarantee this capability, support is required either from the architectural point-of-view and from the protocol viewpoint. These features will be explained in detail both by introducing the proposed communication protocol support and by describing the implementation of the NoC switches and of the proposed routing mechanism.

Communication Protocol Support

This section will present an example of a solution that can be adopted in order to cope with the problems presented in the previous sections. In particular, in order to solve all the issues that arise when designing a reconfigurable NoC it is possible to employ a communication protocol based on *keywords*. Keywords are used to distinguish among different packets, and to disambiguate information carried with them. Each keyword relates to specific information; depending on the packet, this information is to be read either from a network switch or from the destination (the Slave core) to perform controls and checks on the information flow.

Initiator and target components are responsible of generating messages (refer to the Network Interfaces detailed description); switches are only responsible and capable of packet-switched routing during specific communication sessions. The communication protocol is meant to have as low overhead as possible, so that the only necessary information is carried within the packet and only the necessary computations are performed at switch-level.

In this scenario, the communication protocol defines three message classes: header class, data class and control class. Message classes allow the definition of a flexible and extendable communication protocol, as well as a general layered approach to information flow and control. The features of each class are reported in Table 4.

Table 3. Sub-modules contributions to resources requirements

	Input Phase			Output Phase			
	Filter	Buffers	Trigger	FSM	Routing Table	Others	TOTAL
Slices used	2	72	3	57	136	94	364
%Slices (out of 9280)	0.02	0.78	0.03	0.61	1.47	0.01	3.92
Contribution % (out of 364 Slices)	0.55	19.78	0.82	15.66	37.36	25.83	100.00

Table 4. Message classes

Message type	Message usage
Header	It is used to exchange preliminary information about the communication
Data	This contains data that has to be forwarded to the target
Control	Control messages are used by network switches as control information, or by the cores as acknowledge

Information Units Structure

The general structure of a *phit* is composed of two fields: a preamble, and a payload. The preamble is used to encode the specific packet type, with the encoding mechanism as reported in Table 5; on the other hand, the payload contains the actual information, which is interpreted according to the current packet type. The MSB (Most Significant Bit) represents the maximum size of the network packet, and it must be set at synthesis-time. For this reason, it has to be possible, at design-time, to tune the different network parameters, so that each field and the packet size can be tailored to the real application needs. In addition, to support the flexibility of the communication infrastructure, the

packet semantics has not to be affected by these changes; this problem can be solved by fixing the relative positions of the fields and by using the user-defined parameters in order to disambiguate the actual field size.

The keywords mechanism allows for a message to be associated with different information contents. This is useful to decouple the packet type from its real semantics and from the contents. This mechanism is achieved by means of *message specifications*. A *message specification* (or *message option*) is additional information carried by the packet, and related to a given class. Message specifications allow for sub-class messages creation. Table 5 reports a list of message specifications, along with the way they are represented in the packet.

Notice that the specifications are embedded in the packet type, and the reported ones can be applied also to acknowledge part of the communication.

Header Packet

This packet is used to start a Master-initiated communication session, between the Master and the Slave interface module. The field denoted as

Table 5. Message packets and specifications

Message type	phit[0:3]	Message specification	phit specification
header_packet	1000 0010	Initiator-based routing Destination-based routing	
data_packet	0001 0011	Initiator-based routing Destination-based routing	
end_flow	1110 0111	Force routing path Read routing table	
route_update	100x	Destination in new Update existing destination	Phit[3]=1 Phit[3]=0
ack_header_flow	0100 1011	Force routing path Read routing table	
ack_data_flow	1101 0101	Force routing path Read routing table	
ack_tail_flow	1010 0110	Force routing path Read routing table	

request_type contains the identification of the current communication request, i.e. either a write or a read request. This information will be used at the remote end-point of the communication to start a bus-session. The destination identifier, contained within the *destination_ID* field, is used at the end-point module to identify the module (attached to the remote bus) that represents the destination of the current communication. The routing path information, in case the routing is of initiator-based type; otherwise, the destination identifier is here collected. To disambiguate among different initiator cores, the *Sender_ID* field is used, at the end of the packet. Incoming packets with the same starter identifier are considered as belonging to the same *flit*, i.e. a coherent sequence of elementary data-packets. The packet structure is reported in Figure 5 for both initiator-based and destination-based routing.

Data Packet

Data related to the communication request initiated by the Master core is contained in the data class message. There exist two different versions of the data packet: the first containing the route path, and the second containing only the destination identifier information. In the first case, the *routing_path* field contains a list of output ports that the packet has to pass through to get to the target, and this field is followed by the actual token of the sent data, in case a write request is required; as a matter of fact, a write request consists of sending a specific data to the remote bus, via the shared Network-on-Chip architecture. In case a read request is desired, the only single packet is sent with the information related to the destination.

In case the routing path is not required to be forced by the Master network interface, the packet contains the destination identifier, and it is up to the network switches to check the routing tables for the suitable path. In this case, the packet of Figure 6 is employed.

End Communication Flow

The end of the (current) communication session is encoded in the special packet reported in Figure 7. The end flow *phit* marks the tail of the current information flow and it is issued by the initiator. An end-flow packet sets the state of the receiver

Figure 5. Header phit structure with (a) or without (b) routing information

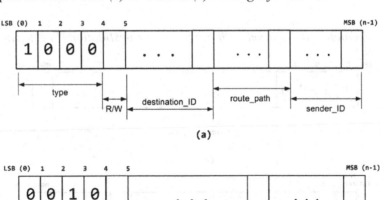

(a)

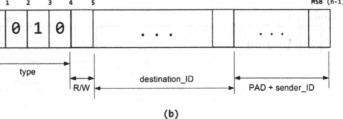

(b)

to the starting state, presenting the received data at the necessary sub-systems connected to it.

Reply Packet

Acknowledge packets belong to the control class messages. They are used by the communication end-points to inform each other about the communication status. The acknowledge message can report either a positive or a negative reply; a reply is considered positive if the communication session ended up successfully (from the target point-of-view). A negative reply, on the contrary, forces the initiator to retransmit the message, due for example to errors in the communication. In case

of a successfully completed communication, the packets flow will contain the desired data (either a result or the data that had to be read from the remote core). The general structure of a reply packet (flow) is the following:

- Reply header, containing the information of the sender, the destination and the path (in case routing has to be forced);
- A set of subsequent data packets, containing one byte of the entire data to be transmitted;
- A reply packet tail, i.e. the end of the current flow of packets.

Figure 6. Data packet

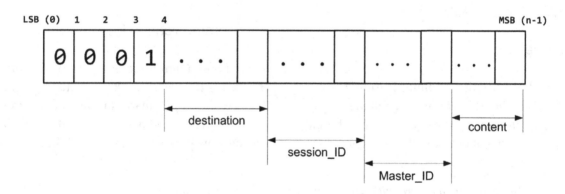

Figure 7. End flow communication packet with routing (a) or destination (b) information

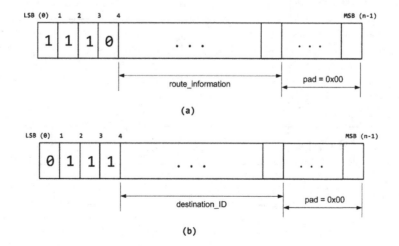

This flow and the related information are necessary since the starting side of the current communication, i.e. the initiator, has to be aware of redirecting the incoming reply data to the module that asserted the communication request.

Route Update and Dynamic Slave Insertion

The packets type presented in this section is necessary in order to support a dynamic routing mechanism. The routing path is initialized at system startup, requiring initial budgeting on the routing tables of the switches at synthesis-time, but it can be changed at run-time (thanks to dynamic routing) accordingly to the network requirements. To support this feature, the *phit* reported in Figure 8 is used. The *destination* field contains the identification code of the module for which the path should be changed. The *new_route* field, instead, contains the entire routing path and it is used hop-by-hop to update the local switch routing tables.

In case the destination identifier relates to a newbie dynamically inserted Slave, the fourth bit of the packet of Figure 8 is set to *1*, stating that a new entry should be added.

RECONFIGURABLE NOCS IMPLEMENTATION DETAILS

This section aims at describing the most relevant implementation issues that arise when designing a reconfigurable NoC. In particular, it is necessary to take into account the placement of the bus-macros, that makes it possible to establish communication channels among switches belonging to different reconfigurable regions and that deeply influences both the sizing of each reconfigurable region (thus the maximum size of each sub-network that has to be deployed on a single reconfigurable region) and the maximum number of communication channels that a single reconfigurable region is able to support.

In addition to this, the second part of this section will deal with the implementation details of the network switches, considering both the input and the output stages, and of the routing mechanism (with particular attention to the routing tables). As previously hinted, these aspects are fundamental when dealing with a NoC which topology can be adapted to the user needs and to the working environment constraints at run-time.

Bus-Macro Placement

As previously hinted, bus-macros placement play a key role during both the definition of the reconfigurable architecture and the NoC implementation, since each signal that has to cross the

Figure 8. Route update packet

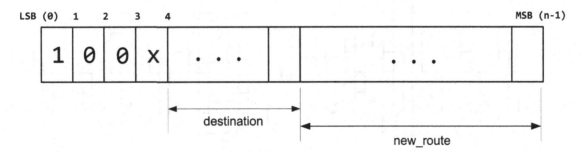

boundary of a reconfigurable region (thus, each signal to/from the others reconfigurable regions and to/from the static part) has to pass through a bus-macro. These hard macros, in fact, have to be placed on the edge between two reconfigurable regions (or between a reconfigurable region and the static part), as shown in Figure 9 (a) for a wide bus-macro. In this way, it is possible to fix the location of the pins of the interface of each reconfigurable region, independently from its internal implementation. This makes it possible to correctly connect all the signals of a new module that is deployed on a reconfigurable region through a reconfiguration process.

When dealing with a reconfigurable NoC, the number of communication channels that can be placed on the 4 edges of a reconfigurable region strictly depends on three main factors: the size of the reconfigurable region itself, the physical placement of the bus-macros and the width of each communication channel. A communication channel, in this context, is the set of signals that are necessary in order to connect a single port of a network switch either to a single port of another switch or to a network interface.

Reconfigurable Regions Sizing

One of the main problems that have to be faced when designing a NoC-based reconfigurable embedded system is the definition of the number and the size of the reconfigurable regions, since this choice will deeply influence both the flexibility and the overall performance of the whole system. In order to obtain a good solution it is necessary to achieve a trade-off between the definition of a fine-grained system (in which a large number of small reconfigurable regions can provide a very high level of flexibility, making it possible to perform very fast reconfigurations when a single reconfigurable regions has to be changed) and the definition of a coarse grained system (in which the performance are usually increased, but that require a higher reconfiguration overhead, due to the large size of each reconfigurable region).

In addition to these metrics, as previously hinted, it is necessary to cope also with the bus-macros placement. The size of each reconfigurable region, in fact, deeply influences the number of bus-macros that can be place on its edges. From this point of view, using a coarse grained approach makes it possible to place a higher number of bus-macros with respect to a fine-grained approach. The main drawback of this choice, in addition to the previously presented ones, consists in the fact that in a coarse grained approach a single reconfigurable region will hold an entire sub-network, which has to be entirely configured even if a single switch has to be changed. On the opposite, using very small reconfigurable regions it is possible to deploy on each reconfigurable region a single switch (or very few switches), making it possible

Figure 9. Placement of a single wide bus-macro (a) and of a set of three wide bus-macros (b) on the edge between two reconfigurable regions

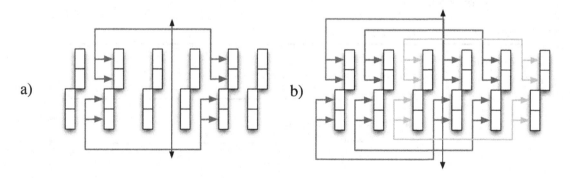

to configure at run-time only the portions of the NoC that have to be changed. Thus, the best choice is to define the reconfigurable regions as the minimum portion of the reconfigurable device that makes it possible to place, on its edges, all the necessary bus-macros to establish all the required communication channels.

Physical Bus-Macros Placement

As previously hinted, bus-macros are the part of the communication channels between the boundaries of two reconfigurable regions or between a reconfigurable region and the static part of the system. As described in the previous section, the placement of the bus-macros can become a problem if the size of the reconfigurable region is so small that it is not possible to place all the necessary bus-macros on its edges. In order to solve this problem, as described by Xilinx Inc. (2006), it is possible to use wide bus-macros and to insert more than one bus-macro on a single row (for left-to-right or right-to-left bus-macro) or on a single column (for top-to-bottom or bottom-to-tom bus-macros), as shown in Figure 9 (b), where three different wide bus-macros are placed on the same row. In this way it is possible to increase the number of bus-macros that can be placed on the edges of a single reconfigurable region by a factor of three.

Communication Channel Width

Since each bus-macro, independently by its direction and its size, is able to route only 8 bits, the number of required bus-macros strictly depends also on the number of bits necessary to implement a single communication channel. A communication channel is the set of signals that are necessary in order to interconnect the ports of two different switch of the NoC. The number and the size of these signals depend on the network protocol that is employed. It is necessary to decrease the width of each communication channels in order to reduce the number of bus-macros necessary to implement it and thus the total number of required bus-macros.

A simple solution can be obtained by reducing the number of bits of a single *phit*, even if this solution obviously increases the latency of each communication across the network, since reducing the number of *phits* brings to the increment of the number of *phits* necessary to build a single *flit*. Thus, this solution has to be employed only if the previously presented techniques have shown to be not sufficient to solve the bus-macros placement problem.

Switch Design

In order to fully support the reconfiguration mechanism, we propose to divide the implementation of each reconfigurable NoC switch into two parts, called *stages*. Figure 10 depicts the switch model, being a connection of two low-level modules: the input stage and the output stage. The first (input) stage is used to realize a simple buffering mechanism, to store the incoming packets. The output stage includes the switch local routing tables, a Finite State Machine (FSM) controller reading the incoming packets and an output controller to forward the message to the desired output.

Input Stage Implementation Details

The purpose of the input stage is to receive the incoming packets and store them (temporarily) into a local buffer. The buffer is used to store the

Figure 10. Switch model implementation

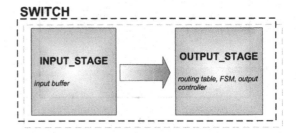

packets during the reconfiguration process, so that they are not lost when the remainder of the system is under reconfiguration process.

There exist a buffer for each input port of the switch. The implementation defines four ports, as reported in Figure 11 (a). In this example, each buffer has 2 slots, so that at most 2 packets can be stored for each incoming port.

The input buffer can be realized as Finite State Machines with 4 states, as reported in Figure 11 (b). The first and the second states reflect the number of packets in the buffer. The third states relate to an error state, in which the buffer is full and no packets can be received unless the buffer flushing is enabled. The *busy* signal is used to inform the connected switches of the status of the input buffers. When this signal is high, the switch connected to the considered port stops sending new data and starts storing the packet within its local buffer, until the *busy* signal value returns low.

The four buffers have to be managed as a whole, and the incoming packets have to be read from the buffer where the data is received. In order to choose the right buffer, a multiplexor is employed. The multiplexor signal is used to short wire the packet that has been received toward the output stage. This is accomplished relying on the *valid_in* bits, by means of an ad-hoc combinatorial network that defines the control signal of the multiplexor. In this way, if a packet arrives at port *j*, then the multiplexor control signal will activate that particular port, enabling the incoming packet to be queued.

Output Stage Implementation Details

The output stage of the switch implementation is reported in Figure 12. The purpose of the output stage is to read the incoming packets, read (or write) the routing tables and control the multiplexor signal at the output ports; in this way, the incoming packet can be routed toward the right port as expressed in the local routing tables.

The routing tables contain the routing path for the destination target in the system. For each of them there exists an entry in the table, containing the output port where the packet should be routed to reach the desired destination. The data of the routing table is used to control the output demultiplexor. The data collected in the routing

Figure 11. Input stage implementation model

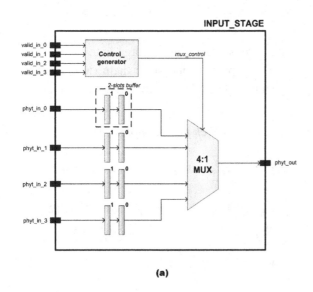

(a)

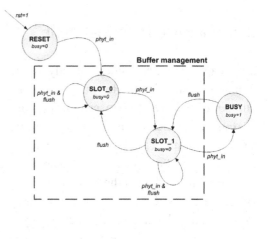

(b)

Figure 12. Output stage implementation model

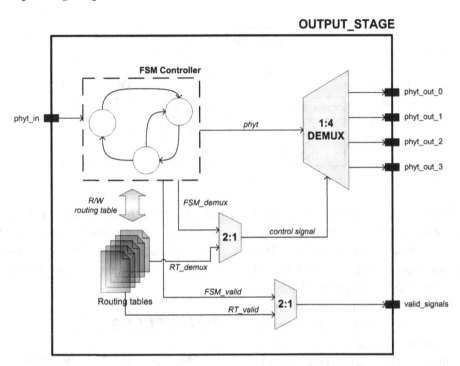

table is controlled by the FSM, depending on the current incoming packet.

The FSM controller, in addition, controls the output valid signals and it can force the routing if required by the packet. The two 2:1 multiplexors are used to choose which signal to effectively output from the switch, depending on the current machine state and on the current incoming *phit*.

Routing Mechanism

As described in the previous sections, the proposed routing mechanism is realized in two ways: either by forcing the routing path in the packets or by reading the routing tables that are stored within the network nodes. In the first case the Network Interface is responsible for generating the required packet with the route field. In the latter case, on the other hand, the information is stored within the routing tables, which are located in the nodes of the network (the switches). The possibility of choosing between these two ways at run-time

makes it possible to cope with the issues presented in the previous sections, such as the dynamic insertion or removal of a single switch (or of a set of switches) in order to change the topology of the NoC architecture at run-time.

Routing Tables

The purpose of the routing tables is to provide the network users with a mechanism to forward the packets, in an easy way. The routing tables must use the least amount of resources as possible, and yet they have to provide a suitable dimension, in terms of the maximum number of Cores that can be addressed.

The routing mechanism can be divided into two functions. The first function, called *table_index_map*, is used to realize logical addressing of the Slave cores. In this function, given a destination identifier the output is the index within the real routing table where the routing path information is stored for that precise target. In this way, for

Figure 13. Routing mechanisms using routing tables

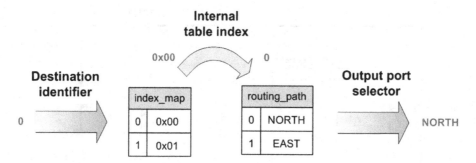

each destination identifier, the table stores internal information in a transparent way with respect to the actual physical location of the module. This is very important when dealing with a NoC-based reconfigurable system, since the physical location of a core can dynamically change over the time. The second function, instead, is used to store the routing path information required for the packet to reach the destination. The function takes as input an index table, and the output is the port number toward which the packet should be forwarded to reach the desired destination. In this way, once a packet has been received, the switch reads from the internal tables the output port to enable for the packet forwarding. This scenario is reported in Figure 13.

CONCLUSION

This chapter described and analyzed several different issues that arise when designing a reconfigurable NoC. As it has been pointed out, it is not possible to utilize a generic NoC infrastructure on a generic reconfigurable system, since the interactions between the two scenarios create a set of difficulties that need to be explicitly faced by the designer.

In particular, this chapter presents and describes the changes that have to be made in the NoC in order to adapt this communication infrastructure to a reconfigurable scenario and the aspects of a generic reconfigurable system that have to be correctly handled in order to obtain an optimized and high-performance reconfigurable NoC.

REFERENCES

Ali, M. Welzl, M., &; Hellebrand, S. (2005, November 21-22)., "A dynamic routing mechanism for network on chip. In *Proceedings of the," 23rd NORCHIP Conference, 2005.* 23rd, (pp. 70-73)., 21-22 Nov. 2005

Bartic, T. A., ; Mignolet, J.-Y., ; Nollet, V., ; Marescaux, T., ; Verkest, D., ; Vernalde, S., & ; Lauwereins, R. (2005, July 8)., "Topology adaptive network-on-chip design and implementation., " *IEE Computers and Digital Techniques,* IEE Proceedings -, vol.*152,* no.(4), pp. 467-472., 8 July 2005

Benini, L., & ; De Micheli, G. (2002, January)., "Networks on chips: a new SoC paradigm., " *Computer,,* vol.*35,* no.(1), pp.70-78., Jan 2002

Bertozzi, D. (2004). Benini, L. (2004)., Xpipes: a network-on-chip architecture for gigascale systems-on-chip. *IEEE. Circuits and Systems Magazine, IEEE, 4*(2), 18–31. doi:10.1109/MCAS.2004.1330747

Bobda, C. (2005, Sept.-Oct.). Ahmadinia, A. (2005, September-October)., Dynamic interconnection of reconfigurable modules on reconfigurable devices. *IEEE Design & Test of Computers, IEEE, 22*(5), 443–451. doi:10.1109/MDT.2005.109

Bobda, C., & Ahmadinia, A. Majer, M.; Teich, J.; Fekete, S., &; Van der Veen, J. (2005, August 24-26)., Dynoc: A dynamic infrastructure for communication in dynamically reconfigurable devices. Field Programmable Logic and Applications, 2005. In *Proceedings of the International Conference on Field Programmable Logic and Applications, 2005*, (pages 153–158)., 24-26 Aug. 2005

Dally, W. J. Towles, B. (2001)., "Route packets, not wires: on-chip interconnection networks., " In *Proceedings of the Design Automation Conference, 2001*. Proceedings, (pp. 684-689)., 2001

Eto, E. Xilinx Inc. (2004, September)., *XAPP290 - Two flows for partial reconfiguration: Module based or Difference based.*. San Jose, CA: Xilinx Inc., September 2004.

Ferrandi, F. Santambrogio, M. D., &; Sciuto, D. (2005, April 4-8)., "A design methodology for dynamic reconfiguration: the Caronte architecture. In *Proceedings of the 19th IEEE International,*" *Parallel and Distributed Processing Symposium, 2005. Proceedings. 19th IEEE International*, (pgp. 4)., 4-8 April 2005

Jovanovic, S. Tanougast, C.; Weber, S., &; Bobda, C. (2007, August)., Cunoc: A scalable dynamic NoC for dynamically reconfigurable fpgas. In *Proceedings of the International Conference on Field Programmable Logic and Applications, 2007*. (FPL 2007). International Conference on, (pages 753–756)., 27-29 Aug. 2007

Kao, C. (2005)., "Benefits of Partial Reconfiguration. In Proceedings of", XCell, (pp. 65-67)., 2005

Keutzer, K. Newton, A. R.; Rabaey, J. M., &; Sangiovanni-Vincentelli, A. (2000, December)., "System-level design: orthogonalization of concerns and platform-based design. In *Proceedings of IEEE Transactions on,*" *Computer-Aided Design of Integrated Circuits and Systems*, IEEE Transactions on, (vol.19, no. 12, pp. 1523-1543)., Dec 2000

Mack, R. J. (1996)., "VLSI physical design automation: theory and practice., " Electronics & Communication Engineering Journal, vol.8, no. (2), pp.56, Apr 1996

Mak, T. S. T. Sedcole, P.; Cheung, P. Y. K.; Luk, W. (2006, August 28-30)., "On-FPGA Communication Architectures and Design Factors. In *Proceedings of the, " International Conference on Field Programmable Logic and Applications, 2006*. (FPL '06). International Conference on, (pp.1-8)., 28-30 Aug. 2006

Pionteck, T. Koch, R., &; Albrecht, C. (2006, August 28-30)., Applying partial reconfiguration to networks-on-chips. In *Proceedings of the International Conference on Field Programmable Logic and Applications, 2006*. (FPL '06). International Conference on, (pages 1–6)., 28-30 Aug. 2006

Pionteck, T. Albrecht, C.; Koch, R.; Maehle, E.; Hubner, M., &; Becker, J. (2007, March 26-30)., "Communication Architectures for Dynamically Reconfigurable FPGA Designs. In *Proceedings of the IEEE International, " Parallel and Distributed Processing Symposium, 2007*. (IPDPS 2007). IEEE International, (pp.1-8)., 26-30 March 2007

Soares, R. Silva, I. S., &; Azevedo, A. (2004, September 7-11)., "When reconfigurable architecture meets network-on-chip. In *Proceedings of the 17th Symposium on, " Integrated Circuits and Systems Design, 2004*. (SBCCI 2004). 17th Symposium on, (pp. 216-221)., 7-11 Sept. 2004

Xilinx Inc. (2006, March)., *Early Access Partial Reconfiguration User Guide*. San Jose, CA: Xilinx Inc., March 2006.

KEY TERMS AND DEFINITIONS

Bandwidth: Amount of information *per unit of time* that a channel can support.

Bus-Macros: Hard macros that provide a means of locking the routing between different reconfigurable regions.

Latency: Amount of time required for an output to be generated, given the inputs to the functional unit.

Packet-Switching: Mechanism by which data is forwarded along the network by means of packets.

Routing: Choice of the output branch toward which an incoming packet should be directed.

Scalability: Degree of adaptability of the system performances as a function of the amount of required workload

Throughput: Amount of data *per unit of time* that is delivered in the entire system.

ENDNOTE

[1] Recall that ad-hoc design methodologies have to be used to enable dynamic reconfiguration in FPGA-based designs (Xilinx Inc. (2004); Xilinx Inc. (2006))

Section 3
High-Level Programming of Reconfigurable NoC-based SoCs

Chapter 8

High-Level Programming of Dynamically Reconfigurable NoC-Based Heterogeneous Multicore SoCs

Wim Vanderbauwhede
University of Glasgow, UK

ABSTRACT

With the increase in System-on-Chip (SoC) complexity and CMOS technology capabilities, the SoC design community has recently observed a convergence of a number of critical trends, all of them aimed at addressing the design gap: the advent of heterogeneous multicore SoCs and Networks-on-Chip and the recognition of the need for design reuse through Intellectual Property (IP) cores, for dynamic reconfigurability and for high abstraction-level design. In this chapter, we present a solution for High-level Programming of Dynamically Reconfigurable NoC-based Heterogeneous Multicore SoCs. Our solution, the Gannet framework, allows IP core-based Heterogeneous Multicore SoCs to be programmed using a high-level language whilst preserving the full potential for parallelism and dynamic reconfigurability inherent in such a system. The required hardware infrastructure is small and low-latency, thus adding full dynamic reconfiguration capabilities with a small overhead both in area and performance.

INTRODUCTION

Networks-on-Chip (NoCs) provide a scalable, efficient and performant communication medium to interconnect complex IP cores. To facilitate interoperation between IP cores, a number of standards have been proposed (e.g. VSIA, OCP/IP) [Kogel et al., 2005] regarding the interface between the IP cores and the communication medium. The purpose of such standards is to facilitate design reuse and as such they are aimed primarily at IP core developers. By their very nature, IP cores are agnostic of the system in which they are deployed. Consequently, the datapaths between the cores are not governed by the IP cores. Thus, a system consisting solely of IP cores and a communication medium is not dynamically reconfigurable. A Dynamically Reconfigurable SoC therefore requires an additional infrastructure to support reconfiguration of the cores and the datapaths. Moreover, to reduce design time,

DOI: 10.4018/978-1-61520-807-4.ch008

it is essential that a Dynamically Reconfigurable SoC can be programmed at high level.

In this chapter we analyse the requirements of such an infrastructure, in particular in view of allowing high-level programmability. We present our architecture of a dynamic reconfiguration infrastructure (DRI), the Gannet framework.

The chapter starts with the background to this work. We then analyse the requirements for NoC-based SoC infrastructures to support dynamic reconfiguration. The subsequent sections discuss the key components of the Gannet framework:

- The Gannet Dynamic Reconfiguration Infrastructure,
- The Gannet language used to program the system,
- The Gannet Machine Model, a formal model used to explain how the Gannet DRI executes Gannet programs.

Section "Examples of Dynamic Reconfiguration" presents examples of dynamic reconfiguration of communication (data path) and computation (IP core). In section "Implementation of the Dynamic Reconfiguration Infrastructure" we discuss the implementation of the Gannet DRI. The chapter concludes with an overview of the current status of the project and avenues for future research.

BACKGROUND

Reconfigurable architecture platforms are gaining increasing popularity as flexible, low-cost solutions for a variety of media processing applications. They contribute to bridging the gap between general purpose processors and application specific circuits. As the application domain for reconfigurable platforms further expands, it becomes imperative that these systems provide a high degree of flexibility and adaptability, while at the same time being extremely cost-efficient. For many applications, fine-grained reconfigurable platforms such as FPGAs are not an acceptable choice because of their high area overhead and reduced performance compared to ASIC solutions. As a result, recent times have seen more focused research on coarse-grained reconfigurable architectures (CGRA), which constitute a middle ground between ASICs and FPGAs in terms of performance-versus-flexibility.

Using the terminology from the review paper on CGRAs by [ul Abdin and Svensson, 2008], the Gannet platform [Vanderbauwhede, 2008, Vanderbauwhede et al., 2008, Vanderbauwhede, 2007] is an Array of Functional Units - as opposed to Hybrid Architectures and Arrays of Processors. The former are typically using a combination of an ordinary processor with a reconfigurable fabric, e.g. [Singh et al., 2000, Mishra et al., 2006] in the latter the reconfigurable fabric consists of fully-featured processors, e.g. [Butts, 2007, Taylor et al., 2004]). Within its category, Gannet bears some similarity to MATRIX [Mirsky and Deon, 1996], Silicon Hive [Cocco et al., 2004] and MORA [Lanuzza et al., 2007, Profit et al., 2008], as it shares with these architectures the local-memory processing model, i.e. every functional unit has its own local memory. However, Gannet distinguishes itself from the other architectures by providing a generic interface layer to third-party processing elements where all other architectures discussed in [ul Abdin and Svensson, 2008] use either dedicated functional units or ordinary processors.

In contrast to most CGRAs, the Gannet platform was specifically designed as a *reconfiguration infrastructure* for SoCs using packet-switched NoCs. From that perspective, Gannet has similar goals as the work done at IMEC [Nollet et al., 2005, Nollet et al., 2004, Marescaux et al., 2004] on operating system control of NoC routing tables and support for run-time reconfiguration of functional units. The two approaches can be considered complementary as Gannet focuses on providing an interface layer which facilitates high-level programming of the interactions between the cores whereas the NoC OS focuses on operating system functionality such as process control and scheduling.

A further similarity to the work at IMEC is the recognition that it is essential to separate control flow from data flow [Nollet et al., 2003] to avoid the performance bottleneck caused by the processor bus in central-memory systems. There is however a fundamental architectural difference as Gannet does not require a centralised control system (e.g. a processor running an operating system). Furthermore, programming the routing tables is a low-level technique. Even when using a distributed programming model such as employed by the Æthereal researchers [Goossens, 2005], the programming model only addresses the NoC communication paths but not the actual functionality of the system, as it governs which slots are in use by which resource. In contrast, the Gannet platform can be considered as a distributed processor, which can be programmed using a high-level language. More precisely Gannet can be categorised as a coarse-grained demand-driven architecture.

Demand-driven (reduction) and data-driven (dataflow) parallel architectures have been extensively studied in the past [Veen, 1986, Treleaven et al., 1982], however, these were fine-grained architectures. The interest in these architectures decreased because they could not match the inexorable rise of the ever more performant von Neumann-style processors. It was generally recognised that communication presented the main bottleneck in fine-grained parallel architectures. However, with the advent of multicore SoCs and the adoption of Networks-on-Chip as an efficient communication paradigm for multicore systems, there is a renewed interest in coarse-grained architectures [Guerrier and Greiner, 2000, Lampinen et al., 2006].

In the specific case of the Gannet platform it is tempting to compare the proposed architecture with the numerous dataflow architectures for executing functional programs [Vegdahl, 1984,

Amamiya and Taniguchi, 1990, Giraud-Carrier, 1994], as Gannet is a demand-driven parallel architecture and it adopts a functional paradigm for high abstraction-level programming. However, the crucial difference is that the aim of Gannet is not to provide an optimal architecture for executing general-purpose functional programs. On the contrary, the aim of our work is to facilitate high abstraction-level design of Dynamically Reconfigurable NoC-based Heterogeneous Multicore SoCs and for that purpose we have adopted concepts from dataflow architectures and functional programming. As a consequence, many concepts in Gannet will be familiar from this earlier work, in particular [Vegdahl, 1984] and [Ashcroft, 1986].

To contrast Gannet with other research on high level programming of dynamically reconfigurable systems, we can consider e.g. the work of [Najjar, 2003] using Single-assignment C (SAC) and the Lime project [Huang, 2008] using Java, in both cases to allow high-level programming of a host processor with attached FPGAs. While these approaches offer indeed high-level programming of dynamically reconfigurable systems, they are targeted at FPGA-based systems whereas Gannet targets NoC-based SoCs. To better illustrate the difference, imagine a NoC-based SoC where some of the IP cores are embedded FPGAs. To create the bit streams for those FPGAs one could use the SAC or Lime approach; however, the reconfiguration decision and management and the communication between the cores in the SoC would be handled by Gannet.

REQUIREMENTS ON SOC INFRASTRUCTURES FOR DYNAMIC RECONFIGURATION

This section presents a number of key observations about Dynamically Reconfigurable NoC-based Heterogeneous Multicore SoCs and derives the requirements for a Dynamic Reconfiguration Infrastructure (DRI).

Characteristics of NoC-Based Coarse-Grained Reconfigurable Architectures (CGRAs)

Assuming a NoC-based System-on-Chip consisting of a heterogeneous set of reconfigurable cores, we can make the following general observations:

- SoCs consisting of large numbers of Heterogeneous IP cores connected over a NoC are inherently parallel and can support large numbers of data flows between large numbers of cores. These data flows can be parallel (independent of each other) or concurrent (dependent on each other). Furthermore, it is in principle possible that a particular core will participate in multiple data paths. In general, data flows in a multicore SoC will not result in purely parallel operation but rather in concurrent operation, i.e. the flows can share resources.

- On the nature of the IP cores, we can observe that they are self-contained, by definition system-agnostic and are typically data processing units.

- Regarding communication, it is obvious that, because of the NoC, all data transfers are packet-based.

- In a dynamically reconfigurable system, it is essential that the data flows are controllable at run time. A system where the functionality of the cores can be reconfigured but the data path remains static would clearly be of limited use.

- In a NoC-based SoC with very large numbers of cores, centralised memory would present a huge performance bottleneck. Hence, IP cores will require local memory. Indeed, several CGRAs [ul Abdin and Svensson, 2008] and multicore processors [Pham and Aipperspach, 2006] have adopted local-memory architectures.

- Streaming data processing is essential in high-performance SoCs. This follows from the previous observation, as without central memory the system must operate in dataflow fashion. The nature of NoC-based SoCs as essentially a connected set of data processing units is optimally suited for this type of operation.

- High abstraction level design of dynamically reconfigurable heterogeneous multicore SoCs is essential to reduce design time. This point is more generic and applies to any type of system design. Traditional low-level design approaches using HDLs are unable to cope with the complexity of today's SoCs. For Dynamically reconfigurable SoCs the reconfiguration of both data paths and core functionality must be expressed at a high level of abstraction and consequently the system architecture must enable high-level programmability.

Requirements on Dynamic Reconfiguration Infrastructures

Based on the above observations, we can derive the requirements for a Dynamic Reconfiguration Infrastructure (DRI) for NoC-based SoCs.

Since neither the cores nor the NoC provides control over the data flows between the cores in terms of deciding where to direct data generated by a core, the DRI must govern the data path. Indeed, the primary purpose of a DRI is to provide dynamic data path reconfiguration capabilities to the SoC. For that purpose, the DRI must provide run-time flow control operations. In doing so, the DRI must support parallel and concurrent operation of cores. As a central controller could present a bottleneck, the DRI should use distributed control as much as possible. Therefore, the DRI must support a local-memory model. For some types of control constructs, a centralised controller is the only option. In that case the DRI must ensure that the data flow is separated from the control flow.

Furthermore, considering the importance of streaming data processing for a very large class of applications combined with the observation that NoC-based SoCs are intrinsically ideally suited for streaming data processing, it is essential that the DRI should support this processing model. Finally, as the provider of dynamic reconfigurability, the DRI must support high-level programmability.

The DRI should also provide support for reconfiguration of the cores. However, as we will see further, this support follows naturally when all the above requirements have been taken into account.

THE GANNET DYNAMIC RECONFIGURATION INFRASTRUCTURE

In this section we propose a functional model for a Dynamic Reconfiguration Infrastructure (DRI) which we call Gannet. The Gannet DRI is a Service-based Architecture (SBA) [Vanderbauwhede, 2006b]. We introduce the Service abstraction and explain how the proposed model meets the DRI requirements set out above. We then present a high-level system overview of the Gannet DRI. Finally, we discuss the operation of a key component, the Gannet service manager.

The IP-Core-as-Service Abstraction

We start by observing that a NoC provides full connectivity between all its nodes. In principle, every node can exchange data with all other nodes in the system. We will call a node consisting of an IP core and a NoC interface a *tile*. From the above discussion it follows that the tile should also contain local memory. The aim of the Gannet DRI is to make the data exchanges between tiles programmable at high level. The choice of the programming paradigm is very important, as the paradigm will have to support all requirements of the DRI as identified in the previous section. The

most popular programming paradigm, imperative programming (with C as the archetypical example) is a poor choice as it does not natively support parallelism or concurrency.

As IP cores are self-contained units with well-specified interfaces and functionality, we can regard them as "service providers", i.e. every core provides one or more services to the system. Typically, services are computational but flow control services including all types of storage are equally possible. We now introduce the service abstraction: a *service* is defined as a unit that consumes data, produces data and can modify its internal state. By "internal state" we mean all writcable memory elements on the service tile. With this abstraction, a service is very similar to a function. Consequently, we can adopt a functional approach to task graph composition.

Functional composition means that all data – apart from constant (hardwired) data (e.g. ROM content) – are the results of calls to other services. This idea is very similar to so-called functional languages such as Scheme, Haskell or ML [Sussman and Steele, 1975, Milner et al., 1990, Hudak et al., 1992]: a program is a tree of nested functions. The key advantage of the functional paradigm is that it natively supports parallelism and concurrency and can easily support pipelined streaming data processing.

By itself, an IP core will not act as a function as it has no control over the source of its data and the destination of the result of the computation. To achieve service-based (i.e. functional) behaviour, every tile of a Gannet SoC contains a special control unit (the Gannet *service manager*, SM), which provides a service-oriented interface between the IP core and the system.

Let us assume for the moment that a service can be modelled by a pure function, i.e. $y=S(x_1, x_2,....,x_n)$. By "pure" we mean that the function has no side effects, i.e. it is a mathematical function. The service manager effectively performs evaluation of the function arguments; the service core (i.e. the IP core) implements the function body.

Figure 1. Gannet system architecture

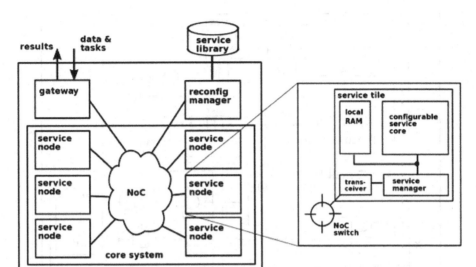

Consider the following pseudo-code for a system with 3 services: DCT(Block(Image(), blocksz)). The service Block takes a block of blocksz*blocksz pixels from Image. blocksz is a constant; Image is a service which obtains an image via I/O, e.g. a camera. The service DCT performs Discrete Cosine Transform on the block of pixels provided by Block. The operation starts at the DCT service: its SM requests data from Block and stores the constant blocksz. As soon as the SM receives a block of pixels from Block, it activates the DCT core. The core computes the transform. The SM will take care of sending the data to whichever service called the DCT. Similarly, the SM of the Block service requests data from the Image service; the SM of the Image service simply instructs the Image core to get an image and returns it to the caller (i.e. Block).

The HDL design and implementation of the SM are discussed in [Vanderbauwhede et al., 2008] and will be revisited in section "Implementation of the Dynamic Reconfiguration Infrastructure". In the following sections we present a high-level architecture of the DRI and the SM.

System Overview

Given a NoC-based System-on-Chip, the Gannet Dynamic Reconfiguration Infrastructure (DRI) consists of:

- The *service managers* acting as an interface between each IP core and the NoC;
- A *gateway* which acts as the interface through which the DRI is configured;
- A *reconfiguration manager* which acts as an interface to the service library, a reconfiguration database which contains configurations for the IP cores.

The high-level view of a Gannet system is a collection of *service tiles* (IP core + SM + transceiver) connected via a NoC. This is illustrated in Figure 1; we call the combination of a service tile and a NoC switch a *service node*.

The operation of the Gannet DRI can be described at high level as follows: thanks to the Gannet service manager, every service tile has the capability to know how it should communicate with other tiles in the system. Tiles with recon-

Figure 2. The Gannet service manager (SM) architecture and its interface with the service core

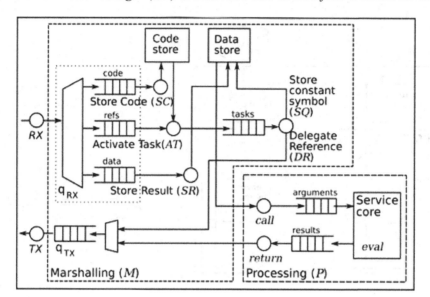

figurable cores also have the capability to know how to reconfigure the core. This capability is provided by the SM; the actual information about the configuration of communication and computation is contained in a program that is executed in a distributed fashion by all tiles in the system. A Gannet system can execute several of these programs (called "task descriptions") concurrently. The task descriptions enter the system through the Gateway. The Gateway has an interface with the outside world, dependent on the overall system in which the Gannet SoC is deployed. For example, if the Gannet SoC would be part of a space probe, the interface would be a radio transceiver; if the SoC would act as an accelerator for a modern desktop PC, it could be PCI-express.

Gannet task descriptions are lists of packets, each of which contains a part of the complete program (called a "subtask"). The Gateway simply transmits the packets on the NoC, which transfers them to their destination Service. Once all services have received all required subtasks, the program is executed; in other words the Gannet SoC runs the specified task.

The dynamic reconfiguration mechanism in the Gannet DRI is based on a library of configura-

tions for the Services. The Service reconfiguration manager is the interface through which the configuration data can be requested from off-chip storage. The actual reconfiguration of the IP core in a Service is handled by the SM. Due to the properties of the Gannet DRI, dynamic reconfiguration is very easily expressed in the task description.

Operation of the Gannet Service Manager

It is clear from the above description that the Gannet service manager circuit plays a crucial role. As all communication between the cores is handled by it, it is important that the SM is a low-latency circuit, in other words marshalling the data for the core should take a minimal amount of clock cycles. Furthermore, the circuit must be small and low-power. As a rule, we could demand that the overhead of the SM should not be more than a few percent for speed, area and power. We have demonstrated in [Vanderbauwhede et al., 2008] that our novel design indeed achieves these goals.

The service manager is a queue-based system (Figure 2) where several queues are processed in parallel, depending on the type of packet in the queue. This pipelined parallel architecture results in very low latency; as the queues are very shallow, the area consumption is very modest. A formal model for the operation of the SM is presented in the next paragraph.

We can consider the Gannet SoC as a machine for running Gannet programs. The Gannet SoC architecture is quite different from the familiar von Neumann-style processor-based SoC architecture: it is a distributed processing system without global memory. There is no program counter, and the program is not executed in a sequential fashion but in a demand-driven fashion [Wilkinson, 1996]. We introduce a more formalised description of the Gannet machine:

- The Gannet machine is a distributed computing system where every computational node *consumes packets* and *produces packets* and can store state information between transactions.

- A Gannet *symbol* is a multi-byte word which encodes operations, code references and constants. A symbol is identified by its *Kind* (*service, reference, constant*), the *Task* and *Subtask* to which it belongs, and its *Name*. Depending on the Kind, symbols can have additional fields to encode e.g. quoting, extensions, processing mode etc. We denote a symbol as a tuple *(Kind, Task, Subtask, Name,...)*.

- A Gannet *packet* consists of a header and a payload. The payload can either be data or instruction code (Gannet symbols). The header consists of the following fields: *Packet Type* (*code, reference, data*), destination address (*To*), return address (*Ret*) and packet identifier (*Id*). We denote a Gannet packet as p(Type,To,Ret,Id;Payload).
  - A *code* packet contains the Gannet code for a subtask, compiled into a list Gannet symbols. We will use the notation $\langle s_1,...,s_n \rangle$ to denote a list of Gannet symbols. A compiled subtask

is called an *instruction* for the Gannet machine. The payload of a code packet is stored in the Code store.
  - A *reference* packet contains a *code reference*, a symbol which acts as an identifier for instruction code stored in the Code store

In the context of the formal model, computational cores such as IP cores will be called "service cores".

The role of the NoC in the Gannet SoC is essential as a medium for transferring packet-based data between services. However, the actual implementation of the NoC will not impact on the functionality of the system – although it will of course influence its performance. Furthermore, the role of the IP cores in the Gannet SoC is purely computational, i.e. IP cores consume data and produce data without awareness of the source or destination of the data.

All the intelligence in terms of the control of the communication between the cores is therefore encapsulated in the service manager. Consequently, the operation of the service manager completely determines the operation of the entire system.

With the above definitions, the operation of a Gannet service can be described in terms of the task code and the result packet produced by the task, using the following set of actions (see *Figure* 2):

- **Receive packet (*RX*):** when a service S_i receives a packet, the packet is stored in one of the queues (*code, refs, data*) dependent on the packet's *Type*. Presence of a packet in the queue triggers the next action:
  - For code packets: **Store code packet (*SC*):** a service S_i receives a *code* packet p(code,S_i,S_j,R_{task};subtask) where *subtask* = $\langle S_i,a_1...,a_n \rangle$. The instruction (compiled subtask code) is stored in the Code store for execution at a later time and referenced by the symbol R_{task}.

193

- ◦ For reference packets: **Activate task (AT):** the service S_i receives a *reference* packet $p(\text{ref},S_i,S_j,R_{id}:R_{task})^1$; the service activates the task referenced by $R_{task}:\langle S_i,a_1...,a_n\rangle$ by transferring it to the *tasks* queue. This results in evaluation of the arguments $a_1,...,a_n$ as follows:
    - ▪ If the argument is a reference symbol: **Delegate by reference packet (DR):** the service manager requests activation of a subtask by sending a *reference* packet, containing the corresponding *reference* symbol, to the corresponding service.
    - ▪ If the argument is a constant symbol: **Store constant symbol (SQ):** all *constant* symbols (e.g. numbers) in the code are stored in the local Data store.
  - ◦ For data packets: **Store returned result (SR):** result data from subtasks are stored in the local Data store.
- • **Process (P):** When all arguments of the subtask have been evaluated, the data are passed on to the service core by pushing them onto the *arguments* queue. The core performs processing on the data; the service produces a result packet p_{res} = $p(\text{Type},S_i,S_j,R_{id}:Payload_i)$ where the $Payload_i$ is the result of processing the evaluated arguments $a_1,...,a_n$ by the core of S_i
- • **Transmit packet (TX):** when the packet resulting from the P or DR action is put in the TX queue, it will be transmitted over the NoC. The packet p_{res} is sent to S_o and there $Payload_i$ is stored in a location referenced by R_{id}.

This operation sequence results in fully parallel execution of all branches in the program tree in an unspecified order governed by the processing time of the packets. We call the set of actions M = $\{SC,AT,DR,SQ,SR\}$ the Marshalling set. Every action in the M set acts on a queue of packets (see Figure 2). Thus the Gannet service manager provides the marshalling of the data and the service core provides the processing.

To simplify the discussion we have omitted state changes in the above explanation. The P action can produce a change in the state of the service, this feature is discussed in detail in section "The Gannet Machine Model".

The implementation of the Gannet service manager is discussed in detail in section "Implementation of the Dynamic Reconfiguration Infrastructure". In the next sections we discuss the Gannet language and we present a model for the Gannet machine. These two components are the foundation on which the Gannet system is built, therefore a detailed discussion is essential for the understanding of the final sections which present the actual implementation and programming of the Gannet DRI.

THE GANNET LANGUAGE

In this section we introduce the Gannet language [Vanderbauwhede, 2006a], and Intermediate Representation language intended as a target for High-Level Language compilers, and we discuss compilation of Gannet programs into bytecode for the Gannet machine.

Gannet programs define the interactions between the services by mapping every service to a *named function* and describing the flow of data between the services in terms of function calls. The Gannet language is an *intermediate representation language* (IR) for the Gannet Virtual Machine, comparable to .NET CIL [Gough, 2001] or PIR, the Parrot Intermediate Representation [Randal et al., 2004]. In other words, Gannet is not a high-level language but a compilation target language. A program written in Gannet syntax can be transformed in machine code in a trivial way.

Language Syntax

Gannet syntax is an S-expression syntax (similar to LISP or Scheme [McCarthy, 1960, Sussman and Steele, 1975]) completely free from syntactic sugar. In Extended Backus-Naur Form (EBNF) [Bray et al., 2008], a Gannet expression must always obey (for simplicity we assume that all expressions include a trailing white space):

```
service-expr::= '(' service-symbol "'"?
arg-expr+ ')'
arg-expr::= service-expr | literal-symbol
```

Consider as a trivial example a SoC with five services: image capture (img) from several cameras, composite image creation (compose), conversion to JPEG format or PNG format (convert), compression (compress) and encryption (encrypt). To obtain a compressed composite of raw images from cameras 1 and 3, the task description would be

```
(compress (compose (img cam1) (img
cam3)))
```

To obtain a JPEG-converted, encrypted image from camera 2 would require:

```
(encrypt (convert jpeg (img cam2)))
```

Language Semantics

An operational semantics for the Gannet language has been presented in [Vanderbauwhede, 2008]. In this section we focus on the key features that make Gannet suitable as an intermediate representation language for a Dynamic Reconfiguration Infrastructure.

Strict Parallel Evaluation: A Gannet service can be considered as a function, though not necessarily a pure function. The *service manager* effectively performs evaluation of the function arguments; the evaluated arguments are passed on to the service core which computes a result based on the argument values and returns it. A key feature of the Gannet language is that the evaluation order is unspecified; the Gannet "machine" evaluates all arguments in parallel. As a result, a Gannet program will automatically exploit the capacity for parallelism present in the system.

Quoting to Defer Evaluation: Quoted expressions are passed unevaluated to the service core. In this way, service cores that implement control structures can evaluate expressions as required.

Custom Control Constructs

Because of Gannet's minimal design philosophy, there is not a unique fixed set of control structures. In principle, every developer can design a custom set of controls. However, for convenience we have defined a small core set, very similar to Scheme's [Kelsey et al., 1998, Vanderbauwhede, 2007]. In Gannet parlance, these are called *control* services as they provide control over the flow of data. The core set consist of the lexical scoping constructs (group, assign, read), the conditional branching construct (if), the function definition and application constructs (lambda, apply) and list operations (list, head, tail, length, cons).

Lexical Scoping

The group construct acts as a block that provides lexical scope to the variables declared inside it using assign, similar to *let* in Scheme, Haskell or ML. Accessing a variable requires an explicit read. Lexical variables are immutable.

```
group-expr::= '(group ' "'"? assign-expr+
"'"? arg-expr ')'
assign-expr::= '(assign ' "'" var-symbol
arg-expr+ ')'
read-expr::= '(read ' "'" var-symbol ')'
```

Conditional Branching

The ubiquitous *if-then-else* construct:

```
if-expr::= '(if ' service-expr "'"? arg-
expr "'"? arg-expr ')'
```

Function Definition and Application

The lambda construct creates a λ-function (which can be bound to a lexical variable). Function application requires an explicit apply.

```
lambda-expr::= '(lambda ' "'" var-symbol+
"'" service-expr ')'
apply-expr::= '(apply ' lambda-expr "'"
arg-expr+ ')'
```

List Operations

List operations are similar again to those in Scheme, Haskell or ML. Lists require an explicit list constructor.

```
list-expr::= '(list ' "'" arg-expr+ ')'
single-list-op::= 'head' | 'tail' |
'length'
single-list-expr::= '(' single-list-
oplist-expr ')'
cons-expr::= '(cons ' list-expr arg-expr
')'
```

Streaming Data Processing

Gannet has provisions for pipelining of operations on streaming data. This feature has no equivalent in languages intended for sequential processors, as pipelining only makes sense if operations can be performed in parallel. In multicore SoCs, pipelining is a key feature.

```
buf-expr::= '(buf ' "'" buf-var "'" ser-
vice-expr ')'
stream-expr::= '(stream ' "'" buf-var')'
```

```
eos-expr::= '(eos ' "'" vbuf-var')'
peek-expr::= '(peek ' "'" buf-var')'
get-expr::= '(get ' "'" buf-var ')'
```

The buf expression stores the return value of *expr* in the buffer *buf-var*. The stream expression returns the buffered value, calls the *expr* bound to *buf-var* and stores the return value in the buffer. The eos expression returns true if the end of the stream is reached.

As an example, a 2-stage pipeline $S1 \rightarrow [b1] \rightarrow S2 \rightarrow [b2] \rightarrow S3$, where b1 and b2 are buffers, can be written as:

```
(let
 '(buf 'b1 '(S1...)
 '(buf 'b2 '(S2 (stream 'b1)))
 '(S3 (stream 'b2))
)
```

Compilation of Gannet Code into Packets

This section explains how a Gannet program is compiled into packets for running on the Gannet machine.

The set of all packets resulting from compilation of a program is called the *bytecode*.

The compilation process is very straightforward:

1. Decompose the nested S-expression into a list of flat S-expressions by replacing the nested expressions by references:

$$e_{\text{root}} = (S_{\text{root}} \ e_1 ... e_i ... e_n)$$
$$e_i = (S_i e_{i,1} ... \ e_{i,j} ... e_{i,n})$$
$$e_{i,j} = (S_{i,j} \ e_{i,j,1} ... e_{i,j,k} ... e_{i,j,m})$$
...

2. Every token in the expression is replaced by a tuple containing the symbol's *Kind* and a

unique byteword representing the Gannet symbol corresponding to the token:

$$e_i \Rightarrow r_i = (reference, n_{ri})$$

$$S_i \Rightarrow s_i = (service, n_{Si})$$

The resulting list of bytewords is called an *instruction*. Instructions are represented using angle brackets: $\langle s_i, r_{i,1}, ..., r_{i,n} \rangle$ is the instruction referenced by r_i. We will use the notation $r_i \Rightarrow \langle s_i, r_{i,1}, ..., r_{i,n} \rangle$.

3. Create code packets: using the notation introduced above, a code packet is represented as:

$p_i = p(\text{code}, n_{Si}, GW, r_i; \langle s_i, r_{i,1}, ..., r_{i,n} \rangle)$ with n_{Si} the name of the service S_i and $_{GW}$ is the "gateway", the interface between the Gannet SoC and the outside world.

4. Create a reference packet to the root task:

$$p_{root} = p(\text{reference}, n_{Sroot}, GW, r_{root}; r_{root})$$

The gateway reads the bytecode and transfers the packets onto the NoC in no particular order.

Compiling High-Level Languages into Gannet

The Gannet language is intended as a compilation target for high-level languages. Because of the functional nature of the Gannet language and system, functional programming languages such as SML or Scheme are ideal candidates for compilation into Gannet. In particular, we have reported on compilation of Scheme into Gannet [Vanderbauwhede, 2007]. Imperative languages such as C and Java are less suitable because they have limited notion of parallelism and concurrency, having been designed for sequential von Neumann-style processors. However, C dialects such as Single-assignment C [Scholz, 2003] and C-syntax dataflow-style languages such as

Mitrion-C and similar languages [Koo et al., 2007] constitute suitable candidates.

THE GANNET MACHINE MODEL

In this section we introduce the Gannet Machine Model, a semantic model to explain the operation of the Gannet DRI [Vanderbauwhede, 2008]. The semantic model is used to demonstrate how a Gannet SoC runs Gannet programs and how the Gannet DRI implements concurrency, global and local flow control and streaming data processing.

Notation and Definitions

Notation

- The notation ● ("followed by") is used to separate a packet from the other packets in the queue: *(p ● ps)* denotes a packet at the head; *(ps ● p)* a packet at the tail.
- The notation * ("don't care") indicates that the value of a field does not influence the operation.
- The notation ... indicates the presence of some non-specified entities. In general, unspecified entities are left out unless omitting them would cause ambiguity.
- The notation _ indicates allocated available storage space
- The notation $e_{\Downarrow} w$ indicates large-step evaluation of e to w, i.e. w is the result of evaluating the expression e.
- We use the following shorthand conventions:
  - expression: e
  - service-symbol: s
  - reference-symbol: r
  - variable-symbol: v
  - argument-symbol: x
  - quoted symbol: prefix '
  - value: w

Definitions

- The Gannet system consists of N service nodes $S_i(...)$, $i \in 1..N$, a packet-switched Network-on-Chip and a gateway to the outside world, $GW(...)$.
- The unit of data transfer in the Gannet Service-Based Architecture is the packet. Depending on the packet's *Type*, the *Payload* can be *data* or an *expression*.
- The packet receive and transmit FIFO queues of the services are represented by q_{RX} and q_{TX}. A received packet is pushed onto the RX queue; a transmitted packet is shifted off the TX queue.
- The RX queue actually consists of four queues multiplexed by the packet's *Type*:
- $q_{RX}(tasks(...),data(...),refs(...),code(...))$
- Thus $q_{RX}(ps \bullet p)$ is actually $q_{RX}(... pt(ps \bullet p)...)$ with $pt \in \{tasks,data,refs,code\}$.
- The *tasks* queue was not mentioned in the high-level overview in section "Operation of the Gannet Service Manager" to simplify the discussion. This queue is intended for activated subtasks: activation of an instruction by reference packet results in generation of a *task* packet which is transferred to the *tasks* queue. The reason for placing the *tasks* queue to the RX queue is to allow the SM to directly receive *task* packets. The instruction in a *task* packet is not stored for later activation but executed immediately.
- Packet receive and transmit FIFO queues: q_{RX} and q_{TX}
- Apart from the RX/TX queues, a service node S_i consists of following entities:
  - The data store $store_d(Label\ data)$. *Label* is a Gannet symbol, *data* is the stored content. Space allocated for data to be stored is denoted by _: $store_d(Label\ \_)$
  - The code store $store_c(Label\ p)$. Here p is the stored packet, *Label* is the Label field from the packet's header.
  - The processing core $core(...)$ which performs the actual processing of the data.

Thus an explicit notation for a service node S_i is:

$S_i(q_{RX}(tasks(...),data(...),refs(...),code(...)),q_{TX}(...)$
$,store_c(...),store_d(...),core(...))$

At any given moment, every service S_i can be performing any number of actions. Actions are data-driven. Furthermore, all services are operating concurrently in a completely asynchronous fashion.

Small-Step Semantics

The semantics expresses an action taken by a service. Actions (indicated with the arrow $\xrightarrow{A}$) are triggered either by arrival of a packet or by completion of a computation by the service core. Actions result in a transition between two states of the system. Every expression in the semantics describes the effect of the action in terms of the state of the service, i.e. of its stores and queues.

Lines above the transition expression define items (e.g. packets) appearing on the left-hand side, lines below the transition expression define items appearing on the right-hand side.

$$p_i = packet(...);...$$
$$S_i(...p_i...) \xrightarrow{A} S_i(...p_j...)$$
$$p_j = packet(...);...$$

Packet Processing by the Services

A service performs a set of actions which result in packets being received from and transmitted to other services.

A subset of actions (the *marshalling* set) is performed by the service manager, which is the generic data marshalling unit through which every service core interfaces with the system. It is important to note that the service manager is generic, i.e. its design and functionality are independent of the design and functionality of the service core. The complementary set of actions (the *processing* set) is performed by the service core.

Packet Transfer between Services

The set of actions to transfer packets between services consists of *TX* (transmit) and *RX* (receive). The semantics are straightforward:

$$p = \text{packet}(*, i, j, *; *)$$

$$S_i(q_{TX}(p \bullet ps)) \xrightarrow{\quad TX \quad} S_i(q_{TX}(ps))$$

$$S_j(q_{RX}(qs)) \xrightarrow{\quad RX \quad} S_j(q_{RX}(qs \bullet p))$$

Both actions carry the implicit assumption that the system's NoC will transfer the packet correctly between nodes S_i and S_j. Note that the actions don't happen synchronously: the NoC is asynchronous and the delay for transmission of the packet is unknown.

Marshalling Action Set M

On activation of a *code* packet, a number of actions can be performed by the service manager, as explained in section "Operation of the Gannet Service Manager". As mentioned there, these actions are grouped in the M ("Marshalling") set. Application of the M set results in evaluation of all arguments of a service call. The actions of the complete set $M = \{SC, AT, DR, SQ, SR\}$ can be expressed as:

$$p_i = \text{packet}(\text{task}, i, *, *; e_i); e_i = \langle s_i, a_1, ..., a_n \rangle$$

$$a_i ::= {'}r_i \mid r_i$$

$$S_i(q_{RX}(p_i \bullet ps), q_{TX}(qs), \text{store}_d(...))$$

$$\xrightarrow{\quad M \quad} S_i(q_{RX}(ps), q_{TX}(qs),$$

$$\text{store}_d(...(a_1 \ wr_1)...(a_n \ wr_n)))$$

$$wr_i ::= w_i \mid r_i$$

This expression describes the evaluation of function arguments by the Gannet service. For the sake of brevity, the actions leading on to the activation of the *code* packet have been omitted. Instead, it is simply assumed that the service manager directly receives a *task* packet. It is easy to show that this is equivalent to applying the $\{SC, AT\}$ action set on arrival of a *reference* packet.

Processing Action Set P

The actions of the service core determine the functionality of the service. This functionality can be defined as the type, destination and payload content of the packet which the service produces based on the values marshalled by the service manager.

The service core implements a function cs_i which takes n arguments with values $w_1 ... w_n$ and produces a result w, optionally modifying the state of the store in the process.

The P set consists of the actions *call, eval, return*: the values are called from the store; the core performs its computation (*eval*) and returns a result.

The processing can be expressed as:

$$S_i(q_{RX}(qs), q_{TX}(ps), \text{store}((s_1 w_1)...(s_n w_n) \ state))$$

$$\xrightarrow{\quad P \quad} S_i(q_{RX}(qs), q_{TX}(ps \bullet p), \text{store}(state'))$$

$$p = \text{packet}(, *, i, *; w); (cs_i w_1 ... w_n) \Downarrow w$$

Hardware Memory Management

An important point to note is that all the memory is fully managed by the SM: it allocates memory for data required by a task and de-allocates it when the task is finished. As a result, the language does not need any memory management constructs and it is not possible for different tasks to impinge on each other's memory. Static memory allocation would offer the same benefits but it would require memory to be allocated for all possible tasks at all times and would thus require a large amount of memory, most of which never to be used. Dynamic memory management introduces a small overhead for managing the address stack but the required area is negligible compared to the memory area for fully static allocation; the run-time overhead is only a single cycle for allocating an address; de-allocation takes a few more cycles but does not add to the run-time overhead as it does not block the processing.

Control and Computational Service Semantics

In this section and the next, we consider actions at the level of the M and P sets without detailing the individual actions in each set. The combined M and P action sets result in the service transmitting a result packet in response to receiving a task packet and potentially modifying the local store. The semantics of the Gannet services can be described completely in terms of the task and result packets and the state of the store. The aim of the next two sections is to illustrate this mechanism.

Computational Service Semantics

Computational services are services of which the core behaviour can be modelled as a δ-application. This type of service includes all third-party IP cores in the SoC, as these cores have no knowledge of the Gannet system.

The resulting packet will be of type *data* and the state of the *store* is not modified by the evaluation.

$$p_{rx} = \text{packet}(\text{task}, i, j, r_j; e_i);$$
$$e_i = \langle s_i, ..., r_j, ... \rangle; r_j \Downarrow w_j$$
$$S_i(\text{store}_d(...))$$
$$\xrightarrow{M} S_i(\text{store}_d(...(r_j w_j)...))$$
$$\xrightarrow{P} S_i(\text{store}_d(...))$$
$$p_{tx} = \text{packet}(\text{data}, j, i, r_j; w_i);$$
$$w_i = \delta(s_i, ..., w_j, ...)$$

Note that this does not mean that the IP core is stateless, only that it does not affect the state of the service manager stores.

Control Service Semantics

Control services provide functional language constructs to the Gannet architecture. Evaluation of a task by a control service can result in *a change of the state of the store* or the creation of a *result packet of type task*. To illustrate these mechanisms, this section presents the semantics for the group, assign, lambda and apply services.

Lexically Scoped Variables

Lexical scoping is implemented through the group and assign services. Variables are bound to an expression by the assign service:

$$p_{rx} = \text{packet}(\text{task}, assign, group, r_a; e_{assign});$$
$$e_{assign} = \langle assign, ..., 'v_j, r_j... \rangle; r_j \Downarrow w_j$$
$$S_{assign}(\text{store}_d(...))$$
$$\xrightarrow{M} S_{assign}(\text{store}_d(...('v_j v_j) (r_j w_j)...))$$
$$\xrightarrow{P} S_{assign}(\text{store}_d(...('v_j w_j)...))$$
$$p_{tx} = \text{packet}(\text{data}, group, assign, r_a; v_j);$$

The read service retrieves the value bound to a variable from the store and returns it:

$$p_{rx} = \text{packet}(\text{task}, read, i, r_r; e_{read});$$
$$e_{read} = \langle read, 'v_j \rangle; 'v_j \Downarrow v_j$$
$$S_{assign}(\text{store}_{assign}(...))$$
$$\xrightarrow{M} S_{assign}(\text{store}_{assign}(...('v_j v_j) (r_j w_j)...))$$
$$\xrightarrow{P} S_{assign}(\text{store}_{assign}(...('v_j w_j)...))$$
$$p_{tx} = \text{packet}(\text{data}, i, *, r_r; w_j);$$

The group service takes as arguments a number of assignment expressions and one or more expressions that may call the assigned variables. The group and assign services have a shared store. The group service returns the result of the last expression and clears all variables resulting from the assignment expressions from the assign store. Consequently, assign-variables are lexically scoped.

$$p_{\text{group}} = \text{packet}(\text{task}, \text{group}, i, r_i; e_{\text{group}});$$
$$e_{\text{group}} = \left\langle \text{group}, \ldots, r_{\text{assign},j}, \ldots, r_k \right\rangle;$$
$$r_k \Rightarrow \left\langle s_k, \ldots, r_j, \ldots \right\rangle \Downarrow w_k; r_j \Rightarrow \left\langle \text{read}, 'v_j \right\rangle \Downarrow w_j$$
$$S_{\text{group}}(\text{store}_{\text{assign}}(\ldots))$$
$$\xrightarrow{M} S_{\text{group}}(\text{store}_{\text{assign}}(\ldots(v_j\ w_j)\ldots(r_k\ w_k)\ldots))$$
$$\xrightarrow{P} S_{\text{group}}(\text{store}_{\text{assign}}(\ldots))$$
$$p_r = \text{packet}(\text{data}, i, \text{group}, r_i; w_k)$$

Lambda Functions

Function definition and application is implemented through the lambda and apply services:

- Functions are defined by the lambda service:

$$p_{\text{lambda}} = \text{packet}(\text{task}, \text{lambda}, *, *; e_{\text{lambda}});$$
$$e_{\text{lambda}} = \left\langle \text{lambda}, 'x_j, \ldots, 'r_\lambda \right\rangle; r_\lambda \Rightarrow \left\langle s_j, \ldots, x_j, \ldots \right\rangle$$
$$S_{\text{lambda}}(\text{store}_d(\ldots))$$
$$\xrightarrow{M} S_{\text{lambda}}(\text{store}_d(\ldots(x_j\ x_j)\ldots(r_j\ r_\lambda)\ldots))$$
$$\xrightarrow{P} S_{\text{lambda}}(\text{store}_d(\ldots))$$
$$p_r = \text{packet}(\text{data}, *, \text{lambda}, *; e_\lambda);$$
$$e_\lambda = \left\langle \ldots, x_j, \ldots, r_\lambda \right\rangle$$

- Function application is performed by the apply service:

$$p_{\text{apply}} = \text{packet}(\text{task}, \text{apply}, *, *; e_{\text{apply}});$$
$$e_{\text{apply}} = \left\langle \text{apply}, r_\lambda, \ldots, 'r_j, \ldots \right\rangle; r_\lambda \Downarrow e_\lambda; 'r_j \Downarrow r_j$$
$$S_{\text{apply}}(\text{store}_d(\ldots))$$
$$\xrightarrow{M} S_{\text{apply}}(\text{store}_d(\ldots(r_\nu\ e_\nu)\ldots('r_j\ r_j)\ldots))$$
$$\xrightarrow{P} S_{\text{apply}}(\text{store}_d(\ldots))$$
$$p_r = \text{packet}(\text{task}, *, j, r_w; e_w); e_w = e_\lambda[x_j\ /\ r_j]$$

As can be seen from this semantics, the apply service does not bind values but rather substitutes code references. The reason for this behaviour is explained in section "Separation of Control Flow from Data Flow".

Multi-Operation Services

It is often desirable for a service to be able to perform more than one operation. In particular for service cores that take up little area, it is desirable to share one service manager amongst several service cores to limit the area overhead of the service manager. For example, for control services the granularity is quite fine and the implementation is relatively simple compared to the actual IP cores; on the other hand, control services need to be very flexible and in particular the sets of services related to lexical variables, lambda functions and list manipulations are closely interlinked and require a considerable amount of memory, although their data throughput is not large. It is therefore most area-efficient to combine control constructs into a single service. We call such a combined service a *multi-operation service*. The *Name* field of the service symbol indicates which operation must be performed for a given call (similar to the opcode in a microprocessor). The only difference with the single-operation services as presented above is the value of the To-field. For example, if we name the combined control service control, an assign task packet now becomes:

$$p_{\text{assign}} = \text{packet}(\text{task}, \text{control}, j, r_j; e_{\text{assign}})$$
$$e_{\text{assign}} = \left\langle \text{assign}, 'v, r \right\rangle$$

The advantages of combining services are less overhead thanks to the shared service manager and local store and the potential to factor out common functionality, thus further reducing the required area. Furthermore, in the particular case of control services, they can be implemented on a small embedded microcontroller for maximum flexibility. The potential disadvantage is due to the lower degree of parallelism and the potentially large number of calls to be handled, i.e. the classic trade-off of speed for area. However, as we will discuss in section "Separation of Control Flow from Data Flow" this drawback can be mitigated by ensuring no control service, either combined or individual, is present in the data path.

Dynamic Service Configuration

Using the concept of a multi-operation service, to achieve dynamic service configuration in Gannet it suffices to add a *reconfiguration service* to every dynamically reconfigurable service core. This configuration service will request the configuration information from a central *reconfiguration manager*. A Gannet program that reconfigures a service S1 to perform a discrete cosine transform (DCT) might look like this:

```
(let
    '(S1-reconf 'dct '(config 'dct 's1))
    '(S1 (block (img '8))
)
```

The key point is that there is no change required to the Gannet DRI to support dynamic service configuration: it is simply a matter of adding the extra services. The global config service (i.e. the reconfiguration manager) is simply a database server which serves the precompiled configurations supported by each service core. The local reconf operation depends on the nature of the actual reconfigurable core. For example, if the core would be a microcontroller, reconf would simply load a different program into memory; if the core would be an embedded FPGA, reconf would load up a bitstream. To indicate to the reconfiguration manager what type of reconfiguration information is required for a given service, the config service takes two arguments: one describing the functionality to be configured and one describing the nature of the service core. The reconf call also takes two arguments, the actual reconfiguration data and a label for the configuration. This label is used to check if the current configuration is different from the new configuration. If the configurations are identical, the current configuration is kept.

In the remainder of this section we use the semantic model to explain two key advanced flow control features of the Gannet system: separation of control flow from data flow and streaming data processing.

Separation of Control Flow from Data Flow

An issue of crucial importance for the performance of SoCs with multiple high-throughput data flows is the separation of control flows and data flows. Although Gannet does not require centralised control, in many cases services will perform flow control tasks. This section introduces a technique used to avoid the potential bottleneck at such control services.

Bottleneck in Central-Memory SoC Architectures

Consider a simple SoC which implements a video webcam. The system captures a video stream, compresses it using a lossy codec, encrypts it and transmits it over a network. The functionality for these four operations is provided by hardware IP cores. Following a conventional SoC design approach, this system will be implemented using a microcontroller which runs an embedded operating system (Figure 3).

The OS communicates with the hardware using device drivers. Data is transferred to and from the central memory either via the microcontroller or via a direct memory access (DMA) controller. Clearly this memory transfer is a bottleneck: if the number of datapaths or the number of operations per datapath would be very large, the memory access time will limit the data rate.

The Gannet architecture allows separation of dataflow from control flow: the data will be transferred directly between the data processing cores (Figure 4), eliminating the bottleneck. Consequently the system will be able to support more

Figure 3. Simple video capture SoC with embedded OS

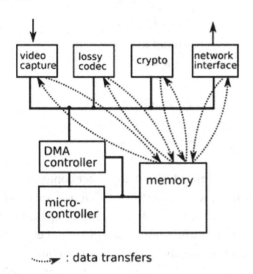

: data transfers

Figure 4. Gannet architecture for simple video capture SoC

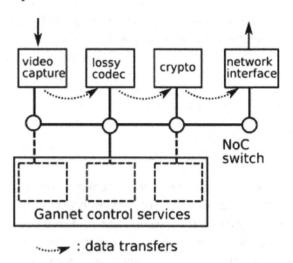

: data transfers

and longer data paths. As the net number of data transfers is lower, the system will also consume less power than the conventional architecture.

Redirection of Data Flows

For optimal performance control flows must be separated from data flows, i.e. data should flow between non-control services while the actual data path is governed by the control service. If this would not be the case, data would have to be copied to and from the control service's local memory, causing a performance bottleneck. The Gannet system as presented solves this issue via a combination of *deferred evaluation* and *result redirection*. To explain this mechanism, we use the control services introduced above as examples.

Consider the expression:

```
(S1 (group
    (assign 'v1...)
    (S2... (read 'v1)...)
))
```

With the semantics as presented in section "Control Service Semantics", the group service

receives the result of the evaluation of the S2 call. It then passes this result on to S1. However, the behaviour is different when the last argument is quoted:

```
(S1 (group
    (assign 'v1...)
    '(S2... (read 'v1)...)
))
```

In this case, evaluation of the last expression is deferred to the group core. The core dispatches a reference packet to S2 but sets the return address to S1:

$$p_{\text{group}} = \text{packet}(\text{task}, \text{control}, S_1, r_i; e_{\text{group}});$$
$$e_{\text{group}} = \left\langle \text{group}...r_{\text{assign},j}...'r_{S_2} \right\rangle;$$
$$r_{assign,j} \Rightarrow \left\langle \text{assign}, 'v_j, r_j \right\rangle$$
$$r_{S_2} \Rightarrow \left\langle s_2,...,r_j,... \right\rangle \Downarrow w_k; r_j \Rightarrow \left\langle \text{read}, 'v_j \right\rangle \Downarrow w_j$$
$$S_{\text{control}}(\text{store}_{\text{control}}(...))$$
$$\xrightarrow{M} S_{\text{control}}(\text{store}_{\text{control}}(...(v_j \ w_j)...('r_{S_2} \ r_{S_2})...))$$
$$\xrightarrow{P} S_{\text{control}}(\text{store}_{\text{control}}(...))$$
$$p_r = \text{packet}(\text{reference}, S_2, S_1, r_i; r_{S_2})$$

Another example is conditional branching as implemented by the if service.

```
(S1 (if (Sp...) (S2t...) (S2f...)))
```

The semantics without quoting (and thus without result redirection) are:

$$p_{if} = \text{packet}(\text{task}, \text{if}, j, r_j; e_{if}); e_{if} = \langle \text{if}, r_p, r_t, r_f \rangle;$$
$$r_{t|f} \Rightarrow \langle s_{t|f}, ... \rangle; r_p \Downarrow w_p^B$$
$$S_{if}(\text{store}(...))$$
$$\xrightarrow{M} S_{if}(\text{store}(...(r_p \ w_p^B)(r_t \ w_t)(r_f \ w_f)...))$$
$$\xrightarrow{P} S_{if}(\text{store}(...))$$
$$p_r = \text{packet}(\text{data}, j, \text{if}, r_j; w_{tf}); tf = w_p?t:f$$

With the above semantics, both branches will always be evaluated. This is obviously undesirable in many cases. If the second and third arguments are quoted, evaluation is deferred to the core which will evaluate only one branch predicated on the value of the first argument.

```
(S1 (if (Sp...) '(S2t...) '(S2f...)))
```

E.g. if the Sp call evaluates to true, the result of S2t will be sent directly to S1.

$$p_{if} = \text{packet}(\text{task}, \text{control}, S_1, r_1; e_{if})$$
$$e_{if} = \langle \text{if}, r_p, 'r_t, 'r_f \rangle; r_p \Rightarrow e_p = \langle s_p, ... \rangle \Downarrow \text{true}$$
$$S_{if}(\text{store}(...))$$
$$\xrightarrow{M} S_{if}(\text{store}(...(r_p \ \text{true}) \ ('r_t \ r_t) \ ('r_f \ r_f)...))$$
$$\xrightarrow{P} S_{if}(\text{store}(...))$$
$$p_r = \text{packet}(\text{reference}, S_{2,t}, S_1, r_1; r_t)$$

Finally, the apply service uses reference substitution for the same reasons. Consider the expression:

```
(S1 (apply (lambda 'x '(S2 x)) '(S3...)))
```

If apply would bind the evaluated arguments to the λ-variables, all results would be copied to the apply service's local store and would have to be requested from there. This excessive back-and-forth copying of potentially large amounts of data is avoided by substituting the λ-variables for code references instead[2]:

$$p_{apply} = \text{packet}(\text{task}, \text{control}, S_1, r_1; e_{apply});$$
$$e_{apply} = \langle \text{apply} \, r_\lambda, 'r_{S_3} \rangle; 'r_{S_3} \Downarrow r_{S_3}$$
$$r_\lambda \Rightarrow e_\lambda = \langle \text{lambda}, 'x, 'r_{S_2(x)} \rangle; r_{S_2(x)} \Rightarrow \langle S_2, x \rangle$$
$$S_{apply}(\text{store}_d(...))$$
$$\xrightarrow{M} S_{apply}(\text{store}_d(...(r_\lambda e_\lambda)...('r_{S_3} r_{S_3})...))$$
$$\xrightarrow{P} S_{apply}(\text{store}_d(...))$$
$$p_r = \text{packet}(\text{code}, S_2, j, r_w; e_w)$$
$$e_w = e_\lambda[x / r_{S_3}] = \langle S_2, r_{S_3} \rangle$$
$$p_r = \text{packet}(\text{reference}, S_2, S_1, r_1; r_w)$$

In a similar fashion any other control service can use the mechanism of deferred evaluation and result redirection to achieve separation of data flow from control flow. Consequently, the Gannet architecture provides fully configurable data paths without incurring the performance bottleneck resulting from repeated transfers of large amounts data to and from a central memory.

Pipelined Streaming Data Processing

Pipelined streaming data processing is a key feature in SoCs, in particular for multimedia processing. In fact most SoCs have traditionally been constructed around a (non-reconfigurable) datapath for streaming data. It is therefore imperative that Gannet should support this processing model.

The semantics refer to the example in section "Streaming Data Processing". It should be noted that (buf 'b1 (S1...)) and (stream 'b1) are syntactic sugar, as the streaming and buffering capability are provided by the SM of service S, not by separate buf and stream services (as this would be extremely inefficient).

$$p_{buf} = \text{packet}(\text{task}, S_1, \text{let}, r_1; e_{buf})$$
$$e_{buf} \Rightarrow \langle S_1^{buf}, 'b, ..., r_{stream}, ... \rangle$$
$$r_{stream} \Rightarrow e_{stream} \Downarrow w_{stream}$$
$$S(q_{RX}(p_{buf} \bullet ps), q_{TX}(qs), \text{store}(...))$$
$$\xrightarrow{M} S(\text{store}(... (r_{stream} \ w_{stream}) (b \ \_)...))$$
$$\xrightarrow{P} S(\text{store}(... (b \ (w, p_{buf}))...)); w = \delta(S_1, ..., w_{stream}, ...)$$
$$p_r = \text{packet}(\text{data}, S_1, \text{let}, r_1; b)$$

The expression (buf 'b1 (S1...)) is translated by the compiler into $\langle S_1^{buf}, 'b,...,e_{stream},... \rangle$; the superscript *buf* indicates that the service manager should buffer the result of the computation rather than return it. The result of the *M* actions is to store the buffer variable and to evaluate all arguments of S1 and store the values. The *P* action performs a δ-evaluation on the arguments; it then stores the result as a tuple (w,p_{buf}) and returns the buffer variable *b*.

The stream call returns the buffered result and reschedules the buffering task:

$$p_{stream} = packet(task, S_1^{buf}, S_2, r_1; e_{buf}); e_{stream} \Rightarrow \langle S_1^{stream}, 'b \rangle;$$
$$S(q_{RX}(p_{stream} \bullet ps), q_{TX}(qs), store(...(b\ (w,p_{buf}))...))$$
$$\xrightarrow{M} S(store(...('b\ b)\ (b\ (w,p_{buf}))...))$$
$$\xrightarrow{P} S(q_{RX}(ps \bullet p_{buf}), q_{TX}(p_w), store(...))$$
$$p_w = packet(data, S_1, S_2, r_2; w)$$

The expression (stream 'b1) is translated by the compiler into $\langle S_1^{stream}, 'b \rangle$; the superscript *stream* indicates that the service manager should return the value from the buffer and reschedule the task p_{buf}. Thus S2 will operate on the previous result returned by S1, i.e. the operations are pipelined.

The buffer/stream mechanism allows interleaving (multiplexing) of streams on every service in the pipeline, i.e. a single service will time-multiplex automatically over streams in a statistical fashion. Apart from the buffer and stream constructs, the buffer can be accessed without restarting the task via a peek instruction. As a consequence, the buffer mechanism also serves to implement a local store mechanism for caching (call-by-need) and accumulation.

EXAMPLES OF DYNAMIC RECONFIGURATION

This section presents examples of Gannet task descriptions to illustrate dynamic (i.e. run-time) reconfiguration of the data path and the service core functionality in a Gannet SoC.

Example of Dynamic Data Path Reconfiguration

Consider a system for periodic image capture, for example for use in space exploration robots such as the Mars Rover. The system periodically captures images of its surroundings and transmits them to a satellite. Because of the orbit of the satellite, at certain times of the day the transmission bandwidth will be high and at other times low. Furthermore, at night images should be captured using an infrared camera rather than a visible light camera. Figure 5 illustrates the system. For clarity, only the paths have been shown.

The Gannet task description that governs the dynamic reconfiguration of the data path is shown below:

```
(let
  `(assign `hi-speed (label `test_bw (>
(bandwidth) TH_BW)))
  `(assign `infrared (label `test_ir (>
(tsr-sensor) TH_IR)))
  `(if infrared
    `(buf `b1i (img-capture-infrared))
  `(buf `b1v (img-capture-visible)))
  `(if hi-speed
    `(buf `b2hi (lossless-codec
  (if infrared' (stream `b1i)' (stream
`b1r))))
        `(let
```

Figure 5. Example dynamic image capture system

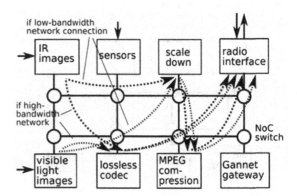

```
   '(buf 'b1s (scale-down
    (if infrared' (stream 'b1i)' (stream
'b1r))))
     '(buf 'b2lo (compress (stream
'b1s)))
     ))
  '(loop
  '(radio-interface
  '(if hi-speed' (stream 'b2hi)' (stream
'b2lo)))
   '(wait SAMPLE_PERIOD)
   '(update 'hi-speed test_bw)
   '(update 'infrared test_ir))
   )
```

The assign statements at the top test the bandwidth and light intensity and set the boolean variables infrared and hi-speed. Depending on the values of these variables, images are captured in infrared or visible light and transmitted uncompressed over a high-speed link to the satellite or scaled down, compressed and transmitted over a low-speed link. The lines with buf statements fill the initial pipeline; stream statements refresh the buffers (i.e. remove the old value and call the corresponding service which refills the buffer; see section "Pipelined Streaming Data Processing" for details). The loop statement at the bottom is the equivalent of a while(1) in C or a forever in Verilog. The system periodically transmits the processed image over the link and updates the

Figure 6. Example dynamically reconfigurable image processing system

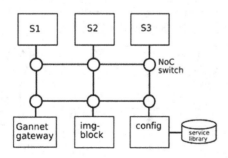

test variables. To understand the program it is important to realise that the description is actually a call tree, i.e. action is initiated by the final service: the radio-interface calls the stream command on e.g. the bufferb2hi (if bandwidth is high), this command results in the corresponding service (lossless-codec) calling the stream command on the bufferb1i (if infrared capture), this results in a call to img-capture-infrared which captures a fresh image. As explained in section "Redirection of Data Flows", the conditional statements simply redirect the requests, they are not part of the data path.

Example of Dynamic Service Reconfiguration

The example in this section is taken from [Noguera, 2006]. The paper reports on dynamic reconfiguration of systems with multiple FPGAs and uses a Sobel filter (used for edge detection in images) as a test case. As Gannet is a SoC infrastructure, we assume a dynamically reconfigurable SoC with three embedded FPGA cores (S1, S2, S3) rather than a discrete system. Furthermore, we assume a fixed core for actual capture of the image block on which the filter acts (img-block). The Gannet system is illustrated in Figure 6.

The original paper uses an embedded processor for IO and to control the dynamic reconfiguration. Our aim is to illustrate how dynamic reconfiguration of service core functionality can be achieved using the Gannet DRI, without the need for an embedded microprocessor. Figure 7 shows the task graph of the Sobel edge enhancement application used in [Noguera, 2006].

The Gannet task description for this Sobel filter *without* run-time reconfiguration is straightforward:

```
(let
  '(buf 'br (RGB2YCrCb (img-block)))
  '(let
  (buf 'bsh (Sobel-Hor br))
```

206

Figure 7. Sobel filter for edge quality enhancement

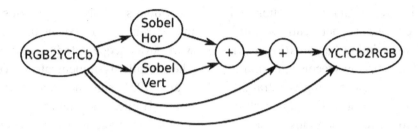

```
(buf 'bsv (Sobel-Vert br))
)
'(buf 'ba1 (add (stream 'bsh) (stream
'bsv)))
'(buf 'ba2 '(add br (stream 'ba1)))
'(loop
  (YCrCb2RGB (stream 'br) (stream
'ba2))
)
)
```

We now consider dynamic reconfiguration assuming a system with three reconfigurable cores consisting of embedded FPGA fabric (Figure 7). The horizontal and vertical Sobel convolutions are implemented on separate cores to allow them to be performed in parallel. A straightforward implementation of run-time reconfiguration, using the commands discussed in the section "Dynamic Service Configuration", is achieved as follows:

```
(let
  '(buf 'br (RGB2YCrCb (img-block)
  (S1-reconf 'RGB2YCrCb '(config 'RGB2Y-
CrCb 's1))))
  '(let
  (buf 'bsh (Sobel-Hor br
    (S2-reconf 'Sobel-Hor '(config 'So-
bel-Hor 's2))))
  (buf 'bsv (Sobel-Vert br
    (S3-reconf 'Sobel-Vert '(config 'So-
bel-Vert 's3))))
  )
  '(buf 'ba1 (add
```

```
(stream 'bsh) (stream 'bsv)
  (S1-reconf 'add '(config 'add 's1))
))
'(buf 'ba2 '(add br (stream 'ba1)
  (S2-reconf 'add '(config 'add 's2))
))
'(loop
(YCrCb2RGB (stream 'br) (stream 'ba2)
  (S3-reconf 'YCrCb2RGB '(config 'YCrC-
b2RGB 's3))
  )
  )
)
```

As can be seen from the code, it is sufficient to simply add the reconfiguration call as an additional argument to the original call, e.g.

```
'(buf 'br (RGB2YCrCb (img-block)
   (S1-reconf 'RGB2YCrCb '(config 'RG-
B2YCrCb 's1))))
```

This works because the Gannet service manager will always evaluate all unquoted arguments of a service call. Consequently, the actual service core will not be called until the reconfiguration is finished. The return value of the reconfiguration command is ignored as the service core only reads the return values of its required arguments. The reconfiguration does not interfere with the streaming data processing because the buffers are managed by the Gannet service manager and are independent of the service core configuration.

Because of the pipelined processing model, requesting and receiving data, computation and reconfiguration can occur concurrently, as illustrated in the following timing diagrams (Figure 8).

The timing diagram shows reconfiguration as a straight, dotted arrow, processing as a straight, full arrow and data request and receipt as short and longer curved arrows. The times of the operations are only indicative. Case (a) shows the behaviour when reconfiguration times dominate; case (b) shows the behaviour when data transfer and processing times dominate, i.e. the block size of the image is much larger in case (b) than case (a). Both cases clearly illustrate the overlap between reconfiguration, data transfer and processing. This example is a nice illustration of the power of Gannet's programming model to support efficient scheduling for run-time reconfiguration.

IMPLEMENTATION OF THE DYNAMIC RECONFIGURATION INFRASTRUCTURE

This section discusses the practical implementation of the Gannet DRI and programming tool chain. We discuss the assembler and compiler. We present results from a System-C cycle-approximate model of a Gannet SoC to illustrate the capabilities. We present a Verilog HDL implementation of the DRI [Vanderbauwhede et al., 2008] which demonstrates that the Gannet DRI combines high performance and low overhead.

Assembler and Compiler

The compiler generates the code packets as explained in section "Compilation of Gannet Code into Packets". Gannet programs are completely static from the point of view of code and variable addressing: all addresses for all instructions and variables are determined at compile time. This allows the SM to use direct lookups of code and

variables. The main complexity is compiling the individual symbols: the symbol's Kind is determined from the context and the compiler needs to work out the scope and addresses for every symbol. The compiler is implemented in Haskell [Hudak et al., 1992] using the Parsec parser combinator library [Leijen and Meijer, 2001]. The design is modular to make it easy to generate back-ends for a variety of platforms. Currently, the compiler exports Gannet bytecode, Gannet assembly language and Scheme.

The Gannet assembly language is the closest possible representation of compiled Gannet that is still human-readable. The Gannet assembler transforms the assembly into bytecode. The assembler (written in Perl) is used to explore new features of the system for which there is no compiler support.

Behavioural Model

We have implemented a behavioural model of the Gannet DRI using SystemC. The model allows cycle-approximate simulation of complete NoC-based SoCs with the Gannet DRI. The model is unclocked and uses the TLM (Transaction Level Modelling) library for transaction-level modelling. Consequently internal pipelining is not simulated accurately, resulting in slightly pessimistic performance. The SM implementation (Figure 9) adheres closely to the Gannet machine model introduced in section "The Gannet Machine Model".

The main differences are explicit arbitration and a separate path for returning requested data, which is required to make streaming more efficient. Other features not explicit in Figure 2 are a separate store for data pertaining to active subtasks i.e. instead of keeping the information in the header of the request packet in the queue, the information is stored in a RAM for faster access and reduced contention. Furthermore, an FSM controls the interface between the SM and the core using a *core status* register: the core can be in one of four states *idle*, *ready*, *busy*, *done*. The ready state indicates that the SM has finished

Figure 8. Timing diagrams for run-time reconfiguration of Sobel filter

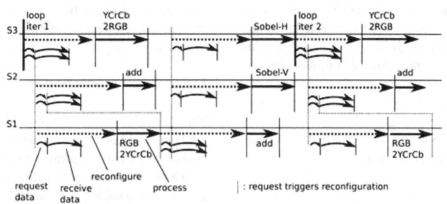

(a) Reconfiguration time is longer than data transfer/processing time

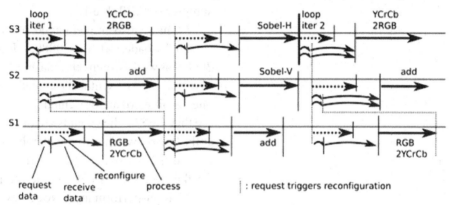

(b) Reconfiguration is shorter than data transfer/processing time

Figure 9. Gannet service manager implementation

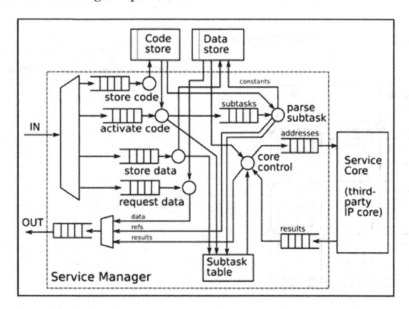

marshalling data; in the *busy* state the *core* is processing; in the *done* state the SM dispatches the result and cleans up the finished task. In the *idle* state the core is ready to receive new data. This process is illustrated in Figure 11 which shows the waveforms of the core status for a simulation of a task running on a 16-core Gannet SoC providing matrix operations (Figure 10).

The task consists of the following matrix operations on 8x8 matrices:

```
(madd
(cross (scale '0.5 (inv
    (if (< (det (a)) '0)
      '(mmult (a) (c))
      '(mmult (a) (d))
    )))
  (tran (a)))
(cross (scale '0.5 (inv
    (if (< (det (b)) '0)
      '(mmult (b) (d))
      '(mmult (b) (c))
    )))
  (tran (b)))
)
```

We can observe that the if control service

requires very few cycles thanks to the deferred evaluation and redirection; we can also observe the parallel computation of the determinant and matrix inversion.

FPGA Implementation

The Gannet DRI is a complex system and has not yet been completely implemented in hardware. However, we have implemented an FPGA prototype of the Gannet service manager in Verilog 2001 targeting the Xilinx Virtex-II Pro XC2VP30 [Vanderbauwhede et al., 2008]. The results are summarised in Table 1.

We have also implemented our Quarc NoC switch [Moadeli et al., 2008b, Moadeli et al., 2008a] as well as a conventional mesh NoC switch as reference. The synthesis results for a flit size of one 32-bit word are presented in Table 2. We can see that the SM slice counts are of the same order as the Quarc switch and considerably smaller than the mesh NoC switch. Note that the slice count for the switch does not include the transceiver. For a high-performance NoC, several virtual channels are required and the flit size will have to be at least four words. Consequently, the area of the SM circuit will generally be a fraction of the NoC switch+transceiver area.

Figure 10. Gannet SoC providing matrix operations

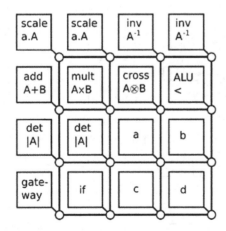

Figure 11. Core status transitions for a 16-core Gannet SoC running an arbitrary task

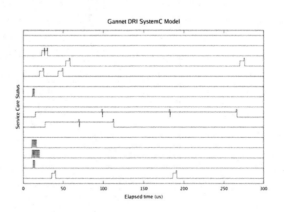

Table 1. Gannet Service Manager synthesis results

Optimisation Goal	Area	Speed
Slice count	822	917
BRAMs	14	8
Max. clock speed	170MHz	242MHz

Table 2. Quarc NoC switch synthesis results

#Virtual Channels	Slice count	
	Quarc	Mesh
0	1,141	1,993
2	1,558	2,663
4	2,401	4,008

The functionally of the design has been extensively verified using the ISIM simulator provided with the ISE toolkit. The simulation results have been presented in [Vanderbauwhede et al., 2008]; the most important observation is that the SM requires a only 20 to 50 cycles (depending on the number of arguments) to activate a task.

Performance Evaluation

The performance of the Gannet DRI was evaluated with a focus on pipelined streaming data processing. We compare the performance with a reference implementation of a single-core system. Our main focus is to evaluate the overhead of the Gannet DRI in terms of reduction of throughput.

Choice of Application Domain

As the Gannet DRI is a generic framework, it is difficult to assess its performance without reducing the results to a particular application domain. We note however that obviously the overhead of the DRI will be determined by the number of cycles required for processing of data by the core and transferring of data by the Network-on-Chip. As our work is concerned with coarse-grained reconfigurable systems, it would be unfair to assess the performance of the DRI using fine-grained cores such as ALUs. On the other hand, for large data blocks and very computationally-intensive cores, the overhead will obviously be negligible.

With these provisos, we have investigated the performance for a key class of applications, i.e. operations on 4x4 and 8x8 matrices of 32-bit floating-point numbers as used in many image processing applications. Image blocks of 8x8 pixels are commonly used in MPEG compressed video streams as the key transform, the 2D discrete cosine transform, works on blocks of that size; the H.264 standard uses an integer transform on blocks of 4x4 pixels. However, we did not assume that a single core performs a complete frame compression or even a complete transform, as these would consume many cycles. Rather, we assumed that the cores perform fundamental matrix operations such as addition, multiplication and transposition. The cycle counts for these operations were obtained from software implementations in C++ compiled using gcc, optimising for speed. We used the following rounded-down numbers:

As Gannet supports variable numbers of argument, the first four operations can have from 2 to 4 arguments and the total cycle counts are adjusted assuming repeated binary operations. It should be noted that the nature of the operations and the exact number of cycles is of limited importance. The aim is to have a realistic set of operations with realistic timings.

Table 3. Cycle counts for floating-point matrix operations

Operation	Matrix size	
	4x4	**8x8**
Addition ($A + B$)	80	240
Subtraction ($A–B$)	80	240
Cross product ($A⊗B$)	80	240
Dot product ($A.B$)	320	1024
Scaling ($a.B$)	80	240
Transposition (A^T)	32	128

Choice of Network on Chip

The Gannet DRI makes no specific assumptions on the Network-on-Chip used in the SoC. For the simulations we have use cycle counts obtained from a Verilog implementation of our own Quarc Network-on-Chip architecture [Moadeli et al., 2008b, Moadeli et al., 2008a]. The Quarc is a NoC with deterministic routing and wormhole switching. The link width is equal to the word size used in the Gannet machine, i.e. 32 bits. The NoC can transfer a single flit with 4 cycles overhead; we have simulated the performance for flits of 4 words (8 cycles/flit) and 1 word (5 cycles/flit). The Quarc NoC has a maximum diameter of N/4, we have assumed N=64.

Design of Experiment

We used a Monte-Carlo approach to evaluate the performance of the system over a wide parameter space. Valid expressions consisting of streaming matrix operations (e.g. (mult (add (a) (tran (cross (b) (c)))) (scale 0.5 (d)))) are generated

Figure 12. Number of operations per expression versus depth of the slowest path

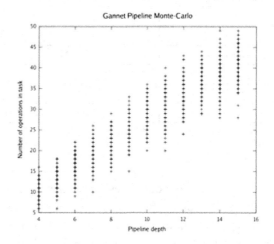

and simulated using the parameter values listed above and the throughput is established. The distances between the cores are chosen randomly. The length of the slowest path is extracted, as is the total number of operations in the expression. The experiment was performed 500 times for both block sizes (4x4 and 8x8) and both flit sizes (1 word and 4 words).

Results

Figure 12 shows the number of operations per expression versus the depth of the slowest path. The figure serves to visualise the distribution of the expressions generated by the Monte-Carlo.

The performance versus the reference implementation is presented in Figure 13. The cycle count for the reference implementation is assumed to be the sum of the cycle counts for all operations in the expression (i.e. no additional overhead for memory transfer etc.).

It is obvious that for small expressions with short paths the advantage of using a multi-core system is small; for more complex tasks however the system performs increasingly well. It should also be clear that the performance of an asynchronous pipeline is determined by the slowest node in the

Figure 13. Performance of Gannet SoC vs serial reference implementation

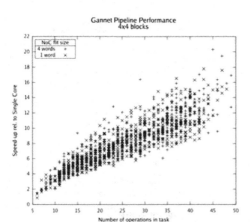

pipeline. To illustrate this point, Figure 14 shows performance results for expressions consisting of operations with a cycle count of 3072, i.e. the slowest operation in the set (a multiplication of four 8x8 matrices).

Finally, the overhead of the Gannet DRI is presented in Figure 15. The overhead is determined by computing the throughput for an asynchronous pipeline without DRI, i.e. a static pipeline, and subtracting this figure from the simulated value. The throughput for an asynchronous pipeline is determined by the slowest process and defined

Figure 14. Performance of Gannet SoC vs serial reference implementation, for slowest operation

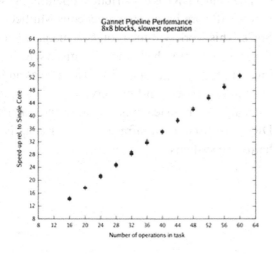

as the size of the processed data block divided by the delay between blocks. An asynchronous pipeline needs a handshaking mechanism to indicate to the upstream process that it can send data. Consequently, the components of the delay are $\Delta t = \Delta t_{req} + \Delta t_{data} + \Delta t_{proc}$. The DRI adds an overhead for activating the task and dispatching the data: $\Delta t_{DRI} = \Delta t_{act} + \Delta t_{dis}$. Thus the relative overhead is given by $\Delta t_{DRI}/\Delta t$. We plot this overhead versus the throughput in millions of operations per second (Mop/s) assuming the system runs at 100 MHz.

It is clear from these results that the overhead of the Gannet DRI is generally very small. Aggregating the data points into a cumulative distribution (Figure 16) we see that for 4x4 blocks, the median overhead is 4.0% with the worst case overhead <10%, whereas for 8x8 blocks the median overhead is 1.5% with the worst case overhead <5%.

CONCLUSION AND FUTURE WORK

In this chapter, we have presented the Gannet Dynamic Reconfiguration Infrastructure, a framework that provides high-level programming for Dynamically Reconfigurable SoCs. Our solution allows IP core-based Heterogeneous Multicore SoCs to be programmed using a high-level language whilst preserving the full potential for

Figure 15. Overhead of the Gannet DRI vs throughput

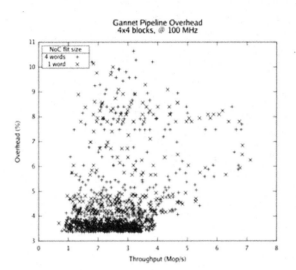

parallelism and dynamic reconfigurability inherent in such a system. We have demonstrated that the Gannet DRI provides full high-level dynamic reconfiguration capabilities with a small overhead both in area and performance.

There are three main directions for our future work. These three avenues will be explored in parallel.

The first strand is to improve the efficiency

Figure 16. Cumulative distributions of Gannet DRI overhead

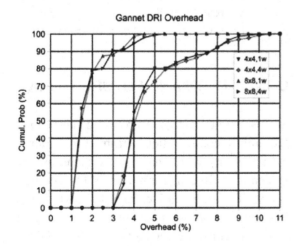

and flexibility of the system. In the short term, this means adding support for multi-threaded cores and implementing the proposed dynamic reconfiguration mechanism. In the longer term this means exploring data-driven operation as an alternative to the current demand-driven control.

The second strand is to realise the integration of a full Gannet system on FPGA, using the Quarc NoC as the communication medium. A particular challenge is the tight integration of the SM with an embedded microprocessor for efficient implementation of global control services.

The third strand is to investigate the potential of the Gannet DRI as a distributed operating system for IP core-based Heterogeneous Multicore SoCs. This work will focus on task management features such as rescheduling, prioritised scheduling, interruption and cancellation as well as adding a policy-based run-time service re-allocation mechanism. These features will allow the Gannet DRI to be used as an infrastructure for adaptive hardware systems.

REFERENCES

Amamiya, M., & Taniguchi, R. (1990). Datarol: a massively parallel architecture for functional languages. In *Proceedings of the Second IEEE Symposium on Parallel and Distributed Processing, 1990. Proceedings of the Second IEEE Symposium on*, (pages 726–735).

Ashcroft, E. A. (1986). Dataflow and eduction: data-driven and demand-driven distributed computation. In J. W. de Bakker, W. P. de Roever, and G. Rozenberg, (Eds.), Current Trends in Concurrency. Overviews and Tutorials. New York: J. W. de Bakker, W. P. de Roever, and G. Rozenberg, Eds. Springer Lecture Notes In Computer Science, vol. 224. Springer-Verlag. New York, New York, NY, 1-50.

Bray, T., & Paoli, J. (2008). *Sperberg-McQueen, C. M., Maler, E., and Yergeau, F. (2008). Extensible Markup Language (XML) 1.0* (5th ed.). Cambridge, MA: MIT Press.

Butts, M. (2007). Synchronization through communication in a massively parallel processor array. *IEEE Micro, 27*(5), 32–40. doi:10.1109/MM.2007.4378781

Cocco, M., Dielissen, J., Heijligers, M., Hekstra, A., Huisken, J., Hive, S., & Eindhoven, N. (2004). A scalable architecture for LDPC decoding. In *Proceedings of Design, Automation and Test in Europe Conference and Exhibition, 2004* (Vol. 3). *Proceedings*, volume 3.

Giraud-Carrier, C. (1994). A reconfigurable dataflow machine for implementing functional programming languages. *ACM Sigplan Notices, 29*(9), 22–28. doi:10.1145/185009.185014

Goossens, K., ; Dielissen, J., & ; Radulescu, A. (2005, September-October)., "AEthereal network on chip: concepts, architectures, and implementations.," *IEEE, Design & Test of Computers, IEEE*, vol.22, no.(5), pp. 414-421., Sept.-Oct. 2005

Gough, J. (2001). *Compiling for the. Net Common Language Runtime (CLR)* (1st ed.). Upper Saddle River, NJ: Pearson Education.

Guerrier, P., & Greiner, A. (2000). A generic architecture for on-chip packet-switched interconnections. In *Proceedings of DATE '00: Proceedings of The conference on Design, automation and test in Europe*, (pages 250–256)., New York: NY, USA. ACM.

Huang, S. S., Hormati, A., Bacon, D. F., & Rabbah, R. (2008, July 7-11). Liquid Metal: Object-Oriented Programming Across the Hardware/Software Boundary. In *Proceedings of the 22nd European Conference on Object-Oriented Programming.* (Paphos, Cypress., July 07 - 11, 2008). J. Vitek, Ed. Lecture Notes In Computer Science, vol. 5142. Springer-Verlag, Berlin, Heidelberg, 76-103.

Hudak, P., Jones, S., Wadler, P., Boutel, B., Fairbairn, J., & Fasel, J. (1992). Report on the programming language Haskell: a non-strict, purely functional language version 1.2. *ACM Sigplan Notices, 27*(5), 1–164. doi:10.1145/130697.130698

Kelsey, R., Clinger, W., et al. (Editors), J. R. (1998). Revised report on the algorithmic language Scheme. ACM SIGPLAN Notices, 33(9):26–76.

Kogel, T., & Haverinen, A. & and Aldis, J. (2005). OCP TLM for Architectural Modeling. (*White Paper.*).

Koo, J. J., Fernandez, D., Haddad, A., & Gross, W. J. (2007, July 9-11). Evaluation of a High-Level-Language Methodology for High-Performance Reconfigurable Computers, Application-specific Systems. In Proceedings of the, *Architectures and Processors, 2007. ASAP. IEEE International Conf. on Architectures and Processors, 2007*, vol., no., (pp.30-35)., 9-11 July 2007

Lampinen, H., Perala, P., & Vainio, O. (2006). Design of a scalable asynchronous dataflow processor. In *Proceedings of the IEEE Design and Diagnostics of Electronic Circuits and systems, 2006 IEEE*, vol., no., (pp.85-86).

Lanuzza, M., Perri, S., & Corsonello, P. (2007). MORA - A New Coarse Grain Reconfigurable Array for High Throughput Multimedia Processing. In *Proceedings of the. International Symposium on Systems, Architecture, Modeling and Simulation,(SAMOS07)*. (LNCS 4599, pp. 159–168)., 2007. Springer-Verlag Berlin Heidelberg 2007

Leijen, D. & and Meijer, E. (2001). Parsec: Direct style monadic parser combinators for the real world. (Technical Report UU-CS-2001-27). The Netherlands: Department of Computer Science, Universiteit Utrecht.

Marescaux, T., Nollet, V., Mignolet, J., Bartic, A., Moffat, W., Avasare, P., Coene, P., Verkest, D., Vernalde, S., and Lauwereins, R. (2004). Run-time support for heterogeneous multitasking on reconfigurable SoCs. *Integration, the VLSI journal, 38*(1),:107–130.

McCarthy, J. (1960). Recursive functions of symbolic expressions and their computation by machine, part i. *Communications of the ACM, 3*(4), 184–195. doi:10.1145/367177.367199

Milner, R., Tofte, M., & Harper, R. (1990). The definition of Standard ML. Cambridge, MA: MIT Press., Cambridge, MA, USA.

Mirsky, E., & DeHon, A. (1996). MATRIX: a reconfigurable computing architecture with configurable instruction distribution and deployable resources. In *Proceedings of IEEE Symposium on FPGAs for Custom Computing Machines, 1996. Proceedings. IEEE Symposium on*, (pages 157–166).

Mishra, M., Callahan, T., Chelcea, T., Venkataramani, G., Goldstein, S., & Budiu, M. (2006). Tartan: evaluating spatial computation for whole program execution. In *Proceedings of the 12th international conference on Architectural support for programming languages and operating systems*, (pages 163–174). New York: ACM. New York, NY, USA.

Moadeli, M., Vanderbauwhede, W., & Shahrabi, A. (2008a). A Performance Model of Communication in the Quarc NoC. In *Proceedings of the 14th IEEE International Conference on Parallel and Distributed Systems, 2008.* (ICPADS'08), (pages 908–913).

Moadeli, M., Vanderbauwhede, W., & Shahrabi, A. (2008b). Quarc: A novel network-on-chip architecture. In *Proceedings of ICPADS '08: Proceedings of The 2008 14th IEEE International Conference on Parallel and Distributed Systems*, (pages 705–712)., Washington, DC: USA. IEEE Computer Society.

Najjar, W. A., Böhm, W., Draper, B. A., Hammes, J., Rinker, R., & Beveridge, J. R. (2003, Aug.). High-Level Language Abstraction for Reconfigurable Computing. *Computer, 36*(8), 63–69. doi:10.1109/MC.2003.1220583

Noguera, J., ; Badia, R. M. (2006, July), System-level power-performance tradeoffs for reconfigurable computing., *IEEE Transactions on Very Large Scale Integration (VLSI) Systems,, IEEE Transactions on*, (vol.14, no.7, pp.730-739)., July 2006

Nollet, V., Coene, P., Verkest, D., Vernalde, S., & Lauwereins, R. (2003). Designing an operating system for a heterogeneous reconfigurable soc. In *Proceedings of IPDPS '03: Proceedings of The 17th International Symposium on Parallel and Distributed Processing*, (page 174.1), Washington, DC: USA. IEEE Computer Society.

Nollet, V., Marescaux, T., Avasare, P., & Mignolet, J. (2005). Centralized run-time resource management in a network-on-chip containing reconfigurable hardware tiles. In *Proceedings of the conference on Design, Automation and Test in Europe* -(volume 1, pages 234–239). Washington, DC: IEEE Computer Society. Washington, DC, USA.

Nollet, V., Marescaux, T., Verkest, D., Mignolet, J., & Vernalde, S. (2004). Operating-system controlled network on chip. In *Proceedings of the 41st annual conference on Design automation*, (pages 256–259). ACM New York: ACM., NY, USA.

Pham, D. C., & Aipperspach, T. (2006). Overview of the architecture, circuit design, and physical implementation of a first-generation cell processor. *IEEE Journal of Solid-state Circuits*, *41*(1), 179–196. doi:10.1109/JSSC.2005.859896

Purohit, S., Chalamalasetti, S. R., Margala, M., & Corsonello, P. (2008). Power-Efficient High Throughput Reconfigurable Data path Design for Portable Multimedia Devices. In *International Conference on Reconfigurable Computing and FPGAs* (Reconfig08), (pages 217–222).

Randal, A., Sugalski, D., & Toetsch, L. (2004). Perl 6 and Parrot Essentials, Second Edition. Sebastopol, CA: O'Reilly Media, Inc.

Scholz, S. (2003, November). Single Assignment C: efficient support for high-level array operations in a functional setting. *Journal of Functional Programming*, *13*(6), 1005–1059. doi:10.1017/S0956796802004458

Singh, H., Lee, M., Lu, G., Kurdahi, F., Bagherzadeh, N., & Chaves Filho, E. (2000). MorphoSys: An Integrated Reconfigurable System for Data-Parallel and Computation-Intensive Applications. *IEEE Transactions on Computers*, 465–481. doi:10.1109/12.859540

Sussman, G. J., & Steele, G. L. (1975). An interpreter for extended lambda calculus. (Technical report)., Cambridge, MA: MIT Press., USA.

Taylor, M., Psota, J., Saraf, A., Shnidman, N., Strumpen, V., Frank, M., et al. (2004). Evaluation of the Raw microprocessor: an exposed-wire-delay architecture for ILP and streams. In *Proceedings of the 31st Annual International Symposium on Computer Architecture, 2004. Proceedings. 31st Annual International Symposium on*, (pages 2–13).

Treleaven, P. C., Brownbridge, D. R., & Hopkins, R. P. (1982). Data-driven and demand-driven computer architecture. *ACM Computing Surveys*, *14*(1), 93–143. doi:10.1145/356869.356873

(in press). ul Abdin, Z., & and Svensson, B. (2008). Evolution in architectures and programming methodologies of coarse-grained reconfigurable computing. [Uncorrected Proof:–.]. *Microprocessors and Microsystems*.

Vanderbauwhede, W. (2006a). Gannet: a functional task description language for a service-based SoC architecture. In *Proceedings of the. 7th Symposium on Trends in Functional Programming (TFP06)*.

Vanderbauwhede, W. (2006b). The Gannet Service-based SoC: A Service-level Reconfigurable Architecture. In *Proceedings of 1st NASA/ESA Conference on Adaptive Hardware and Systems* (AHS-2006), (pages 255–261), Istanbul, Turkey.

Vanderbauwhede, W. (2007, September 30). Gannet: a Scheme for Task-level Reconfiguration of Service-based Systems-on-Chip. In *Proceedings of the 2007 Workshop on Scheme and Functional Programming* (pages 129–137)., September 30th, 2007, Freiburg, Germany., pages 129–137. N/A.

Vanderbauwhede, W. (2008). A Formal Semantics for Control and Data flow in the Gannet Service-based System-on-Chip Architecture. In Proceedings of *The International Conference on Engineering of Reconfigurable Systems and Algorithms, ERSA 2008*. N/A.

Vanderbauwhede, W., Mckechnie, P., & Thirunavukkarasu, C. (2008). The Gannet Service Manager: A Distributed Dataflow Controller for Heterogeneous Multi-core SoCs. In Proceedings of the *NASA/ESA Conference on Adaptive Hardware and Systems, 2008.* (AHS'08). *NASA/ESA Conference on*, (pages 301–308).

Veen, A. (1986). Dataflow machine architecture. [CSUR]. *ACM Computing Surveys*, *18*(4), 365–396. doi:10.1145/27633.28055

Vegdahl, S. (1984). A survey of proposed architectures for the execution of functional languages. *IEEE Transactions on Computers*, *100*(33), 1050–1071. doi:10.1109/TC.1984.1676387

Wilkinson, B. (1996). *Computer architecture: design and performance* (2nd ed., pp. 434–437). Upper Saddle River, NJ: Prentice-Hall.

KEY TERMS AND DEFINITIONS

Coarse Grained Reconfigurable Architecture (CGRA): A System-on-Chip (SoC) architecture which allows to reconfigure parts of the system at a coarse granularity, i.e. at the granularity of several instructions or gates.

Demand-Driven Computational Model: A model of computation based on reduction of the call tree of a functional program. This model supports parallel and concurrent computations.

Dynamic Reconfiguration Infrastructure (DRI): A System-on-Chip infrastructure to allow dynamic (i.e. run-time) reconfiguration of data paths and computational functionality in Heterogeneous Multicore Systems-on-Chip.

Functional Programming: A programming paradigm that treats programs as trees of functions, emphasising function application rather than sequential imperative statements. Examples of functional languages are Haskell, ML, Scheme.

Gannet: The Gannet (Morus Bassanus) is a seabird and is the largest member of the gannet family, Sulidae. It can be found around the coasts of Scotland, with the largest colonies on the Bass Rock.

Heterogeneous Multicore System-on-Chip: A System-on-Chip (SoC) with multiple non-identical processing cores.

Network-on-Chip: An on-chip communication medium using a network of switches to route data between components of the system.

ENDNOTES

[1] The order of arrival is actually irrelevant: if the reference arrives earlier, activation will occur as soon as the code arrives.

[2] To be precise, the apply service performs symbol substitution: the mechanism is not limited to reference symbols.

Section 4
Simulation Framework for Fast Reconfigurable NoC Emulation

Chapter 9
Dynamic Reconfigurable NoC (DRNoC) Architecture:
Application to Fast NoC Emulation

Yana E. Krasteva
Universidad Politécnica de Madrid, Spain

Eduardo de la Torre
Universidad Politécnica de Madrid, Spain

Teresa Riesgo
Universidad Politécnica de Madrid, Spain

ABSTRACT

The aim of this Chapter is to present a highly flexible reconfigurable NoC solution for commercial FPGAs on one side, and on the other side, to provide an innovative approach for fast NoC emulation. The reconfigurable on-chip communication solution that is proposed in this chapter is capable of being reconfigured by means of adapting routers, network interfaces and cores themselves. The main distinguishing characteristic of the presented on-chip communication approach is that it permits to distribute the available on-chip communication resources among different communication topologies and thus, independent and application specific communication strategies can coexist and run in parallel. Furthermore, the proposed solution is not limited to NoCs and it permits to build a variety of on-chip communication. The proposed method in this Chapter for fast emulation provides a rapid way of validating different communication alternatives. The emulation method is based on the original idea of hard core re-usability through the exploitation of partial reconfiguration capabilities of some state of the art FPGAs. Both aspects have been tested and validated using a proof of concept approach and are discussed along this Chapter.

INTRODUCTION

Reconfigurable Systems on Chip (RSoCs) bring flexibility and adaptability to meet the demanding requirements of the electronic market. The most common topology for reconfigurable systems is the combination of an Field Programmable Gate Array (FPGA) and microprocessor(s), either embedded on the same device or separated in different pack-

DOI: 10.4018/978-1-61520-807-4.ch009

ages. Some FPGA devices include Digital Signal Processing (DSP) related blocks, like embedded multipliers, memories and microprocessor cores, which facilitate the mapping of complex functions. Even more, some FPGAs are able of being partially reconfigured, that is, to change the configuration memory of a part of the FPGA while the rest is kept active. This feature takes the powerful benefits of reconfigurability to a much higher level and permits to overcome restrictions related to long configuration times and performance losses when reconfiguring the entire device.

Partial run-time reconfigurable systems permit to keep different cores (tasks) running on the hardware and change/update them when needed without affecting other cores in the FPGA.

In a reconfigurable system, tasks may be loaded, unloaded or relocated to other FPGA areas without previous knowledge of what will be the future combination of cores and dependencies among them. Therefore, considering this evolving situations, on-chip communication solutions that are suitable in a given moment may not be the most adequate in other conditions (due to traffic changes, application updates, etc.). In such changing environment, communications reconfiguration is an important issue that may lead to better exploitation of on-chip resources and increased device performance. The communication reconfiguration process should be enabled and carried out without affecting other cores data exchanges, and data losses must be avoided even for the part that is being reconfigured.

The communication reconfiguration problem can be tackled from several points of view, focused on solving different system design problems. From higher levels systems modeling, trough system flexibility control usually based on extending operating systems, to system architecture and low level reconfigurability control. In this chapter, the problem is tackled from the system architecture and the system reconfigurability points of view and provides real implementations of the proposed solutions that have been tested with core models.

First, a brief state of the art is included along with a problem statement, and then, a solution that goes beyond NoCs by providing an on-chip communication adaptation solution is described in detail. The proposed solution covers several aspects: a communication architecture definition, along with its implementation on an FPGA and the definition of a router and a packet format that support system reconfigurability. Additionally, while describing the adopted solution, several reconfigurability aspects, performed tests and practical implementation issues will be presented. Conclusions, analysis and discussions of the discovered problems and achieved reconfiguration results are also included in the chapter. After that, in the second part, the adopted architecture and design solutions are integrated in an emulation framework, which is based on the innovative idea of exploiting hard core reusability through partial reconfiguration, for fast emulation of RSoCs, is described in detail. Three use cases, based on core models, have been implemented for proving the emulation approach and achieved speedup reports are included at the end of the Chapter.

BACKGROUND

NoCs for Reconfigurable Systems

The topic of Networks on Chip (NoCs) is quite new and challenging. It has appeared in the beginning of this century (Benini & De Micheli, 2002; William & Brian, 2001; Guerrier. & Greiner, 2000; Kumar et al., 20002) to solve problems related to on chip communications of highly integrated Systems on Chip (SoCs), where a large number of heterogeneous IP Cores, running at different clock frequencies, have to be interconnected. The communications paradigm adopted is based on the Globally Asynchronous, Locally Synchronous (GALS) communication scheme, where IP Cores interconnections are created by circuit or packet switching.

A few years after the first appearance of NoCs, the community working in reconfigurable and adaptable systems started studying the use of NoCs for on-chip communication in reconfigurable systems like in (Nollet et al., 2003) that is one of the first publications in this topic.

The selection and design of optimal communication strategies in reconfigurable systems is a great challenge because the system application might not be known at device design time. As a result, the on-chip communication of a reconfigurable system has to be prepared to cover "unknown" requirements.

A direct and general solution for this problem is to have an over-sized, fixed, on-chip communication infrastructure, where the routers and the communication resources are fixed in the reconfigurable device or even melt in the silicon. For instance, a general purpose NOSTRUM NoC instance is a 128 (256 wires) bit mesh NoC (Millberg et al., 2004). On the other side, there are ad-hoc NoCs that are preferred when a specific application or a set of applications are targeted. For instance, the design flow presented in (Srinivasan et al. 2007) targets to minimize the number of used network routers and NoC links. Additionally, it is known that NoCs can be considered an option when more cores have to be interconnected. Therefore, hierarchical NoCs have appeared, like HiNoC presented in (Hollstein et al., 2006), where buses are used for local connections and NoCs for global connections. These NoCs attenuate problems related to NoC high area overhead and latency when the number of locally connected cores is low.

The design of general or customized NoCs for fine grain reconfigurable systems will be part of the RSoC design process in the near future. NoCs will be either loaded in the array as it is done nowadays with system buses in FPGA based RSoC design flows, or be a hardwired part of the system. Such hard wired soution is proposed in (Hetch, R. et al, 2005), where are regular mesh architecture is described and simulation experiments are

presented and also in (Goossens K., et al., 2008), where the hardwired NoC is composed of some hard routers, but in this case the system can be further extended with additional soft routers and Network Interfaces (NIs) by means of extending a Linux operatin system.

The work presented in this Chapter searches innovative solution that target a higher level of system reconfigurability and flexibility on one hand, and on the other hand, searches for the actual realization of the proposed ideas on real hardware. In this work, the actual implementation is considered an important issue as it shows the deficiencies of the current technology to afford the future trends.

NoCs for Runtime Reconfigurable Systems

The on chip communication challenge, mentioned in the previous section, is even harder in runtime reconfigurable systems, where the system changes over time and new cores are loaded and/or removed from the system while it is running. For such time-dynamic systems, flexible and adaptable on-chip communications are necessary, and aspects like the allocation of cores of different sizes in the reconfigurable array, which might result in over-writing some on-chip communication resources, cores reconfiguration and core relocation along the architecture become very important.

The problem of interconnecting constantly changing modules in NoC based reconfigurable systems was firstly addressed by Bobda et al. in (Bobda et al., 2004). The main problem addressed by Bobda et al. is the placement of cores of different size and to achieve a routing strategy that could cope with this problem. Their solution is called DyNoC - Dynamic Network on Chip that is a mesh of interconnected routers that have a defined position in the reconfigurable system, and therefore they can be overwritten during the reconfiguration process if more room is needed for loading a new core. To solve this problem, the overwritten routers are deactivated and packets are routed through the routers that are surrounding the newly loaded module, using a modified XY routing algorithm. When the execution of a module is finished, and the occupied area becomes free, the routers are activated again. The architecture proposed by Bobda et al. follows a set of requirements that are defined with the aim of achieving dynamic changes and better logic/ programming elements (PE) ratio in the systems. The most important requirements for PEs are that: i) they have to be coarse grain in order to attenuate the wasted area due to the NoC, ii) all PEs should have access to the NoC and iii) it is stated that "the NoC has to be flexible enough to be used within the module to which it belongs" (Bobda & Ahmadinia, 2005). The DyNoC architecture and the surrounding routing algorithm have been validated with a light controller control system in a 3x3 DyNoC mapped on a Virtex II FPGA (Bobda et al., 2005).

The same problem is addressed in (Pionteck, Koch, & Albrecht, 2006), but the approach selected by Pionteck et al. is based on the idea that switches can be also removed and or added to the network, just like cores or modules. The proposed solution, called Configurable Network on Chip – CoNoChi, consists of switches with four full-duplex equal links. The links can be used either for connections to adjacent switches or for connecting a hardware module to a switch. As a result, depending on the number of non local links, a linear network or a 2D grid can be instantiated and reconfigurable modules of different sizes can be allocated by inserting and removing NoC switches using partial reconfiguration. In order to map CoNoChi to an FPGA, it is partitioned into a grid of rectangular modules that can be configured as a switch, as a horizontal or a vertical communication line or as part of the hardware module. In the CoNoChi approach, the relative location and the resources used by switches and cores are not restricted, and therefore, the flexibility provided is very high.

However, the main drawback is that the length between switches might dynamically change and thus the link delay. Also, the defined grid FPGA distribution implies that all the unused areas of the grid have to be configured to connect at least default communication lines. Pionteck et al. also focus on problems related to the routing policy during the adaptation process. The presented solution is based on routing table updates by sending new table contents and control data using the same NoC. The method used is as follows: i) first data in the network is re-routed to free the connection link where the new router is going to be included. This permits data in the NoC to be transmitted using switches not involved in the reconfiguration, then ii) switches related to the link under reconfiguration are disabled to avoid glitches. Finally, after partial reconfiguration, iii) affected routers routing tables are sent through the network and updated.

A different problem related to controlling the reconfigurability of cores connected to the NoC was addressed in (Möller et al., 2007), where a NoC called ARTEMIS is proposed. For controlling the core reconfiguration process, control packets are sent through the NoC and used to enable or disable a special structure allocated in the core NI to NoC border. This special structure is composed of *bus macros* that include enable and disable (Möller et al., 2006) control lines. Bus macros are special structures that guarantee routing in runtime reconfigurable systems and, therefore, are used to pass signals among reconfigurable modules and among reconfigurable modules and the fixed, non reconfigurable, area. A 2x2 ARTE-MIS NoC has been mapped in a Xilinx Virtex II FPGA, along with a use case application which is a reconfigurable Arithmetic Logic Unit (ALU) where switching between a divider, a multiplier and a square root is possible.

A slightly different approach, which is based on a NoC, but targets a more general on chip communication adaptability in runtime reconfigurable systems, has been presented in (Hübner et al., 2008). There, a multilayer approach to interconnect modules is proposed. By exploiting fine grain partial reconfiguration, switch multiplexers are reconfigured in order to connect different modules. These reconfigurable switches are used to change the topology of the on-chip communications by interconnecting the available network signals. The system has been mapped on a Xilinx Virtex II FPGA, where a one dimensional architecture distribution is allocated. The architecture is composed of: i) a set of fixed network signals implemented by bus macro structures cross the entire FPGA width, and ii) reconfigurable modules that span the entire FPGA height. As an example, a mix bus-ring communication has been mapped in the architecture by switching four bit multiplexers implemented with four input Look Up Tables (LUTs). LUTs are the basic reconfigurable element of FPGAs. A previous system from the same research group has been presented in (Becker, Donlin & Hübner, 2007). There, instead of the small grain reconfiguration of the switches' multiplexors, a higher grain reconfiguration of switches has been used.

On-Chip Communications Adaptability in Reconfigurable Systems

As it was mentioned before, one of the main aspects in reconfigurable systems is that the final application is "unknown" at design time, or it may change during device utilization. Even more, each core to be loaded in the system has its own communication requirements that have to be granted by the underlying communication architecture, and it has been clarified that flexibility, adaptability are searched in order to meet this challenge. The solutions presented in the state of the art cover different aspects, like the placement of cores of different size, the placement of new routers to the system, or the reconfiguration of switches' multi-

plexors. Differently, in this Chapter an enhanced and general solution, where the reconfigurability of the on-chip communications is tackled at different levels, is presented.

The characteristics that are being adapted or reconfigured could be referenced or associated to the OSI protocol stack communication layers. In (Benini & De Micheli, 2002), the OSI protocol stack has been reassembled for NoCs, and is called the micro-network stack. This stack is composed of several layers: i) a software layer that corresponds to the system application, and includes the application itself, the presentation and the session OSI layers, ii) the architecture and control layer that includes the transport, network and data link OSI layers and iii) the physical layer that defines the physical connection links. According to (Dehyadgari et al., 2006), where a slightly different OSI layer grouping is presented, NoCs lower layers (transport, network and data link) are fully implemented in HW.

In the following sections, the possibility of providing higher flexibility and reconfigurability through partial reconfiguration in all the NoC layer implemented in hardware will be analyzed and evaluated for a fine grain commercial FPGA.

The approach is based on the idea of defining a general on-chip communication architecture, called Dynamic Reconfigurable NoC (DRNoC), that represents as a template which is customized afterwards for a given application. Furthermore, each application occupies a different DRNoC communication layer, composed of a set of communication resources, allowing that several independent communication schemes coexist in the same system and run in parallel. Moreover, communication schemes for an application might not be restricted to a single communication type - homogeneous approach, but they could rather be a combination of point to point (P2P), point to multipoint (P2M) and/or NoC communication elements – heterogeneous or hierarchical approach.

DYNAMIC RECONFIGURABLE NOC (DRNOC) ARCHITECTURE

This section describes in detail, on one hand, the DRNoC architecture and its reconfigurability, and on the other hand, the mapping of the architecture on a Xilinx FPGA along with reconfiguration tests, performed for evaluating and validating the defined reconfigurations. Test results and the derived restrictions, which mainly result from the current technology, are also discussed. However, the main ideas and approaches, as well as the higher level implementations presented in this section, can be reused for any other reconfigurable system.

Apart from the architecture, its mapping and the reconfiguration, a set of design resources will be described in the following subsections. The focus is put to the NoC communication scheme because it is the most complex one and includes: the definition of a NoC packet format and a router that are intended to support system adaptability. A summary of the architecture features and an implementation comparison with the state of the art can be found at the end of this part of the Chapter.

DRNoC Architecture

A general and technology independent view of the Dynamic Reconfigurable NoC (DRNoC) architecture can be seen in Figure 1.

The Dynamic Reconfigurable NoC (DRNoC) is composed of a set of heterogeneous elements connected through Reconfigurable Network Interfaces (RNIs) to an on-chip mesh of Reconfigurable Routing Modules (RRMs). Each DRNoC element has a specific position and occupies a defined amount of predefined resources in a RSoC, aspects that are defined at RSoC design time. Differently, the specific configuration of each element and the amount of used resources, from the ones assigned to each element at design time, are defined at runtime. This approach leads to on-chip communications customization and also permits to define independent communication strategies if different applications are running concurrently in the same RSoC.

The defined DRNoC mesh also includes diagonal links and is referred as *Xmesh*. These diagonal links, which are indeed "S"- like when mapped on a device, are not common in NoC designs. Diagonal links have been included in this solution in order to provide the feature of independent on-chip connectivity for each application that might

Figure 1. DRNoC architecture general view

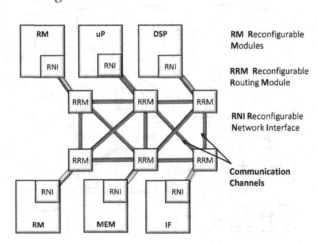

exist on the device, and, also, to give support for a broad set of communication schemes.

For instance, Figure 2 shows a 6x6 DRNoC Xmesh with four fully different communication schemes mapped on it: i) a regular mesh NoC that uses two diagonals on the top-middle part, ii) an irregular mesh NoC on the left, with a "Large core" allocated in the top, iii) one P2P connection in the bottom side and iv) one heterogeneous hierarchical communication scheme on the right, where a P2P link is used locally in the upper side, and a star NoC is used for the rest.

A floorplanning like this, where all nodes are distributed in fully independent communication schemes, would not be possible if the diagonal links were not available. An alternative to the Xmesh diagonal links is to concatenate some links in an "L"- like connection, but in that case latency will not be equal and resources from other links will be required, which will result in a more complex mapping procedure.

The main elements of the DRNoC system are described next.

- **Reconfigurable Network Interfaces (RNIs).** RNIs define a standardized interface between cores and the network. They are in charge of appropriate data formatting. The main Network Interface (NI) configurations that can be loaded in RNIs could be: serializers, deserializers, packetizers, depacketizers, multiplexers, demultiplexers, buffers, feed-throughs and empty designs. All these elements provide reconfigurability at the transport and network NoC layers.

- **Reconfigurable Routing Modules (RRMs).** Each RRM has nine communication channels used to build the Xmesh. RRMs define the application communication strategy. Routing modules could be: buffers, any type of NoC router with up to eight ports (apart from the local one) that fit in the reserved RRM area, switches, multiplexers, feed-throughs and empty designs. Depending on which aspect is modified, these elements provide reconfigurability at the transport, network and/or data link layers. For instance, according to (Dehyadgari et al., 2006), virtual channels are part of the transport and network layers, while a

Figure 2. A 6x6 DRNoC architecture with four independent communication schemes mapped

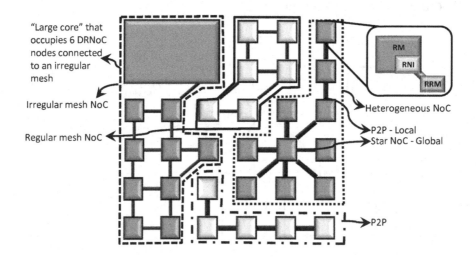

switching communication protocol belongs to the data link layer.

- **Communication Channels** define the communication physical links. They are composed of a fixed number of wires with specific characteristics and position. To reduce problems related to link delays, these are the only elements that cannot be changed in the system (in a custom RSoC system, they will be silicon metal wires). Communication channels are divided into subchannels that are composed of an equal amount of wires. All channels are equal except the local one, which connects RNIs with RRMs. The local channel size restricts the amount of independent communication schemes that a core can simultaneously take part in. Since each RRM is connected through communication channels to all its eight neighbors, the ideal size of a local channel is eight times the number of a communication channel wires. Further, the number of subchannels in a RRM channel defines the maximum amount of independent communications that can coexist in the system.

- **Switch Matrixes (SMs).** These elements are indeed part of RRMs, but are listed here separately because they may be reconfigured independently if the architecture mapping is design to permit that. SMs define which communication channel will be used and thus provides reconfigurability to the physical layer in the sense that the overall communication scheme topology can be modified.

The amount of resources available in a RRM and the number of available communication channels limit the amount and type of communication elements (routers, switches, etc) that can be used, and thus, the type and number of independent communication schemes that can coexist in the system. Every communication scheme mapped into DRNoC can use any amount of the available communication channels (1, 2,..) and an integer amount of subchannels. Thus, each independent communication in the device uses a communication layer that is composed of some of the available communication resources (RRMs logic resources, channels and subchannels). For instance, a simple link composed of two wires could be defined by loading feed-throughs in a RNI and a RRM using one subchannel of each communication channel. The remaining RRM resources (area, channels and subchannels) can be used by a different communication scheme. On the other hand, the RNI available resources and the number of wires in the local channel define the amount of simultaneous communications that a core can take part of (for instance in heterogeneous NoCs, a bus-to-NoC bridge takes part in two communications schemes).

DRNoC Addressing

This section presents the DRNoC communication system addressing and packet format.

The main goals of the DRNoC approach are to permit independent communication schemes to coexist and that, the system has to deal with cores that can be allocated in any position in the defined architecture.

The direct addressing approach in an embedded communication network is to have a unique address for each element and to use this address for data transfer. However, on dynamically changing systems, logic and physical addresses have to be differentiated. In DRNoC, three different addressing layers, described next, (one physical needed for system control and two logical) have been defined and are graphically represented in Figure 3 along with the packet format that is introduced in the next section.

- **Slot Address**. A unique physical address is assigned to each DRNoC node and to each node element (Core, RNI and RRM). A common row-column address is assigned

Figure 3. DRNoC addressing on the left and the packet format on the right

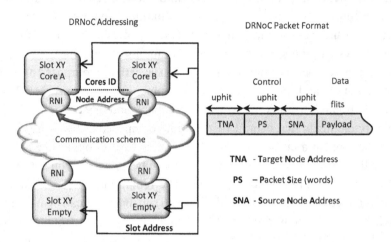

to each DRNoC node and two additional bits are used to select the target node element. For instance, in a 4x4 DRNoC architecture, 6 bits are needed to address all node elements. The 4 most significant bits are related to node row-column coordinates, while the last two are used to select the node element. For example, the address of Core:11 is "111100", while RNI:11 has the address "111101" and for RRM:11 it is "111110". Slot addresses are intended to be used by the control system to manage the node elements state and control the reconfiguration process.

- **Node Address**. It is a unique number that identifies a node in a mapped Communication Task Graph (CTG). In DRNoC, node addresses are related to the implemented communication scheme that might involve several, but maybe not all, DRNoC nodes. This permits to have a reduced number of addressing bits. For instance, in a 10x10 DRNoC, 7 bits are required for addressing all nodes, independently from the number of nodes the communication scheme requires. Differently, with node addresses, if the mapped

communication scheme, in the 10x10 DRNoC, is a 2x2 mesh, then only two bits are required for addressing.

- **Core ID**. A unique number that is assigned to each core that can be mapped in the system. This is the address that is used by cores to communicate between each other. Furthermore, in a dynamic reconfigurable system, there might be more cores in the local library than currently loaded in the reconfigurable system. Some bits of this ID number are related to the core functionality and others to the mapping. The use of this addresses give the possibility to abstract the core to core communication from the current application it is involved in, and also from the selected communication scheme.

DRNoC NoC Communication Scheme Packet Format

A packet format is usually specific for a given NoC and differs from each other in the header part. Some of them, for instance, include special data that give routing priority to a packet, while others include all the routing path the packet has to follow. Anyway, the common characteristic is that a header always includes the destination address and the size of the packet, just like the DRNoC NoCs packet format. However, differently from other packet formats in DRNoC a specific unit called *Phit Unit - uphit* is introduced to provide dynamic NoC data width adaptability.

The DRNoC NoC communication scheme packet format can be seen on the right side of Figure 3 and, as usual, it is composed of a header and a payload part. The header part (that contains the control data) includes the target node address, the amount of data to be sent, measured in words, and the source address. The packet size is measured in words instead of flits because data transactions between two cores are usually organized in words. On the other hand, like in any other NoC, flit size in DRNoC is equal to phit size. The flit values used in the current implementation are 4, 8 or 16 bits, thus for the worst case where flits = 4, only 16 flits (4 words of 32 bits) could be sent in a packet, while if the measurement is directly in words, 16 words will be sent.

Further, in order to provide support for the adaptability of the NoC bandwidth, that is, in order to change the NoC phit size, the data path communication scheme is separated from the control path. To achieve this, a parameter called Phit Unit - *uphit* is introduced (see Figure 3). The entire packet header is transmitted in uphits and routers switch control is based on them. As a result, when the phit/flit size is changed, the header part of the packet that is directly related to the control remains unchanged. Contrary, due to the uphit introduction, phit sizes are restricted to be: word size < phit < uphits. For instance, if uphit=4 and the word size is 32 bits, then the NoC phit can be 4, 8, 16 or 32. Additionally, uphit size restricts the number of nodes that can be involved in an independent communication scheme. For the previous example, where uphit=4, the maximum number of nodes that can be assigned to a communication scheme is 16 (four if the communication scheme is a mesh with XY routing).

DRNoC Reconfigurability

Two types of partial reconfiguration have been distinguished and could be supported by DRNoC:

- **Intra-Core reconfiguration** is defined and used to change only a certain parameter of a communication element (NI, routing modules or even emulation-specific cores like traffic generators and traffic receivers, as it will be seen later in the application example). This type of reconfiguration is usually based on changing come bits of the system configuration and is related to the network and/or transport layers.

- **Inter-Core reconfiguration** is used to define, build or modify the overall communication strategy. This reconfiguration type is linked mostly to full cores reconfiguration and affects the protocol layers implemented in HW (in NoC configurations, this reconfiguration can affect the data and/or control paths). For instance, channel width can be modified by changing RRM configuration.

DRNoC Mapping to FPGAs

In this section, the mapping of the presented DRNoC to FPGAs will be described, and more specifically a real implementation on a partially reconfigurable Xilinx FPGA, a Virtex II. Additional implementations for the other Xilinx FPGA, Virtex 4 and Virtex 5, are also included.

As it has been mentioned, each element of the DRNoC communication architecture has a limited amount of possible locations in the system, and is composed of specific RSoC resources. In FPGAs these resources are the array programmable elements and more specifically, in Xilinx FPGA they are Configuration Logic Blocks (CLBs).

For mapping DRNoC on a commercial fine grain FPGA, a *Virtual Architecture (VA)* has to be designed. In runtime reconfigurable systems, a virtual architecture defines the distribution of the device resources into reconfigurable slots and the on-chip slot interconnections. In order to achieve the highest possible flexibility, a set of steps have to be followed during the virtual architecture design (Krasteva et al., 2008). A general virtual architecture is shown in Figure 4, along with a DRNoC mapping.

In order to define a virtual architecture, a resource division phase requires the FPGA to be divided into a reconfigurable and a fixed area. The fixed area is usually in charge of controlling FPGA communication with peripheral devices and does not change. During this resource division, all elements that do not have a homogenous distribution in the device need to be included in the fixed area. In this case, for Virtex II FPGAs, the fixed area occupies the leftmost, rightmost and the middle CLB columns of the FPGA, while for Virtex 4 and Virtex 5 the distribution is more complex due to their higher heterogeneity. Contrary, the reconfigurable area, that has to be as homogeneous as possible, is divided into reconfigurable slots (Krasteva et al., 2005). Slots can be distributed following a one dimensional (1D) or two dimensional (2D) model and have to

be equal in size and resources. In this case, they are composed only of CLBs, and a 2D distribution has been used (see the left side of Figure 4). Additionally, depending of the target device specific features, some elements might be composed of additional components, for instance the available in Virtex 5 on-chip FIFOs might be part of a slots whenever homogeneity is preserved.

With respect to the internal communication, fixed communication channels are allocated along the architecture in order to guarantee access to the on-chip communication for every slot (vertical arrows on the figure shown on the left side of Figure 4). Like in any partially reconfigurable system, communication channels are implemented with bus macros.

To summarize, the main properties derived from the use of virtual architectures are that slots can be grouped to allocate bigger cores and that a core can be loaded in any slot position.

Each *DRNoC Reconfigurable Node (RN)* is composed of four elements: cores, RNIs, RRMs and SMs. For achieving the highest adaptability level, the best mapping of the DRNoC architecture to a virtual architecture will be to separate one slot for each DRNoC node element and thus,

Figure 4. General two dimensional virtual architecture (left) view and DRNoC mapping example (right)

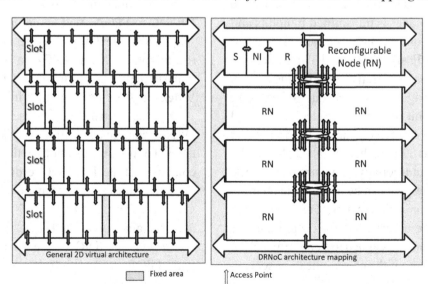

four slots will be needed. Estimating that the area needed for state of the art routers with three inputs is around 150 CLBs, and that part of the available CLBs will be used for the communication channels implementation, the best slot size, with respect to RRMs, is higher than 200 CLBs. However, in order to save some area, SMs are grouped with RRMs and every reconfigurable node is split into slot, network interface and routing module as shown in the upper left node on the right side of Figure 4, where RRMs and SMs slots are marked with R, RNIs with NI, and cores with an S. From here after, the cores to be mapped in S will be referred as slot/core to be differentiated from the general term "slot".

Regarding the Xmesh communication channels implementation, they have been entirely created with bidirectional vertical bus macros that pass four wires from a RRM to the main horizontal communication channel and four wires from the horizontal communication channel to a RRM.

Apart from all the described connections, there are some extra wires individually routed to each slot and used to pass control signals (run, reset and reconfigure) to each node element and also, for the specific case of our emulation application example, a special bus used for measured data extraction from cores and network interfaces has been added to every DRNoC row. Reconfigurable routing modules are not connected to the measuring systems in the current implementation in order to save area. All these modifications of the

original 2D virtual architecture, made to map the DRNoC, have not affected the original architecture properties and cores can be allocated in any slot/cores position, NIs in any RNI position and routing elements in any RRM position and, DRNoC nodes can be grouped to allocate bigger cores (cores that need to use more than one DRNoC node).

The selected DRNoC connectivity for DRNoC mapping examples on FPGAs has the characteristics listed next and two layouts can be seen on Figure 6:

- The width of a communication channel is 80 wires, 40 per direction, divided into four subchannels of 10 wires each. Thus, a maximum of four independent communications can coexist in parallel if there is enough free area in RRMs.
- Local channels, between RRMs and RNIs, are composed of two subchannels with a total of 160 wires, 80 per direction. This permits a core to be part of two independent communication schemes of a maximum of 40 wires per direction each.
- Access points, between slot/cores and RNIs, have 152 wires, 76 per direction. This number for instance is enough to implement a full AMBA master-slave duplex communication.
- Each slot/core, RNI and RRM has four control wires.

Figure 5. Four example DRNoC implementations on Xilinx FPGAs. From top to bottom: one Virtex 5, one Virtex 4 and tow Virtex II

DRNoC Size FPGA (array size)	Used slot/core (%CLBs)	Used RNI (%CLBs)	Used RRM (%CLBs)	Real slot/core Size (CLBs)	Real RNI Size (CLBs)	Real RRM Size (CLBs)	DRNoC node No.bus. macros	Total FPGA (%)
2x2 XC2V3000 (64x56)	13.1	27	17.3	125	105	238	108	12
4x4 XC2V8000 (112x104)	15.8	32.5	41.6	101	81	120	158	21.7
4x4 XC4VLX100 (160x68)	14.8	30.4	39	109	90	156	158	23.2
6x6 5VLX330 (240x109)	9.8	20.3	26.1	173	153	284	158	21.7

The previous list does not include the measuring system connectivity (an additional AMBA bus) since it is not properly considered as a part of the DRNoC architecture.

It is important to know the real area that is available for logic implementation after placing this huge connectivity in the FPGA. A summary of all the required resources for mapping DRNoCs of different size and with the previously listed connectivity on several FPGAs can be found in Figure 5. The table entries are as follows: a 6x6 DRNoC mapping on a Virtex 5, a 5VLX330 FPGA with a slot size of 192 CLBs, a 4x4 DRNoC for a Virtex 4, an XC4LX100 with slot size of 125, another 4x4 DRNoC but for a Virtex II with slot size of 120, and finally, a 2x2 DRNoC for a Virtex II FPGA, an XC2V3000 with slot size of 144 CLBs.

Worst cases for RRMs have been considered for all FPGAs: for the 2x2 DRNoC implementation, the worst case is a RRM with three channels, while for the reaming 4x4 and higher it is with eight channels. The table includes the percentage of the CLBs needed from each slot assigned to a DRNoC node element, the total amount of the needed communication bus macros in each DRNoC node (*DRNoC node No.bus. macros)* and the total percentage of CLBs needed for the entire DRNoC connectivity.

The most important values from the Table are the real slot/core, real RNI and real RRMs sizes, as these values define the available area to load hard cores and thus, the limit of the number of independent communication schemes that can be defined. These values are in the range of 100 to 170 CLBs for cores and NIs to be loaded in slot/core and RNIs and 20-70% more for routers to be loaded in RRM.

Additionally, as example, two FPGA Editor images of real DRNoC implementations layouts can be seen on Figure 6. On the bottom part, the one used along this Chapter for validating the proposed ideas, a 2x2 DRNoC on an XC2V3000, and on the top part of the same figure a 4x4 DRNoC layout on an XC2V8000 FPGA.

Figure 6. Two DRNoC implementations on Xilinx FPGAs. On the top 4x4 DRNoC on an XC2V8000 FPGA and on the bottom a 2x2 DRNoC on an XC2V3000

4x4 DRNoC for an XC2V8000 FPGA

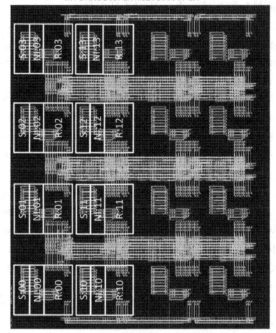

2x2 DRNoC for an XC2V3000 FPGA

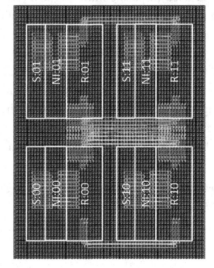

Having into account the problems related to the current Xilinx Place and Route (PAR) tools versions (it is not possible to define routing restriction, for example a bounding box) and the need of bus macros, the available area for the presented implementations is quite low and this restricts the communication schemes that can be loaded and thus the applications used for testing. Therefore, due to the restricted area the use cases presented in the second part of the Chapter will be based on traffic receivers and traffic generator models and not on real cores.

Anyway the use of bus macros might not be required in future reconfigurable systems design flows and this will permit to overcome the area consumption related deficiencies. Nevertheless, the truly implemented use cases on the Virtex II 2x2 DRNoC are a proof of concept of the DRNoC architecture and the remaining innovative ideas presented along the next sections.

DRNoC Routers

As it has been explained the DRNoC solution is not restricted to NoCs, however a brief definition of a router for the DRNoC system can be found in this section as it is one of the most complex elements that can be found in a communication scheme.

The main distinguishing characteristic of the DRNoC routers is that they have separate data and control path. This decision permits to dynamically change the NoC flit width. To achieve this, the control part of the router is defined to work with uphit units, while the data part deals with flits. Furthermore, DRNoC routers FIFO buffers are allocated at the router input and data buffering is prior to routing decisions. This is important for reconfigurable systems as in this case, buffers do not have to be freed if changes have to be made in the routing tables or the routing strategy. Also, as data and control are split, the control data have to be buffered in a separate buffer whit uphit size data width. A consequence of the introduction of data path reconfigurability is that an entire packet has to be kept in the buffer in case a reconfiguration is triggered in the middle of a packet transmission.

If this happens, the packet that is being transmitted is deprecated from all buffers where it is not complete, and retransmitted starting from the point a complete packet is available.

Furthermore, the router has also to have, as far it is possible, the buffering, arbitration and control parts well defined in order to enable changing each module independently. Any router that has a modular design and a behavioral or register level description, where buffers, switches and data control are independent components could be adapted to follow the division between control and data path approach of DRNoC. Anyway, the main restriction of routers to be used in the architecture is that they have to fit into the RRM available area, shown in Table1 for several example DRNoC mappings.

As example, the HERMES network (Moraes et al., 2004) router that covers part of the requirements previously mentioned and was kindly provided by the Pontificia Universidade Catolica do Rio Grande do Sul (PUCRS) has been adapted to a DRNoC router. Indeed, the HERMES router is not fully modular, the routing and the arbiter are combined in one module in order to reduce latency. This results in the restriction of no independent routing algorithm and arbitration control reconfiguration. In order to adapt the HERMES router to a DRNoC one, the input buffers have been redesigned to follow separate data and control paths, and the FIFOs have been implemented in Block RAM memories (BRAMs). Dynamic changes in data width are executed with Intra-Core reconfiguration by keeping data buffered in the memory untouched and changing the BRAM interconnections configuration with partial reconfiguration. However, on board tests have shown that, with the Virtex II memories this technique does not always work properly. In more than half of the performed tests, the expected data to be read from data memories after flit width changes was not correct. Although this, packets with modified fit width were successfully router to destination nodes.

Regarding the routing algorithm, two types have been defined: i) the original HERMES XY routing, but instead of dealing with phits it has been modified to deal with uphits, and ii) table-based routing, that is needed for irregular NoCs. Tables in the current implementation are mapped to distributed memory (CLB-LUTs). Intra-Core reconfiguration for changing the routing tables can be used and have been successfully tested for small table sizes (up to 4 entries of 4 bits width) by generating partial configuration files with just the LUTs content with a tool called BITPOS (Krasteva et al, 2005).

DRNoC Design Resources

Apart from the DRNoC router that is the main element of a NoC communication scheme, other design resources, described in this section, are required during the on-chip communication design and validation process.

• Traffic Generators

Traffic Generators (TG) transmit a certain amount of data to a receiver core and are used during NoC design to emulate the behavior of real cores. TGs to NIs have a standardized interface that in the current version is based on the AMBA APB protocol. OCP has also been studied as an option, but it requires much more wires than APB, and thus, more bus macros, that would result in cores with less available real (usable) area. In TGs, Intra-Core reconfiguration can be used and has been tested to modify TGs target node addresses. There are two types of TGs, regular and Pareto TGs. Regular TGs generate constant, regular traffic to a single or to multiple receivers. In these TGs, the traffic generation is regulated by changing the time interval between two sent packets and the numbers of words that are included in each packet. On the other side, Pareto TGs are common for modeling non uniform traffic. Several studies have shown that traffic in multimedia applications is burst based and self similar (Varatkar & Marculescu 2004). According to (Pande et al, 2005) such type of traffic can be modeled by mixing ON and OFF periods. The time a traffic generator spends in either the ON or the OFF state is selected according to a distribution which exhibits long range dependence. In the same work, the Pareto distribution has been presented as the most appropriate one for defining the ON and OFF periods, along with suitable values for modeling MPEG 2 video traffic. Base on these data, a simplified VHDL implementation, valid only for MPEG 2, has been designed and included in the DRNoC design resources. Intra-Core reconfiguration in traffic generators has been tested and validated to change the target node address, but can also be used to modify the traffic regulation related parameters.

• Traffic Receivers

Traffic Receivers (TRs) implement DRNoC traffic measuring points (MPs). TRs MPs include all the needed measuring registers. Two types of traffic receivers have been distinguished depending on the selected measuring scheme: either to perform online measuring (keep the current value), or to keep only max/min values into the internal measuring registers. Each register of an MP is related to a parameter to be measured: latency, the time a packet header enters the NoC, the time a complete packet has been received in the TR and the number of received data packets. Data from these registers is used further to calculate latency, injection rate and throughout. Each measurement point tracks data related to a single TG. There may be more than one MP if data for multiple sources have to be tracked. Intra-Core reconfiguration in traffic receiver can be used and have been tested and validated to modify tracked TG node address.

- **Network Interfaces**

Several models have been designed for network interfaces (NIs). On the TGs side: i) one that generates packets for XY network routing, and translates core IDs to XY node addresses and ii) common NIs, that translate core IDs to common node addresses. Regarding the TR side, NIs do the inverse address translation and, additionally, include measuring points that are the same as the ones previously described for TRs. For NIs, Intra-Core reconfiguration could be used to change the address translation table when a core is moved to a different slot. However, this feature has not been tested so far.

- **Other Resources**

Some general DRNoC resources are needed to create direct links between two or multiple cores. Several resources have been designed: i) for resources to be loaded in RNIs, different feed-troughs, multiplexers/demultiplexers, buffers and dummy logic, ii) while for resources to be loaded in RRMs, feed-troughs of 2, 3 and 4 inputs/outputs, buffers and dummy logic, have been designed.

DRNoCs Design Resource Area Requirements

The main idea behind this subsection is to evaluate the area requirements of some DRNoC design resources for being included, following a hard core design flow presented in (Krasteva et. al, 2005), into a hard core library.

Synthesis results for different design resources can be found in Figure 7. An important value in the table is the slot usage percentage (*% Slot*). This value provides an idea of how much resource from the host slot area are going to be used when this core is loaded in the DRNoC system and thus, how much area is left free for loading elements that belong to different communication schemes. All results included in this section have been obtained from the Integrated System Environment (ISE) SW provided by Xilinx after the synthesis process using area optimization.

The first input on the Table is a Pareto ON/OFF traffic generator, which generates traffic to a single TR (*TG-Single-Pareto*) and is the biggest TG, because it includes, apart from the common logic, all the logic needed for random value generation and Pareto periods calculation. The next

Figure 7. DRNoC design resources area requirements and target slot percentage occupation

Core Type	Slices	FlipFlops	LUTs	% Slot
TG-Single-Pareto	200	210	290	40
TG-Single-Regular	107	142	177	21.4
TG-M2-Regular	112	143	180	22.4
TG-M4-Regular	117	143	182	23.4
TR-Single-Max-Min	275	208	490	55
TR-Single-online	286	290	395	57.2
TR-M2-online	489	350	781	97.8
TR-M4-online	810	613	1293	162
NI-TR-Single-Max-Min-8b	260	208	479	61.9
NI-TR-Single-online-16b	244	326	450	58.0
NI-TR-Single-online-8b	232	328	426	55.2
NI-TR-M4-online-8b	879	613	1293	209
NI-TG-XY-8bflit	230	320	389	54.7
NI-TG-XY-16bflit	241	337	405	57.3
R-XY-8b2C	262	183	505	27.5
R-XY-8b3C	300	245	630	31.5
R-XY-16b2C	360	228	690	37.8

entry is a regular TG that generates uniform traffic for a single receiver (*TG-Single-Regular*). After it, the number of associated TRs has been doubled for the same type of TG (*TG-Mx-Regular* on the table, where *x* is the number of associated traffic receivers). TRs synthesis results are presented on the same table, after the TGs. The table includes TRs of both available types: i) one follows the max/min value scheme, marked as *TR-single-max-min* in the Table that includes a single measuring point (tracks data from one TG) and ii) the remaining TRs follow an online measuring scheme but have a different amount of measuring points (*TR-Mx-online*). For instance, *TR-M4-online* tracks data from four TGs and this core cannot be loaded in the DRNoC current implementation as it occupies more than 100% of the available RRM area. Regarding NoC NIs and routers, the table includes results for an 8 and a 16 bit flit XY NoCs. The number of flits in the TR NIs are indicated in the name. For instance, a *NI-TR-Single-online-8b* corresponds to a network interface traffic receiver that tracks online data from a single source in an eight bit flit NoC. The same naming has been applied for traffic generators NIs that are included next in the Table (for instance *NI-TG-XT-8bflit* is a traffic generator NI for an 8 bit flit XY NoC). Finally, DRNoC XY routers for an 8 bit flit NoC with different number of used channels, two and three, are included in the Table, along with a two channel DRNoC XY routers for a 16 bit flit NoC (*R-XY-16b2C* is an XY router for an 16 bit flit NoC that uses two communication channels).

Apart for the area, another important value is the occupied by the cores communication wires, that is, the number of occupied channels and subchannels. This value that has not been included in Figure 7, is accounted again, like the area, in a percentage of the number of wires available in the target DRNoC implementation. Afterwards, these percentage values are used to calculate cost parameters for both, area and communication resources (channels and subchannel wires), in order to know how costly is to arrange a given

communication scheme into the system and to account the available on-chip resources.

DRNoC NoC Model Generation Tool: DRNoCGEN

A SW tool called DRNoCGEN was developed in order to automatically generate synthesizable VHDL models for the DRNoC NoC communication scheme.

This approach is similar to other, like for instance, the environment presented in (Ost et al, 2005) that generates HERMES NoC models, and the XeNoC system presented in (Joven, 2008) that generates complete and synthesizable NoCs from XML descriptions. The main DRNoCGEN difference is that the generated NoC systems, apart from being synthesizable are mapped to a defined DRNoC template. For instance, a user can map a 2x2 mesh NoC based on XY routing, or a star NoC based on table routing on a defined 8x8 DRNoC architecture.

The tool also permits to create new DRNoC virtual architectures for any FPGA by specifying some parameter values: i) the uphit size, ii) the channel width, iii) the number of subchannels and iv) the number of slots, value directly related to the number of nodes in the architecture. Based on these parameters, the tool generates architecture templates that, afterwards, are used as a base for building NoCs by specifying the NoC elements: i) NoC nodes (TGs/TRs and NIs), ii) NoC phit size, defined as an integer number of uphits, iii) NoC routers and iv) the buffer size. Most of these elements are technology independent at this level, even the DRNoC router as the selection of the implementation of the FIFOs in BRAMs belongs to further design steps. However, as currently the only FPGAs that provide partial reconfiguration capabilities are the Xilinx one, most of the designs have been validated for them.

The tool generates VHDL NoC models that are mapped to the defined DRNoC architecture following a specific hierarchy and system templates.

The generated top VHDL file includes all NoC required resources mapped to a DRNoC slots: cores to slot/cores, NIs to RNIs and all routers mapped to RRMs.

In order to synthesis the automatically generated system for a given FPGA, a constraint file the bus macros (for Xilinx FPGAs) have to be provided (this step is done manually).

Reconfigurability & Current Technology Restrictions

During the previous subsections, some reconfigurability restrictions, resulting from preliminary tests and analysis performed, have been pointed out; next, a summary of what has been practically validated in the currently available DRNoC implementation can be found:

- For Intra-Core reconfiguration the currently supported reconfigurations are: i) TR and TG target/source node address and in a given NoC communication schemes, ii) change router buffers size and iii) change routing tables content. Due to router restrictions, currently the routing strategy is linked to the arbitration and cannot be reconfigured separately. Another restriction is that currently a router data buffer can occupy only a single BRAM.
- For Inter-Core reconfiguration, some problems related to dynamic data path reconfiguration due to FPGAs Virtex II BRAM technology restrictions have been reported and therefore data path reconfiguration of the flit width has been restricted in the current DRNoC implementation. This problem may be overcome with the Virtex 4 and Virtex 5 built in FIFO logic. The Inter-Core reconfiguration actually supported is to independently modify complete NIs and RRMs,

Anyway, independently from the used FPGA, the main technology restriction, which derives in restrictions of the hard cores (partial configuration files) that can be loaded in the system, results from the currently available placed and route tools that do not permit to enclose the routing in a bounding box (slot borders). However, in relatively short time, more sophisticated design flows might be available and it will be possible to implement all the reconfigurability features that DRNoC provides.

NoCs for Runtime Reconfigurable Systems Feature Summary

This subsection includes in Figure 8 a features summary of the NoC solutions described in this chapter state of the art, along with the proposed DRNoC. The same table also includes a summary of implementation results that have been reported.

Some aspects of the presented approach are similar to others briefly described in the introduction part of the Chapter. In DyNoC, for instance routers have also a fixed position, but there they are not reconfigured, and similarly to CoNoChi, a switch can be loaded in the system, but it's position in DRNoC is restricted to a set of possible positions. Additionally, DRNoC permits topology changes like the Hübner et al. NoC. However, DRNoC was designed to provide a complete reconfigurable solution: bring reconfigurability to the NoC protocol stack lower layers and also, to permit to create concurrent and independent, ad-hoc, communication strategies.

From Table it can be noticed that DRNoC has the widest reconfigurability support compared to the remaining approaches. DRNoC permits to reconfigure Cores, routers and also NIs. Moreover, DRNoC permits both, to load large cores in the system and to relocate cores along the architecture. Further, from the implementation point of view, it can be noticed that the currently used DRNoC implementation has an average size and number of reconfigurable nodes, 2x2 and 4

Figure 8. NoC solutions for partial runtime reconfigurable systems features and implementation summary

Architecture		Reconfigurability Support					Implementation		
Name	Type	Cores	Switches	NIs	Relocate Cores	Allocate Big Cores	Size	FPGA	Mapped Application
GENKO (Nollet et al., 2003).	Mesh NoC	Yes	No	No	No	No	3x3 with 2 reconf. nodes.	Virtex II	Multimedia
(Hübner et al., 2008)	Switches connected to a horizontal routing channel	Yes	Switch Matrices	No	Yes	Yes	NS	Virtex II	Run on an embedded device
DyNoC (Bobda et al., 2005)	Mesh NoC with reconf modules	Yes	No	No	NS	Yes	3x3 with one reconf. module	Virtex II	Light Controller
ARTEMIS Möller et al., 2007	Mesh NoC with reconf modules	Yes	No	No	Yes	NS	2x2 with two reconf. module	Virtex II	Reconfigurable ALU
CoNoChi (Pionteck et al., 2006)	Mesh of four port switches	Yes	Yes	No	Yes	Yes	7 switches with 16 bit links	Virtex II	NS
DRNoC (Krasteva et al. 2007)	Xmesh NoC	Yes	Yes	Yes	Yes	Yes	2x2 with 4 reconf. modules and 40 bit links	Virtex II	Emulation

NS –Not Specified

reconfigurable nodes, but large connectivity, 40 bit links, compared to other real implementation of this type of communication solutions.

DRNoC Emulation Framework: Fast Emulation Based on Partial Reconfiguration.

One of the main challenges in NoC design is to find the optimal NoC solution for a given application. Several methods and design flows that permit to perform design space exploration, at different abstraction levels have been proposed. Higher abstraction levels permit to rapidly evaluate different mapping and NoC implementation options without paying the time cost of long simulations. For instance in (Mahadevan, S., Virk,K. & Madsen. J, 2006), a framework for MPSoC NoC systems modeling, simulation and evaluation, based on System C models is presented. Systems are generated matching application and platform models. Other frameworks are based on object-oriented languages, like the presented in (Coppola et al., 2004), where a C++ library is built on top of SystemC, or based on the Matlab simulation environment, like (Samira et al., 2007). Other approaches generate NoC topologies getting the application graph representations or application descriptions as starting point, using analytical and/or heuristic methods. Examples of these are SUNMAP (Murali & De Micheli, 2004) that, based on the Xpipes (Jalabert et al., 2004) NoC generator, creates topologies modeled in System C. These approaches are very suitable for system design early stages as they permit to have the fastest design space exploration. Further, there are other HDL or mixed (System C and HDL) solutions for NoC modeling, NoCGeN (Chan & Parameswaran, 2004), where the NoC topology can be selected. Lower level, VHDL or RTL, permit to have more accurate results, but are time consuming. Usually, at this abstraction level, traffic generators and traffic consumer models that simulate real core behavior are used in order to reduce simulation time.

On the other side, there are FPGA based emulation solutions proposed to drastically reduce simulation time, and therefore, overall system evaluation time. For instance, in (Genko et al., 2005), a HW-SW FPGA based emulation framework is presented and combined with the Xpipes environment. Four orders of magnitude of speedup are reported in that work. Emulation, depending on the FPGA available area, may permit to test NoCs using real applications instead of traffic models. In (Ogras et.al.,2007), four real applications mapped into a NoC and prototyped in an FPGA are presented.

Nevertheless, the main disadvantages of emulation based solutions are:

- Synthesis time. Every time a system parameter needs to be changed, the system has to be re-synthesized, re-placed and re-routed (from here after this process will be referred simply as synthesis). This is not a real problem if a system has to be synthesized once, but if the goal is to come up with the optimal system implementation, many combinations have to be synthesized and emulated. An approach to overcome the synthesis time problem is to have all the system options implemented in the FPGA and switch between them. This forces cores used during emulation to be different from the final ones, and also it increases the area overhead.

- Measured data extraction from the FPGA: the FPGA has much more limited resources in this sense in comparison with a simulation approach.

- The available FPGA area permits only to emulate relatively small systems. In (Wolkotte et al., 2007) a solution for this problem is proposed. There, sequentially, parts of a parallel system are loaded into an FPGA. Speedups of 80 to 300 in comparison with System C simulation are reported. Anyhow, each new FPGA generation has more logic available and thus permits to emulate bigger systems.

In the following sections, a fast emulation approach is presented. In the solution hard core re-usability through partial reconfiguration is proposed as a possible solution to reduce synthesis time, compared to a non-partial reconfiguration emulation flow. In this approaches NoCs to be emulated are built and/or modified by loading reusable hard cores in the system using partial reconfiguration. To achieve this, the presented in the previous sections, DRNoC architecture,

Figure 9. DRNoC emulation design space exploration working flow

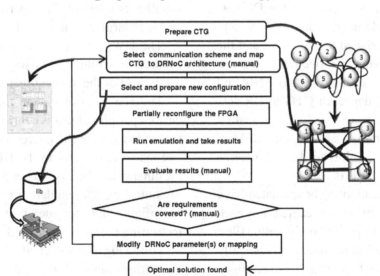

has been used. To demonstrate the feasibility of the proposed solution, three use cases have been built and emulated. The previously mentioned 2x2 DRNoC implementation has been used as a proof of concept of the DRNoC based fast emulation proposal. Apart from the synthesis, emphasis has also been put on the measured data extraction and the configurations administration.

Emulation Working Flow

A general view of the proposed working flow for design space exploration, based on partial reconfiguration is presented in Figure 9. In some aspects, the flow it is similar to other NoC design space exploration interactive flows, like (Genko et al., 2007.), but here, hard core re-usability is exploited and the system emulation is not restricted to NoC communication schemes.

The flow begins with a mapped application Communication Task Graph (CTG), where application tasks are assigned to available designed resources. For each node of the CTG, a suitable DRNoC resource model, traffic receivers (TR) and/or traffic generators (TGs), are assigned if they are available in the hard core library, or generated if they are not.

Afterwards, in the first step of the interactive flow, the new CTG is mapped to the DRNoC architecture available in the emulation system. TGs and TRs are assigned to slot/cores, and the communication scheme is defined. NIs, routers, P2P and P2M links are assigned to RNIs and RRMs. From here after, each CTG to DRNoC mapping, along with an assigned communication scheme will be referred as a configuration. Also, in this step, measuring points are defined in terms of number and tracked TGs.

After all configurations to be emulated have been setup, the emulation is executed. For each DRNoC configuration, partial configuration files, available in the hard core library or manually provided to the emulation system, are arranged and loaded in the FPGA. It is important to remark that if several configurations are defined, then for switching from the current configuration to the next one, only the differences are reconfigured.

All configurations are emulated consecutively, and results related to each measuring point are independently saved in a host PC. Later, results can be plotted, analyzed by a user and, if needed, new DRNoC configurations can be added by modifying the CTG to DRNoC mapping and/or the communication schemes definition. This process continues until the best communication scheme and CTG mapping has been found.

DRNoC Emulation Framework

A general view of the emulation framework is presented in Figure 10. The framework is distributed in three platforms: i) An FPGA board which holds the DRNoC implementation, where measuring points are allocated in TRs and receiving nodes NIs, ii) a measuring system HW control, currently mapped to an XUP board with an XC2VP30 FPGA and iii) a host PC that runs the DRNoC emulation SW.

It has been necessary to allocate the HW control in a separate platform in order to have an entire FPGA for the DRNoC. The XUP board has been selected because building the control system on the PPC is relatively fast, permits easy updates and is re-targetable to other FPGAs.

The HW control system is described in this subsection, while the DRNoC emulation SW is the subject of the next subsection.

The HW control system main element is a custom peripheral that is connected to the FPGA embedded Power PC microprocessor (PPC) on one side and directly to the DRNoC FPGA on the other side. The custom peripheral is in charge of controlling the emulation process (run, stop, reconfigure and continue), pulling data out from measuring points registers and data buffering.

Following (Grecu et al, 2007) the system measuring points are allocated in TRs and in the associated TRs NIs. Data is extracted from these

Figure 10. DRNoC emulation framework

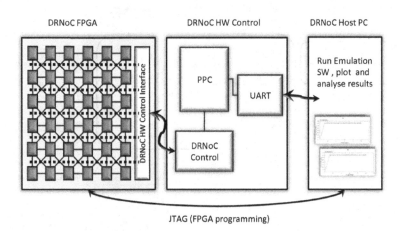

measuring points using an AMBA APB bus interface that shares the same FPGA area with the Xmesh communication. There is one AMBA APB bus for each DRNoC row.

As the PPC system runs at 100 MHz and the DRNoC FPGA runs at 25 MHz, an asynchronous four phase communication interface has been used for connecting both systems and a bridge has been included on the DRNoC FPGA fixed area to multiplex the internal APB buses.

In the current HW control implementation, twelve lines have been separated for DRNoC slot addressing and two for register selection. As a result, the same DRNoC control system (HW and SW) can be used with up to a 32x32 DRNoC implementations simply by changing the DRNoC FPGA. Also, it can be easily adapted to other internal communication standards (not AMBA based ones) by simply changing the communication bridge. Data is pulled out from each register consecutively and buffered in a FIFO allocated in the custom peripheral. If online measuring has been selected, data are constantly pulled out from the registers and buffered in the FIFOs from where it is constantly read by the PPC and sent to the Host PC. When the emulation process ends, the DRNoC FPGA generates an interrupt. The DRNoC HW control makes a final read of all registers

and, after that, it interrupts the PPC. Differently, if the design registers store max/min value, then appropriate TRs and NIs have to be included on the DRNoC. In this case, data are pulled out from the measuring registers only when the DRNoC FPGA interrupt has been received. After all data are pulled out, the interrupt is retransmitted to the PPC that triggers the host PC DRNoC SW to read all data from the DRNoC HW control.

For having a time reference in the measuring system, a timer has been included in each TG, NI and TR. All timers are synchronized with the run command. The limit of the emulation system time is defined by these timers (in the current implementation they use 32 bits), which corresponds to 4G clock cycles. The DRNoC FPGA runs at 25 MHz resulting in 172 seconds total emulation time.

Regarding the data pulled from the measuring registers, the limit factor is the XUP board (HW control) to Host PC connection that is currently implemented by a serial link (RS232). The parallel DRNoC FPGA to XUP board link has a throughput of 65 Mbps and the PPC FIFO depth is 16K words, but the Host PC to XUP serial link is 9600 bps. Therefore, the system can pull and save in the host PC up to 10000 measured values before it runs out of FIFO space. This problem can be easily solved by switching from the serial

Figure 11. DRNoC emulation SW

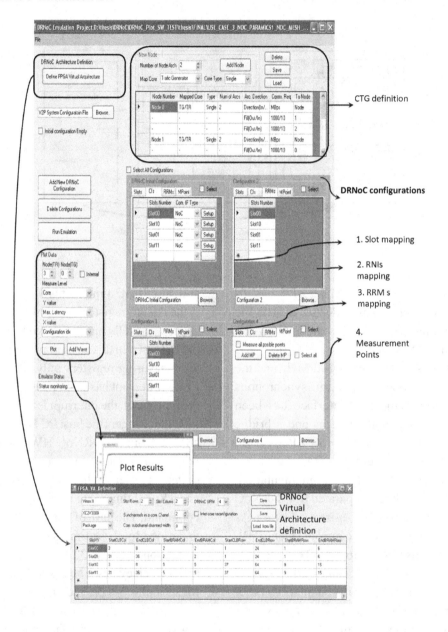

link to the Ethernet connection available in the XUP board.

DRNoC Emulation SW

A DRNoC emulation control tool has been designed and runs on the measuring system host PC. The tool is in charge of controlling the entire emulation process following the previously described emulation working flow and to administrate measured data and configurations. A screen shot of the software interface can be seen in Figure 11, where main functions are highlighted.

The tool main features are listed next:

- The SW main function is to organize all configurations and data involved on the design space exploration process.

- To define the application CTG and the DRNoC virtual architecture. This information is then used to properly allocate/relocate hard cores in order to be loaded into the system.
- To define DRNoC configurations which are composed of a CTG mapping and a communication scheme definition.
- To control the measuring process with the commands: run, stop, reconfigure and continue. The control is performed through the DRNoC HW control.
- To prepare raw data, pulled out from the DRNoC system, to be plotted (organize pulled data, calculated throughput and injection rate).
- To prepare FPGA partial bitstreams. Partial configuration files are technology dependent and therefore for adapting the emulation software to other DRNoC FPGAs, this part has to be modified. For the further presented use cases, that are based on a 2x2 DRNoC, the BITPOS and pBITPOS (Krasteva et al, 2006) tools have been integrated in the emulation SW. BITPOS is used when the system full FPGA configuration includes hard cores. In this case, BITPOS is executed to extract hard cores needed for Intra-Core reconfiguration. Differently, pBITPOS is used to properly allocate and relocate hard cores during the DRNoC configurations loading process.
- The tool prepares all the scripts needed to configure both FPGAs and calls the FPGA programming tool (iMPACT) to program the DRNoC FPGA through the JTAG port. In this case the FPGA internal configuration access port (ICAP) has not been used, because additional FPGA resources will be required for that and also because the main point here is not the systems self-reconfigurability but the emulation framework.

The tool works with hard cores that are held in the library. To generate hard cores for the library, a hardware core design flow has been followed. For instance, if a new NoC router is going to be added, first a DRNoC with the proper options has to be generated with the DRNoCGEN tool. The generated model includes more than a router, apart from the TGs, TRs and NIs. For having fast synthesis and since only one router is needed, an appropriate router is selected and isolated. As the code generated by DRNoCGEN follows a series of system templates, this step is quite easy. Only the top and the user constraint file have to be modified (all but the router and the bus macros have to be commented). Then, a partial configuration file containing only the router hard core is extracted with BITPOS and added in the library along with a related description file. In both cases, synthesis times are reduced compared to synthesizing the complete system.

DRNoC Validation Through Use Cases

The objective of the three use cases briefly described in this subsection is not to come up with the conclusion of which is the best communication scheme for each modeled applications CTG. Differently, the use cases have been selected in order to test the proposed working flow, the DRNoC architecture reconfigurability and to evaluate the feasibility of the proposed, partial reconfiguration based, emulation approach. However, performance parameters, used for NoC characterizing, latency and throughput, have been calculated to evaluate the different DRNoC NoC communication schemes. Furthermore, cost parameters are applied to evaluate and compare different implementation options (configuration) options.

As it has been mentioned several times, the implemented DRNoC on the available XC2V3000 FPGA is 2x2 with a uphit size of 4 bits and it has been used for the use cases described in this section.

As example, only the first use case will be explained in detail, while the two following ones are simply overviewed.

- **Inter - Core Reconfiguration for Changing the Communication Scheme - UC1**

The first use case is intended to test the switching between two different communication schemes within an application design space exploration by using Inter-Core reconfiguration. In this case a NoC communication scheme will be substituted by direct point to point links that do not use routing. For this aim, partial reconfiguration will be applied to the network interfaces and the routing modules, RNI and RRM respectively in DRNoC. The complete emulation working flow can be seen in Figure 12.

The selected application is a pipeline like application where a task generates irregularly data that are saved in an intermediate FIFO from where it is pulled out immediately (contentions are not foreseen).The application is modeled by three tasks assigned to tree nodes of a Communication Task Graph (CTG) (see Figure 12, top left side), and a specific DRNoC design resource is assigned to each node:

a. The node that generates traffic to the FIFO (node0) has been modeled with a single Pareto traffic generator (*N0:TG-S Pareto* in the top right side of Figure 12.)

b. The FIFO and the data pulling task, assigned to node 1, have been modeled with a TR and a TG (*N1:TR+TG-S Pareto* in Figure 12), the TG generates traffic with the same distribution and throughput as node0, but with some initial delay

c. Node2 is a traffic receive, a TR-MP, that receives data and measures traffic (*N2:TR+MP* in Figure 12.).

This is the first step of the working flow and is defined with the SW tool user interface (see Figure 11.- CTG definition).

For this use case, each traffic generator has been configured to send 10000 packets of 10 words each (1 word is 32 bits) in an MPEG 2 burst distribution. Afterwards, two more words are added to the 10 words packets in the NI. One word corresponds to the packet header (that apart from the defined DRNoC packet format -target address, source address, and packet size, also includes a last packet flag used for the measuring system) and the second one is a time stamp, also used by the measuring system.

Figure 12. Inter - Core reconfiguration to change the communication scheme - UC1

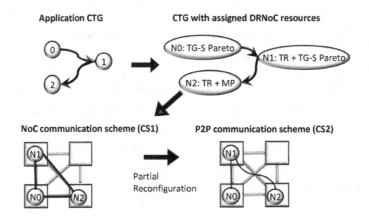

Once the CTG has been modeled with DRNoC design resources, the configurations are defined and the designed resources are mapped to the reconfigurable DRNoC architecture (DRNoC Configuration region of the SW tool shown in Figure 11.).

Two configurations with the same slot mapping to the DRNoC architecture (node0 to slot00, node1 to slot01 and node2 to slot10), but with different communication schemes have been defined:

1. The first configuration communication scheme (*CS1*) is an 8 bit flit NoC (see Figure 12.). The NoC is built by:
   a. Two 8 bits XY NoC network interfaces loaded in RNI00 and RNI01.
   b. One network interface that measures traffic on the receiving node RNI10 and
   c. Three two channels eight bits XY routers loaded in RRM00, RRM01 and RRM10.

Notice, that in the NoC configuration, the RRM01 and RRM10 (N1 to N2 link in Figure 12) switch matrixes have been modified to redirect respectively the south and north router port and take advantage of the Xmesh diagonal. Otherwise and additional router had to be loaded in RRM11 and this will increase NoC latency and total area overhead.

2. The second communication scheme (*CS2*) is a pure P2P (see Figure 12.) built by:
   a. One bidirectional feed-though loaded in RNI01 that passes the core AMBA OPB master and slave signals to the router and uses 146 wires of the 160 available in the core to NI interface.
   b. Two unidirectional feed-through loaded in RNI00 and in RNI10
   c. Three channels short-circuits loaded in RRM00 and RRM10.

For switching from CS1 to CS2 and vice versa, following the proposed working flow, four Inter-Core partial reconfigurations are required: two to modify RNIs and two more to modify RRMs.

Maximum and minimum latency and throughput values can be found for both configurations in Figure 13.

From the table it can be noticed that the CS1 min. and max. latency measured at core level, is higher than the NI latency (almost 30 cycles). This is because 6 cycles are spend in both, the TG core to NI-TG interface and the data serialization process at the source node and, 24 cycles are needed in the NI-TR receiver. The NI-TR first has to remove the packet header and the time stamp, putted by the NI-TG and then it has to receive and prepare a complete word previously to transmit it to the TR core.

Regarding the CS2, the latency is constant and the max. throughput value is 0.3 words/cycle because one cycle is waited between two consecutive AMBA transmissions that have 2 cycle latency. At this stage, if only performance is taken into account during the design space exploration, then CS2 is a better choice.

Figure 13. Communication scheme CS1 (NoC) and CS2 (P2P) configuration for UC1

Communication Scheme	Core Latency (cycles)		Core Throughput (words/cycle)		NI Latency (cycles)		NI Throughput (words/cycle)	
	max.	min.	max.	min.	max.	min.	max.	min.
CS1(NoC)	216	47	0.09	0.0077	190	19	0.4	0.036
CS2 (P2P)	2	2	0.3	-	-	-	-	-

Apart from the performance parameters, in dynamic reconfigurable systems it is important to know how much resources are required to build a communication scheme, For this aim a set of cost parameters that are directly related to number of communication channels, subchannels and wires that are needed, as well as the required slot area have to be defined. For instance the channel wires cost (Ccwij) is a relative value that represents to the portion of wires in a single channel (Nwires) used by a given element of the communication scheme divided by the total number of available wires in a single channel. For the presented use case, Ccwij is 1.25 for UC1 and 4.5 for UC2, while the area cost for UC1 is 2.95 and for UC2 is cero (only feed-throughs are required).

In this situation, if wires related cost are the leading parameter for selecting the communication scheme, then CS1 is a better choice, while if area is more important, CS2 is the best choice.

To conclude, the performance results along with the implementation costs have shown that the best option (between the two emulated configurations) for this use case is the P2P communication scheme (CS2). Of course if there are enough free wires and area to load it.

- **Inter-Core Reconfiguration for Modifying a NoC Communication Scheme - UC2**

The second use case (UC2) is oriented to modify a NoC communication scheme, more specifically to change its topology. This process involves Inter-Core reconfiguration, where RRMs are reconfigured. Additionally, in this use case, Intra-Core reconfiguration has been used to change the measuring point tracking node in the traffic receiver node. The task that has been modeled is a common situation where tree sources, traffic generators, try to access the same media, a traffic receiver.

Again, like in all use case, two configurations have been defined, emulated and analyzed.

The first configuration has been an eight bit flit MESH NoC communication scheme (CS1) and the second one an eight bit flit STAR NoC (CS2). For switching from one configuration to the other one, partial reconfiguration of the four RRMs has been used.

The performances parameters, resulted from the emulation process, have shown that CS2 is a better choice for this application as it has lower latency and better throughput. Furthermore, although the star router (CS2) requires three links, and thus more area than the mesh one (CS1) which requires 2 links, it needs only one router, feed-throughs are loaded in the remaining RRMs.

- **Intra-Core Reconfiguration for Tuning a Communication Scheme Parameter – UC3**

The last use case (UC3) is oriented to test Intra-Core reconfiguration for tuning a specific NoC communication scheme parameter. In this case, the buffer size is changed and again, Intra-Core reconfiguration has been used to switch between different measuring point tracking nodes.

The modeled application is composed of four traffic generator traffic receiver nodes where data are transmitted in the NoC without following a specific traffic pattern.

The first communication scheme that has been emulated was a 16 bit flit NoC with 64 entries buffer, while the second communication scheme was the same NoC but with 256 entries buffer.

To pass from one communication scheme to another, two RRMs Intra-Core small bit reconfigurations were required.

Evaluation of the DRNoC Emulation System

To evaluate the presented solution for NoC emulation, the use cases described in the previous section have been compared with a flow that is composed of the same steps but, instead of using partial reconfiguration it is fully based on synthesis (complete synthesis and place&route flow). Speedup results have been included in Figure 14., along with simulation, emulation and synthesis times for each configuration of the three previously discussed use cases.

Three evaluation aspects have been differentiated are described next:

- **System emulation to system simulation speedup** (*total emulation to simulation* row in Figure 14.). This value gives the speedup which results from using emulation instead of simulation that is usually reported in the emulation systems related literature. Here, this value represents the sum of the time needed for the simulation of each system configuration for a given case (*full sys. sim. time* row in the table), divided by the total time that is spent in complete use cases emul ation (*DRNoC total emul. time*) that is in the minutes range (this time is the one spend for the emulation of all configuration).

It is worthy to remark that the total emulation time includes also the data retrieving from the emulation system by the DRNoC emulation software running in a host PC. Therefore this time is much higher than the FPGA emulation time, *DRNoC emul. time* row in the table, that is in the ms range. Furthermore, if the last time is taken into account, then the speedup achieved is four orders of magnitude.

- **Partial reconfiguration synthesis time to full synthesis** (*PR synthesis to synthesis* row in Figure 14.). This value represents the speedup, which is gained with the proposed solution, as only a piece of the entire system has to be re-synthesized (when a hardcore has to be included in the library). For this aim, the time needed for the entire systems synthesis (*Full sys. synt. time* in the table) has been compared to the worst case individually system elements synthesis time (Worst case synt. time in the table). The speedup achieved, compared to a no partially reconfigurable system, is in the range of several units.

Figure 14. DRNoC emulation system evaluation

| | Parameters | Use Cases Results | | | | | |
| | Use Case | UC1 | | UC2 | | UC3 | |
	Communication Scheme	CS1	CS2	CS1	CS2	CS1	CS2
General	Full sys. sim. time (hours)	33.3	0.12	48.3	1.2	66.6	66.6
	Full sys. synt. Time (min)	12	3	16	9	15	15
	DRNoC emul. Time (ms)	80	20	92	45	125	125
	DRNoC total emul. Time (min)	4		9		9	
	Worst Case sim. Time (min)	2.5	2	2.5	2.5	2.5	2.5
	Worst Case PR synt. Time (min)	10	7	10	3	10	0.5
Speedup	Total emulation to simulation Speedup = Emul./sim	501		247		888	
	PR synthesis to synthesis Speedup = PR synth./synth.	5.2	1.5	6.4	4	6	6
	PR emulation to synthesis Speedup = PR emul./synth.	144		281		400	

- **Partial reconfiguration emulation to full system synthesis** (*PR emul. to synthesis* row in Figure 14.). Gives the speedup of the partial reconfiguration emulation flow compared to the time spend on following the same steps but using full system synthesis that is the case of traditional emulation systems. For both approaches, the worst case has been assumed. For the proposed flow, the worst case is when all cores have to be loaded using partial reconfiguration and contrary, the worst case for the synthesis based approach, assuming that the emulation system is exactly the same, is when a new synthesis is required for each parameter to be changed.

For instance, for the use case 2 (UC2), a total of 16 seconds are required for all partial reconfigurations. Initially, 10 seconds are needed to load each slot/core, NI and router in the DRNoC FPGA that corresponds to CS1. Then, 1.5 seconds are required for the measuring point reconfiguration. After that, 3 seconds to pass to CS2 and finally 1.5 seconds more, again for the three needed measuring point reconfigurations. Differently, for the synthesis based approach, 3 re-synthesis of each configuration will be required, that is, 48 minutes for CS1. The same reasoning has been followed for UC3 and UC1.

The achieved speedup results are in the range of hundreds of times (being 400 times the best time reduction). Nevertheless, for performing a more complex design space exploration, more configurations have to be synthesized and tested and speedup will increase (notice that for each use case two configuration has been defined – only two points of the design space). Furhter, reconfiguration times can be drastically reduced if other programming interfaces are used, like the FPGA internal configuration port (ICAP) or the parallel programming port (SelectMAP). In addition, it is important to remark that the synthesis times reported in the use cases are relatively low (less than 20 minutes). Differently, for instance, in (Genko et al., 2005) it is mentioned that each re-synthesis takes hours.

FUTURE RESEARCH DIRECTIONS

Having a look ahead and thinking in the shortest time to market and lowest device cost, it is clear that chips adaptability after device deployment, following the software idea of patching, will be a reality in future systems. Moreover, this adaptability will not be restricted to the systems cores and it will also include the on-chip communications. However, in order to successfully provide complete (cores and communications) adaptable solutions suitable to be consumed by the industry, research efforts have to be put in finding optimal adaptation methods and adaptation control systems that will permit the design of reliable and highly flexible adaptable system.

Furthermore, taking into account the technology advances in system design and in the state of the art of embedded systems, it is clear that in the near future reconfigurable SoC providers will include NoCs in their design flows and in the silicon die as a bus alternative for the on-chip communications. Regarding NoCs, apart from the adaptability related challenges, main problems are NoCs area and power requirements, as well as, the design complexity that is added due to the inclusion of NoCs. More specifically for the FPGA world, apart from the mentioned challenges, an important aspect is the selection of the best adaptation grain. Currently available FPGA architectures could be perceived as fine grain NoCs composed of configurable logic blocks and switch matrices and, at this level, adaptability can be applied by reconfiguring switch matrices. On the other hand, taking into account the huge connectivity and the constantly increasing resource diversity of FPGAs, a coarse grain solution seems to be more suitable. However, coarse grain adaptation in state of the art FPGA is not an easy task due to technology restrictions that result in high area overhead and flexibility restrictions. Nevertheless, reconfiguration technologies in FPGAs are being constantly improved and this will permit to provide more qualified solutions

Regarding the presented in this Chapter solutions and ideas, considering the fast improvements of electronic systems technology, very soon real-live application will run on reconfigurable NoC, like the one proposed in this Chapter. Furthermore, the proposed reconfigurable NoC solution is suitable for the RSoC and FPGAs and also, it could be integrated in ASSPs.

CONCLUSION

In this chapter, a solution for on-chip communications adaptability, called Dynamic Reconfigurable NoC (DRNoC), that permits to improve the system reconfigurability and flexibility, has been presented. The main, distinguishing, characteristics of the presented solution are that: i) it permits to define a broad set of ad-hoc communication strategies, heterogeneous, homogeneous, point to point, point to multipoint and/or NoC and ii) permits several communication strategies to independently coexist in the system. The proposed reconfigurable NoC architecture has been mapped to several FPGAs and mapping results have been included in the first part of the Chapter along with a comparison with the described background work.

Further, an innovative emulation solution which exploits the partial reconfiguration technology to achieve hard core reusability and thus exclude the synthesis step from the emulation flow has been presented in the second part of the Chapter. The solution includes an emulation workflow supported by a framework and a control software. For the validation of the emulation approach, a real implementation on a Xilinx FPGA, a 2x2 DRNoC on an XC2V3000 FPGA, has been used for running three use cases. The main advantage of the proposed emulation solution is that it permits to

have faster design space exploration with respect to a no partial reconfigurable approach, as it have been reported by the presented speedups. Also, another advantage is that the partial reconfiguration based solution might permit to use more "real" models, because it reduces the need of having diverse logic grouped in a single model in order to easily switch between a set of features.

On the other hand the main drawbacks of both solutions, the reconfigurable NoC architecture and the fast emulation approach, derive from restrictions of the partial reconfiguration technique available nowadays; that is the high area overhead and the more complex cores design flow that does not permit to implement larger NoCs. Nevertheless a tendency of improving the partial reconfiguration capabilities in the newest FPGAs can be noticed and this will permit in a near future to test the proposals in real-life applications with larger NoCs.

To summarize, the overall conclusion of this chapter is that the proposed DRNoC architecture provides connectivity flexibility and scalability to reconfigurable system compared, while the proposed emulation approach permits to speedup both, system emulations and design space exploration.

REFERENCES

Becker, J., Donlin, A., & Hübner, M. (2007). New tool support and architectures in adaptive reconfigurable computing. In *Proceedings of the IFIP International Conference on Very Large Scale Integration VLSI-SoC,* (pp. 134–139).

Benini, L., & De Micheli, G. (2002). Networks on chips: A new soc paradigm. *IEEE Computer*, *35*(1), 70–78.

Bobda, C., & Ahmadinia, A. (2005). Dynamic interconnection of reconfigurable modules on reconfigurable devices. *IEEE Design & Test of Computers*, *22*(5), 443–451. doi:10.1109/MDT.2005.109

Bobda, C., Ahmadinia, A., Majer, M., Teich, J., Fekete, P. S., & van der Veen, J. (2005). Dynoc: A dynamic infrastructure for communication in dynamically reconfigurable devices. In Tero Rissa et al. (Ed.), *Field Programmable Logic and Application FPL,In Proceedings of the 15th International Conference on Field Programmable Logic and Application* (FPL), (pp. 153–158). Washington, DC: IEEE Press.

Bobda, C., Majer, M., Koch, D., Ahmadinia, A., & Teich, J. (2004). A dynamic NoC approach for communication in reconfigurable devices. In Becker et al. (Ed.), *Field Programmable Logic and Application FPL, 14th International Conference volume 3203 of Lecture Notes in Computer Science* (pp. 1032–1036). New York: Springer.

Chan, J., & Parameswaran, S. (2004). Nocgen: A template based reuse methodology for networks on chip architecture. In *Proceedings of the 17th International Conference on VLSI Design* (VLSID), (pp.717-7120).

Coppola, M., Curaba, S., Grammatikakis, M., Locatelli, R., Maruccia, G., & Papariello, F. (2004). Occn: a noc modeling framework for design exploration. *Journal of Systems Architecture, 50*(2-3), 129–163. doi:10.1016/j.sysarc.2003.07.002

Dehyadgari, M., Nickray, M., Afzali-Kusha, A., & Navabi, Z. (2006). A new protocol stack model for network on chip. In *Proceedings of the Annual Symposium on VLSI, ISVLSI* (pp 440–44). Karlsruhe, Germany: IEEE Computer Society.

Genko, N., Atienza, D., De Micheli, G., & Benini, L. (2007). Feature - noc emulation: a tool and design flow for mpsoc. In *Circuits and Systems Magazine, 7(4), 42 – 51.* IEEE.

Genko, N., Atienza, D., De Micheli, G., Benini, L., Mendias, J. M., Hermida, R., & Francky, C. (2005). A novel approach for network on chip emulation. In. *Proceedings of the IEEE International Symposium on Circuits and Systems, 3,* 2365–2368.

Goossens, K., Bennebroek, M., Young, J. H., & Wahlah, M. A. (2008). Hardwired Networks on Chip in FPGAs to unify Data and Configuration Interconnects. In *Proceedings of the Second Symposium ACM/IEEE International Symposium on Networks on Chip NOCS* (pp.45-54).

Grecu, G., Ivanov, A., Pande, P. P., Jantsch, A., Salminen, E., Ogras, U., & Marculescu, R. (2007). Towards open network-on-chip benchmarks. In *Proceedings of the NOCS,* (pp. 205). Washington, DC: IEEE Computer Society.

Guerrier, P., & Greiner, A. (2000) A generic architecture for on-chip packetswitched interconnections. In *Design, Automation and Test in Europe Conference and Exhibition DATE* (pp. 250–256).

Hetch, R., Kubisch, S., Herrholtz, A., & Timmermann, D. (2005). Dynamic Reconfiguration With hardwired Networks-on-Chip on Furure FPGAs. In Tero Rissa et al. (Ed.), *Field Programmable Logic and Application FPL, 15th International Conference* (pp. 527- 530). Washington, DC: IEEE.

Hetch, R., Kubisch, S., Michelsen, H., Zeeb, E., & Timmermann, R. (2006). A distributed object system approach for dynamic reconfiguration. In *Parallel and Distributed Processing IPDPS* (p. 8). Washington, DC: IEEE.

Hollstein, T., Ludewig, R., Zimmer, H., Mager, C., Hohenstern, S., & Glesner, M. (2006). Hinoc: A hierarchical generic approach for on-chip communication, testing and debugging of socs. *VLSI-SOC: From Systems to Chips, 200*(4), 39–54. doi:10.1007/0-387-33403-3_3

Hübner, M., Braun, L., Göhringer, D., & Becker, J. (2008). Run-time reconfigurable adaptive multilayer network-on-chip for fpga-based systems. In *Proceedings of Parallel and Distributed Processing IPDPS* (pp. 1–6). Washington, DC: IEEE.

Jalabert, A., Murali, S., Benini, L., & De Micheli, G. (2004). xpipesCompiler: A tool for instantiating application specific Networks on Chip. In Proceedings of Design, Automation and Test in Europe DATE.

Joven, J., Font-Bach, O., Castells-Rufas, D., Martinez, R., Teres, L., & Carrabina, J. (2008). xenoc - an experimental network-on-chip environment for parallel distributed computing on noc-based mpsoc architectures. In Proceedings of Parallel and Distributed Processing (pp. 141–148). Washington, DC: IEEE Computer Society.

Krasteva, Y., Jimeno, A. B., de la Torre, E., & Riesgo, T. (2005). Straight Method for Reallocation of Complex Cores by Dynamic Reconfiguration in Virtex II FPGAs. In *Proceedings of the IEEE International Workshop on Rapid System Prototyping RSP* (pp. 77-83). Washington, DC: IEEE.

Kumar, S., Jantsch, A., Millberg, M., Öberg, J., Soininen, J. P., Forsell, M., et al. (2002). A network on chip architecture and design methodology. In *Proceedings of the IEEE Computer Society Annual Symposium on ISVLSI* (pp. 117–124). Washington, DC: IEEE Computer Society.

Mahadevan, S., Virk, K., & Madsen, J. (2006). Arts: A systemc-based framework for modelling multiprocessor systems-on-chip. *Design Automation for Embedded Systems*, *11*(4), 285–311. doi:10.1007/s10617-007-9007-6

Millberg, M., Nilsson, E., Thid, R., Kumar, S., & Jantsch, A. (2004). The nostrum backbone - a communication protocol stack for networks on chip. In *Proceedings of the International Conference on VLSI Design* (pp. 693–696). Washington, DC: IEEE Computer Society.

Möller, L., Grehs, I., Calazans, N., & Moraes, F. (2006). Reconfigurable Systems Enabled by a Network-on-Chip. In *Proceedings of the International Conference on Field Programmable Logic and Application FPL, 16th International Conference* (pp.1-4). Washington, DC: IEEE

Möller, L., Grehs, I., Carvalho, E., Soares, R., Calazans, N., & Moraes, F. (2007). A noc-based infrastructure to enable dynamic self reconfigurable systems. In *Proceedings of the Reconfigurable Communication Centric System on Chip ReCoSoC* (pp. 23–30).

Moraes, F. G., Calazans, N., Mello, A., Möller, L., & Ost, L. (2004). Hermes: an infrastructure for low area overhead packet-switching networks on chip. *Integration (Tokyo, Japan)*, *38*(1), 69–93.

Murali, S., & De Micheli, G. (2004). Sunmap: a tool for automatic topology selection and generation for nocs. In *Proceedings of the 41st annual conference on Design automation* (pp.914–919). ACM.

Nollet, V., Coene, P., Verkest, D., Vernalde, S., & Lauwereins, R. (2003). Designing an operating system for a heterogeneous reconfigurable SoC. In *Proceedings of the Parallel and Distributed Processing Symposium* IPDPS (pp. 7).

Ogras, U., Marculescu, R., Lee, H. G., Choudhary, P., Marculescu, D., Kaufman, M., & Nelson, P. (2007). Challenges and promising results in noc prototyping using fpgas. *IEEE Micro*, *27*(5), 86–95. doi:10.1109/MM.2007.4378786

Ost, L., Mello, A., Palma, J., Moraes, F., & Calazans, N. (2005). Maia: a framework for networks on chip generation and verification. In Tang, T.-A. (Ed.), *ASP-DAC* (pp. 49–52). New York: ACM Press.

Pande, P., Grecu, C., Jones, M., Ivanov, A., & Saleh, R. (2005). Performance evaluation and design trade-offs for network-on-chip interconnect architectures. *IEEE Transactions on Computers, 54*(8), 1025–1040. doi:10.1109/TC.2005.134

Pionteck, T., Albrecht, C., & Koch, R. (2006). A dynamically reconfigurable packet-switched network-on-chip. In Georges, G., & Gielen, E. (Eds.), *European Design and Automation Association DATE* (pp. 136–137). Leuven, Belgium.

Pionteck, T., Koch, R., & Albrecht, C. (2006). Applying partial reconfiguration to networks-on-chips. In *Proceedings of the 2006 International Conference on Field Programmable Logic and Applications FPL* (pp 1–6). Washington, DC: IEEE.

Samira, S., & Ahmad, K. Mehran & Armin. (2007). Smap: An intelligent mapping tool for network on chip. In Proceedings of Signals, Circuits and Systems (ISSCS), (pp. 1–4, 13-14_.

Srinivasan, K., Chatha, K. S., & Konjevod, G. (2007). Application Specific Network-on-Chip Design with Guaranteed Quality Approximation Algorithms. In Proceedings of the *Asia and South Pacific Design Automation Conference ASP-DAC* (pp.184-190).

Varatkar, G., & Marculescu, R. (2004). On-chip traffic modeling and synthesis for mpeg-2 video applications. *IEEE Trans. VLSI Syst., 12*(1), 108–119. doi:10.1109/TVLSI.2003.820523

William, J. D., & Brian, T. (2001). Route packets, not wires: On-chip interconnection networks. In *Proceedings of the Design Automation Conference DAC* (pp. 684–689). ACM.

Wolkotte, P. T., Holzenspies, P. K. F., & Smit, G. J. M. (2007). Fast, accurate and detailed NoC simulations. In Proceedings of Network on Chips.

ADDITIONAL READING

Bouldin, D. (2006). Enhancing electronic systems using reconfigurable hardware. *IEEE Circuits & Devices Magazine, 22*(3), 32–36. doi:10.1109/MCD.2006.1657847

Gokhale M., & Graham. P. S. (2005). *Reconfigurable Computing: Accelerating Computation with Field-Programmable Gate Arrays*. New York: Springer Books.

Habay, G., Butel, P., & Rachet, A. (2004). *Managing partial dynamic reconfiguration in virtex-ii pro fpgas. Xcell Journal Online*. San Jose, CA: Xilinx Press.

Hartenstein, R. (2006). Proposal of a new Magazine by Reneir Hartenstein. In TU Kaiserslautern.

Hartenstein, R. (2007). The von neumann syndrome. In *Proceedings of the Future of Computing Symposium* (invited paper).

Hauck, S. DeHon, A (Ed.). (2007). Reconfigurable Computing: The Theory and Practice of FPGA-Based Computation (Systems on Silicon). Amsterdam: Elsevier Science & Technology Books.

Hübne, M., Ullmann, M., Braun, L., Klausmann, A., & Becker, J. (2004). Scalable application-dependent network on chip adaptivity for dynamical reconfigurable real-time systems. In Becker et al. (Ed.), *Field Programmable Logic and Application FPL, 14th International Conference volume 3203 of Lecture Notes in Computer Science* (pp. 1032–1036). Springer, Leuven, Belgium.

Kao, C. (2005). *Benefits of Partial Reconfiguration*. San Jose, CA: Xilinx Press.

Krasteva, Y. (2009). *Reconfigurable Computing Based on Commercial FPGAs. Solutions for the Design and Implementation of Partially Reconfigurable Systems*. Published Ph.D. dissertation, Universidad Politécnica de Madrid, Spain.

Krasteva, Y., Criado, F., de la Torre, E., & Riesgo, T. (2008). A Fast Emulation-Based NoC Prototyping Framework. In *International Conference on ReConFigurable Computing and FPGAs ReConFig* (pp. 211-216). Washington, DC: IEEE.

Krasteva, Y., de la Torre, E., & Riesgo, T. (2006). Partial reconfiguration for core reallocation and flexible communications. In Proceedings of ReCoSoC, (pp. 91–97).

Krasteva, Y., de la Torre, E., & Riesgo, T. (2007). Reconfigurable Heterogeneous Communications and Core Reallocation for Dynamic HW Task Management. In *Proceedings of IEEE International Symposium on Circuits and Systems* ISCAS (pp.873-876). Washington, DC: IEEE.

Krasteva, Y., de la Torre, E., & Riesgo, T. (2008). Virtual Architectures for Partial Runtime Reconfigurable Systems. Application to Network on Chip based SoC Emulation. In *Proceedings of IEEE Annual Conference of the IEEE Industrial Electronics Society IECON* (pp. 187).

Kubisch, S., Cornelius, C., Hecht, R., & Timmermann, D. (2007). Mapping a Pipelined Data Path onto a Network-on-Chip. In*Proceedings of the Second International Symposium on Industrial Embedded Systems* (pp. 527-530). Washington, DC: IEEE.

Kumar, S., Jantsch, A., Millberg, M., Öberg, J., Soininen, J.-P., Forsell, M., et al. (2002). A network on chip architecture and design methodology. In *Proceedings of the IEEE Computer Society Annual Symposium on VLSI ISVLSI* (pp 117–124). Washington, DC: IEEE Computer Society.

Mangione-Smith, W. H., Hutchings, B., Andrews, D., DeHon, A., Ebeling, C., & Hartenstein, R. (1997). Seeking solutions in configurable computing. *IEEE Computer*, *50*(12), 38–43.

Millberg, M., Nilsson, E., Thid, R., Kumar, S., & Jantsch, A. (2004). The nostrum backbone - a communication protocol stack for networks on chip. In *Proceedings of the 17th International Conference on VLSI Design VLSID* (pp 693–696). Washington, DC: IEEE Computer Society.

Peattie, M., & Lim, D. (2004). *Two Flows for Partial Reconfiguration: Module Based or Small Bit Manipulation XAPP 290 (v1.0)*. San Jose, CA: Xilinx.

Pionteck, T., Albrecht, C., Koch, R., Maehle, E., Hubner, M., & Becker, J. (2007). Communication Architectures for Dynamically Reconfigurable FPGA Designs. In *Proceedings of the Parallel and Distributed Processing Symposium* (IPDPS), (pp.1-8). Washington, DC: IEEE.

Rodrigez-Andina, J.J: Moure, M.J. & Valdes. M.D. (2007). Features, design tools, and application domain of fpgas. *IEEE Transactions on Industrial Electronics*, *54*(4), 1810–1823. doi:10.1109/TIE.2007.898279

Salminen, E., Kulmala, A., & Hämäläinen, T. D. (2008). Survey of Network on Chip Proposals. In OCP-IP white paper (Tech. Rep.). Finland: Tampere University of Technology.

Scrofano, R. Gokhale. M. B., Trouw, F., & Prasanna, V. K. (2008). Accelerating molecular dynamics simulations with reconfigurable computers. *IEEE Trans. on Parallel and Distributed Systems*, 19(6), 764–778.

Shiflet, E., Dorairaj, N., & Goosman, M. (2005). *PlanAhead Software as a Platform for Partial Reconfiguration (Tech. Rep.)*. San Jose, CA: Xilinx Inc.

Srinivasan, K., Chatha, K. S., & Konjevod, G. (2007). Application specific network-on-chip design with guaranteed quality approximation algorithms. In *Asia and South Pacific Design Automation Conference ASP-DAC*, (pp 184–190). New York: ACM Press.

Wahlah, M. A., & Goossens, K. (2009). Modeling reconfiguration in a FPGA with a hardwired network on chip. In *Proceedings of the Reconfigurable Architecture Workshop* (RAW).

Xilinx (2003). Xilinx application notes, *Virtex Series Configuration Architecture User Guide, XAPP 151 v1.6* (Tech. Rep.). San Jose, CA.

Xilinx (2006). *Early access partial reconfiguration user guide* (Tech. Rep.). San Jose, CA: Xilinx Inc.

Section 5
State-of-the-Art Reconfigurable NoC Designs

Chapter 10
Dynamically Reconfigurable NoC for Future Heterogeneous Multi-Core Architectures

Balal Ahmad
Government College University, Pakistan & University of Edinburgh, UK

Ali Ahmadinia
Glasgow Caledonian University, UK & University of Edinburgh, UK

Tughrul Arslan
University of Edinburgh, UK

ABSTRACT

To increase the efficiency of NoCs and to efficiently utilize the available hardware resources, a novel dynamically reconfigurable NoC (drNoC) is proposed in this chapter. Exploiting the notion of hardware reconfigurability, the proposed drNoC reconfigures itself in terms of switching, routing and packet size with the changing communication requirements of the system at run time, thus utilizing the maximum available channel bandwidth. In order to increase the applicability of drNoC, the network interface is designed to support OCP socket standard. This makes drNoC a highly re-useable communication framework, qualifying it as a communication centric platform for high data intensive SoC architectures. Simulation results show a 32% increase in data throughput and 22-35% decrease in network delay when compared with a traditional NoC with fixed parameters.

INTRODUCTION

Traditional bus based communication medium will prove to be a bottleneck in achieving the high processing requirements in future SoC, especially with the on-chip communication requirements rising over 9 GB/s. NoC is considered as a solution to this communication bottleneck. The NoC architecture is generated by choosing a network topology, one of the routing and switching schemes and fixing a packet size. These network parameters are chosen at the generation time, mostly to deal with the worst case communication scenario. However, in the real world, the cores connected are not utilizing all the available bandwidth, and some cores utilise more bandwidth then others. This results in some of the

DOI: 10.4018/978-1-61520-807-4.ch010

allocated resources being not fully utilised, causing waste of resources. In an invited paper by (Atienza et al. 2007), it is suggested that 74% improvement in power can be achieved by using an application specific NoC against the traditional NoC approach.

In this chapter a dynamically reconfigurable NoC (drNoC) architecture is proposed as a high data throughput communication centric platform for future multi-core (Ahmad, Ahmadinia and Arslan, 2005, 2008; Ahmad, Erdogan, and Khawam, 2006). Exploiting the notion of reconfigurability, the proposed drNoC allocates the resources and parameters suitable for the desired application domain at run time, thus, not having to deal with effects of unfavourable selection of parameters. In order to develop a generic multi-domain communication centric platform, the proposed NoC has been given an OCP socket based interface to allow for ease of integration.

The chapter begins with the review of the work done in the field of NoC by different research institutes. The proposed drNoC is then presented followed by the description of different layers and the router design. The layered structure makes it easy for designer to understand the different aspects of proposed architectures. Simulation results are finally discussed to establish the effectiveness of the proposed network.

HISTORY OF DRNOC

NoC has been under the spotlight since it was first introduced and many research groups are working on different aspects of NoC design. (Hemani et al., 2000) proposed a packet switched architecture with switches surrounded by six resources and connected to 6 neighbouring switches. The architecture was called honeycomb due to the hexagon based pattern of switches and resources. The concept of packet switching re-appeared in other consecutive approaches but the topology simplified in most proposals to a mesh of resources and switches (Guerrier and Greiner, 2000).

(Dally and Towles, 2001) proposed replacing global wiring with a general purpose on chip interconnection network at an area overhead of 6.6%. A 12mm×12mm chip in 0.1μm CMOS technology was developed. Dally concluded that there can be up to 6,000 wires on each metal layer crossing each edge of a tile(3mm×3mm). It is quite easy to achieve over 24,000 'pins' crossing the four edges of a tile. By effectively choosing the network topology these abundant wiring resources can be converted into bandwidth.

(Benini and Micheli, 2002) proposed a layered design methodology borrowing models, techniques and tools from the network design field and applying them to SoC design. Several open problems at various layers of the communication stack were addressed and a basic strategy was given to effectively tackle them for energy efficient design. Xpipes compiler was also presented as a tool for automatically instantiating an application specific NoC for heterogeneous multi-processor SoCs.

(Kumar et al. 2002; Sun, Kumar and Jantsch, 2002) constructed a model of NoC using a public domain network simulator NS-2 and evaluated design options for a specific NoC architecture which has a two dimensional mesh of switches. S. Kumar analysed the series of simulation results to determine the relationship between buffer size in switch, communication load, packet delay and packet drop probability. The results are useful for the design of an appropriate switch for the NoC.

(Shin and Daniel,1996) proposed a hybrid switching scheme that dynamically combines both virtual cut-through and wormhole switching to provide higher achievable throughput values compared to wormhole switching alone. (Hu and Marculescu, 2004) proposed a smart NoC, which combines the advantages of both deterministic and adaptive routing schemes in a NoC environment.

NoC is relatively a new concept but it has been rapidly accepted in academia with much industrial interest in it. A comprehensive survey on current research and practices of NoC can be found in (Bjerregaard and Mahadevan, 2006).

PROPOSED DYNAMICALLY RECONFIGURABLE NOC

Keeping in view the success of busses and increased communication requirements in multi-core SoC, a dynamically reconfigurable NoC (drNoC) has been proposed that combines the advantages of bus based systems and the advantages of NoC. The network changes its characteristics with the changing communication requirements of the system. The design inspiration comes from the fact that different cores connected in a system have different bandwidth requirements. e.g., in an advanced PDA with 3G communication capabilities, the on-chip communication can vary depending on the user application, thus a network with fixed communication design parameters is not the optimal solution in terms of power and data throughput.

The allocation of network resources depends on network topology, switching and routing decisions. Packet size also affects the network performance both in data throughput and overall system energy consumption (Benini and Micheli, 2001). In the proposed design, the network configures itself in terms of its routing, switching and packet size to maintain the QoS requirement of the system.

This intelligent network has its kernel in the form of a micro network stack called smart network stack (SNS) of the node processor. Depending on the data about to be transferred, the SNS makes the decisions about the packet size, switching and routing, required for the data and includes this information in packet header. This information is read by the router and packets are processed as desired. Thus, in the case of a processing core with high bandwidth requirements, packet size would be increased, also switching would change from packet switching to circuit switching. These

changes increase data throughput and decrease the switching power and timing delays. Distributed control of a network has the advantage that if a single node failure occurs, the network continues to perform its functions.

The router is the main building block of the network. It serves two main functions. Firstly, it acts as the interface between the core and network. To allow for heterogeneous component integration, the router is built with a bus like interface based on OCP socket standard which allows cores to integrate with less effort. Secondly, the router routes the data packets and control signals to the right path.

SNS is a modified version of the micro network stack proposed by (Benini and Micheli, 2002). The SNS is composed of five layers, application, transport, network, data link, and physical layer (shown in Table 1).

The application layer provides the user interface to the communication system. Thus the applications running on the processor core does not have to worry about the complex communication network facilitating its communication with other applications connected to other nodes.

NETWORK ARCHITECTURE AND CONTROL

The network architecture refers to the topology and physical organization of the interconnect network. The protocols specify the use of these network resources during system operation. Network control is taken care of by transport, network and data link layer.

Table 1. Smart network stack

Software	Application Layer
Architecture & Control	Transport Layer Network Layer Data Link Layer
Physical	Wiring

Transport Layer

Unlike the bus based system, the data transported over NoC is in the form of small packets. This packetization of data is dealt with by the transport layer. All packets have same format, "packet header" followed by the "payload". Packet header carries information necessary for the routing of packets to its destination.

Some important subfields of the packet header include:

- *Type*: Indicates the start of packet and specifies if the packet is a broadcast packet (meant for all connected cores) or is meant for only a specific destination.
- *Destination Address*: Indicates the destination address. In case of broadcast packets, the destination bits are not read by the router for routing the packets.
- *Switching Type*: Instructs the router to switch between packet switching and circuit switching (set by the Network Layer).
- *Packet type*: Specifies if it's a data packet, acknowledgment of received packet, or request for retransmission in case of not received or receipt with error packet.
- *Source Address*: Source address is read by the destination node and is used in replying to the packets.
- *Data offset*: This indicates where the data begins in the packet. It can also be referred to as the sequence number of packet useful in assembly of data from the received packets at the destination.
- *Checksum*: Used for error detection.

Header bits are followed by the payload. The size of the payload depends on the type of data transmitted. As mentioned earlier, the size of packet changes for different types of communication. Three types of data thresholds are defined, normal, medium and high. Normal threshold refers to any communication where data throughput is under 100 Kbits/sec e.g. interrupt handling etc., high data threshold refers to any communication above 10Mbits/sec e.g. CPU cache to main memory, compressed or uncompressed video. Medium data threshold is referred to as any communication between these two thresholds.

In the prototype, these values are fixed. However, data threshold values can be added by users when defining the network parameters. Figure 1 shows a conceptual data packet structure.

Network Layer

The network layer deals with switching and routing aspects of the packetized data. Switching can be packet switching and circuit switching. If only a small amount of data is to be transmitted, then setting up a circuit to transfer data is inefficient. Thus, the network layer transmits data by packet switching. The advantage of packet switching is that the network is only used when there is information to be sent, also a single path can be used by packets from different sources at the same time thus utilising bandwidth more effectively.

In the proposed network, circuit and packet switching co-exist in the same network and only the paths needed for high data throughput are converted to circuit switching, the rest of the network keeps working on packet switching and excludes the circuit switched paths from their routing deci-

Figure 1. drNoC data packet structure

Type	Destination Address	Switching Type	Packet Type	Source Address	Data Offset	Payload	Checksum

sion. The decision to change switching depends on the amount of data to be transmitted. A data threshold value defined in section 4.1 is used to determine the type of switching for a specific path. This is also coupled with an increase in the packet size by transport layer for maximum data throughput. The routers are notified about the decision by the "switching type" in data packet header. Thus the routers in the path modify themselves for the new switching type. The circuit switched path formed between the two nodes can be looked upon as an existence of bus as the two communicating cores are connected via a dedicated path for the length of data transmission. Figure 2 shows the formation of a circuit switching path between node 6 and 11. Once the circuit switched path is established, it appears invisible to neighbouring nodes and cannot be used as part of their packet switching.

Routing Algorithm for Packet Switching

Routing is performed in the proposed drNoC based on concept of wormhole routing. In the wormhole routing technique (Dally, 1992), a message is divided into flits. If a communication channel transmits the first flit of a message, it must transmit all the remaining flits of the same message before transmitting flits of another message. In this method, the message latency is proportional to the sum of the number of cycles spent in waiting for suitable channels to route message flits, number of hops, and message length. In the example

above with a b-flit message transverse a path of length d, the first flit does not wait for the rest of the message. It therefore arrives at its destination after d steps, and the last flit of the message arrives after $d+b-1$ steps. The difference in time is due to a better utilisation of network edges by the wormhole router. In addition to reduced latency, wormhole routing also has the advantage that it can be implemented with small and fast switches.

Wormhole routing performance is prone to deadlock and live-lock issues (Su and Shin, 1996). A deadlock occurs when a message waits for an event that will never happen; a live-lock keeps a message moving indefinitely but not letting it reach the destination. To solve this problem, dimension-order-routing (Duato, 1994; Wu, 2002) in which packets are routed in only one dimension till they reach the destination row, and then switch to other dimension until the destination, is used or virtual channel approach (Duato, 1994; Dally, 1992) is used in which one physical channel is split into several virtual channels. However, using virtual channel require the use of more buffer space.

Three virtual channels per port are employed in the proposed design in order to deal with live-lock and dead-lock situations. The proposed adaptive routing algorithm works in two steps:

Step I: Output paths are selected for control flits by choosing the most profitable channel from the available free channels. Profitable paths correspond to ones that will take the packet closest to its destination. The path, once established by the control flit, is utilised by the data flits since

Figure 2. Change of packet switching to circuit switching between node 6 and 11

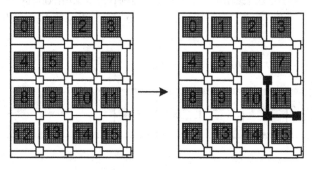

Figure 3. Eight neighbours and ports of a router

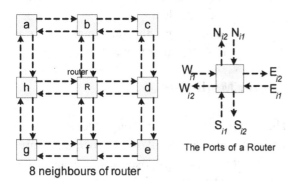

8 neighbours of router

The Ports of a Router

it is based on wormhole routing. If all channels are busy or there is a faulty neighbouring router, the packet is stored in the buffer. The packets following the blocked/stored packets are stopped at the destination by a backward signal from router to source.

Step II: Once the output channel is available for routing, packets are injected from the buffer or the processing core into the network. If the channel is not available, packets are kept stored in the buffer. This process is repeated for all the packets in the buffer before any other packet can be injected into that channel. Once the entire packet from buffer is sent, packets from other incoming ports are injected into the network.

To explain the routing further, consider the network in Figure 3. Router "R" has eight neighbours represented by a, b, c, d, e, f, g. Router communicates with its neighbours through its input ports W_{il}, E_{il}, N_{il} and S_{il} and output ports W_{i2}, E_{i2}, N_{i2} and S_{i2}. Let the virtual channels be VC_1, VC_2, and VC_3. Each router tests the state of its neighbours to establish the availability of channel.

The message routings according to the current router R (x_s, y_s) and the destination (x_d, y_d) are divided into four types: WE, EW, SN and NS routings, where WE (from west to east) routing is taken if $x_s < x_d$, EW (from east to west) routing if $x_s > x_d$, NS (from north to south) routing if $x_s = x_d$ and $y_s < y_d$, and SN (from south to north) routing if $x_s = x_d$ and $y_s > y_d$. In presence of an unblocked outgoing channel, the router directs the header flit on the shortest path according to Table 2. However, if the output channel is busy, then router checks for path with penalty of two more packet hops. In absence of even router's second choice of output channel, the packet is stored in the buffer till the shortest path or the second choice path becomes available.

Data Link Layer

As mentioned earlier, the data link layer abstracts the physical layer and treats it as a medium with a non-zero probability of errors in the transmitted bit stream. Error correction with retransmission is employed in the proposed network to keep silicon cost low. A checksum is calculated (one's complement sum of all 16 bit words in the header and payload) at the transmitter and the checksum bit is sent with the packet to the receiver where the checksum is calculated again and compared with transmitted checksum to detect any error. At the present stage of network implementation, the checksum is implemented but is not used in the simulations due to the processing overhead associated with it.

Physical Layer

Keeping in view the abundance of wiring resources, separate wires are used for control signals and data transfer. Using separate wires for control signals facilitates transmission of information between routers to avoid network congestion. More details of signals are discussed in next section.

SYSTEMC MODELING OF DRNOC

Modeling Level

The proposed systemC modeling layer lies between transfer level and transaction level providing cycle accurate models compatible with

Table 2. Available channel for message routing

Routing Type	Channels available
WE routing	In VC_1
EW routing	In VC_1 or VC_2
SN routing	In VC_1, VC_2 or VC_3
NS routing	In VC_1, VC_2 or VC_3

read() write() functions and signals to maintain bus cycle accuracy. Clock cycle accuracy allows accurate communication space exploration and modeling at the boundary of the transaction layer provides faster simulation times. The system thus implemented follows a specific communication protocol in order to explore its characteristics and for system verification.

Data interface - The data transfer supports passing along pointers as well as explicitly transferring the data to the channel. The channel contains a pointer that is shared between the master and the slave. The data pointer is set to the internal buffer in the master that is transmitting the data. In case of read request, the pointer is pointed to the buffer in the slave. In case of the data copy method, the data has to be copied to and from the channel. A mixture of both is also possible where one copies and other points to the buffer in during the same initiated communication.

Control interface – The control interface is needed for the synchronisation and sequencing mechanism used with the data interface to complete the transaction. The control flow is monitored closely by the controller. As it is cycle driven, the control signals can only be called at clock edges. The channel is implemented as clocked and stores state information between clock cycles. Some of the synchronisation mechanisms supported include,

- A request mechanism that the master core has data available to transmit.
- A response notification that the slave core has data for the master core.
- An acknowledgement that the master core has processed the data.

- Acknowledgement that the slave core has processed the data.
- Flag to show that the slave/master core is busy in some other transaction.

Router Modeling

The router is the main building block of NoC. It acts as an interface point of the processing cores to the system. In the proposed design, each router is connected to four neighbouring routers and the interfaced core. The proposed NoC is also given a bus-like interface as mentioned above, which helps to eliminate the use of wrappers and any component designed to be integrated to a bus based platform can be integrated in the proposed NoC in plug and play fashion. The proposed router can be divided into four components (shown in Figure 4).

The Input Controller that manages the routing table also known as look-up-table (LUT) and determines the fate of arrived packets after header inspection. The input controller of a router is connected to the input controller of its neighbouring routers. This connection is to update routing tables and pass control signals. Thus when a router is instructed to change mode to circuit switching, it informs its neighbours to exclude the specified path from their routing tables, and remembers the path established till it receives the end of transmission packet. This passage of control signal to neighbouring routers makes it possible to avoid the need for big buffers to store the packets as once a packet is blocked, a control signal is sent backwards till it reaches the source to stop packet injection. The input controller checks the arrived packets at each input port in a round robin pattern to establish the output path.

Input port that is the point of entry of the incoming packet, it has a buffer to store one packet that is getting inspected for its header contents. Information extracted from an incoming packet includes its destination address and type of switching.

Figure 4. Conceptual model of drNoC router with OCP interface

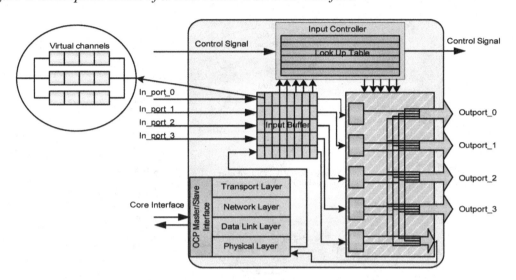

An *Interface module* has been added to deal with the packetization of data. Each router has one interface module and is present at the port where the component is connected to the router. Its main function involves dealing with handshaking with the connected component and once data has arrived, packet assembly with the correct destination and source address. De-packetization of packet is also done by interface module. The layered SMS responsible for control of the dynamically reconfigurable NoC is implemented in the interface module.

Finally the *Switching* Logic that connects the input ports to the output ports depending on the instructions from the Input Controller.

The implementation of the system started with modeling a basic NoC router and adding the reconfigurable features into the system. The modelled basic NoC also served in evaluating the performance of the system.

The node is implemented as an FSM with different states representing different tasks carried out by the interface node or router. Figure 5 shows the FSM of the first stage of implementation where a simple router is designed. At the start, the router waits for a packet to arrive at its input nodes. On packet arrival the destination

address is read by the input controller. The LUT is then checked to determine the output port for the packet. Every packet can be routed to any of the four output ports, with each route carrying a penalty in terms of packet hop or packet delay. The entries in LUT are according to the routing algorithm explained in 4.2.1.

If the output port is available for transmission of packet, the packet is routed to the neighbouring router connected to that port, otherwise, the packet is stored in the buffer and a control signal is sent to the packet source to stop data production. The first packet is sent to determine the path and is called a control packet. Once a complete path is discovered, the following packets (data packets) follow the same route.

After implementation of the basic structure of NoC router, the novel features of the proposed NoC router were added in the model. Figure 6 shows the FSM of the proposed drNoC router. It starts like the normal NoC waiting on packet arrival. However, once the packet has arrived and its address is decoded, the router checks the data threshold set in the packet. Data threshold is the option included in the proposed drNoC that determines the packet size, switching and routing of the packet and is described in section 4.1.

Figure 5. FSM showing basic NoC flow

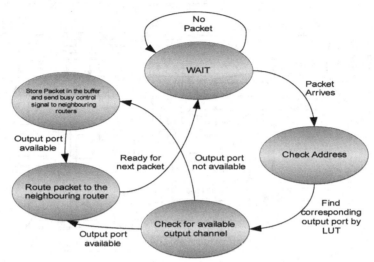

Figure 6. FSM showing drNoC flow

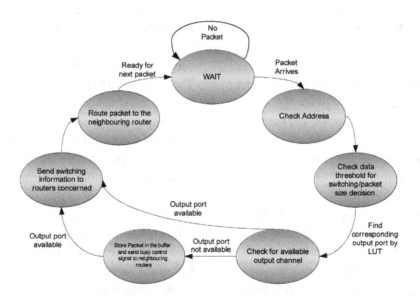

The input controller, after establishing an output channel, acts on the data threshold and sends control signal to neighbouring routers about the routing/switching information. If the core is connected through the OCP socket, all the features of SMS are performed by the router. However, in case the packets are being generated by the connected core, issues like packetization, routing, threshold decisions are made by the core.

When implementing the complete NoC system, different topologies are considered on the basis of degree, maximum distance between two nodes, average distance between nodes and wire cost. 2D-mesh is the most simple 2D network structure consisting of a grid of horizontal and vertical lines with nodes placed at their intersections. 2D Mesh is considered as a preferred topology for NoC architectures due to its simple addressing scheme and predictable inter-node delay. (Zho, 2005) argues that torus theoretically outperforms the 2D-mesh. However due to the complexity involved in keeping the wire lengths the same, the minor performance improvement of torus can be neglected compared to the profit of placement optimization. Network properties for 16 node 2D mesh can be given by (Bijlsma, 2005);

Degree = 3-5

$$\text{Average Distance} = \frac{2}{3}\sqrt{N} = 2.6$$

$$\text{Wire Cost} = 2\left(N - \sqrt{N}\right) = 24$$

A prototype router design is implemented in Verilog. Each input port has a fixed link width of 32 bits. When analysed under 0.18μm technology, router area is found to be 66362.3 μm². Area of drNoC is considerably higher than a router for the circuit switched presented by (Phi-Hung, Kumar and Chulwoo, 2006), but when compared with the 32-bit link router for packet switched network in (Kim, Kim, and Sobelman, 2005), it is found to be only 0.9% more in area. This increase is area is due to the complex circuitry involved in the implementation of smart network stack.

Hardware Reconfigurability

As mentioned above, novelty of the proposed NoC architecture lies in its ability to change its routing, switching and packet size dynamically with the changing communication requirement of the system. The concept of hardware reconfigurability is employed for co-existence of the desired packet and circuit switching in the network. Once the switching is decided, the routing algorithm is chosen at the software level. The reconfiguration bits for switching is included in the packet, and once received at the router, performs the reconfigu-

ration of router making it invisible to non circuit switching links (in case of establishment of circuit switching path). This hardware reconfigurablilty also saves power as parts of network not getting utilize are switched off (no switching activity) hence, a low power consumption.

Different to other approaches, where the complete reconfigurable area or the reconfigurable modules is implemented as buses, the reconfigurable modules remain in their original scope and the surrounding design is slightly changed. The topology of a static design remains unchanged if some parts are made reconfigurable. The straight-forward way to achieve this is to intercept the communication between static and reconfigurable modules at the channels, which interconnect those parts. The current configuration is simulated by forwarding channel events only to the currently active modules by using dynamic switches. Additionally, all channel accesses of inactive modules are suppressed. If the modules are either sensitive to certain events or make use of blocking channel accesses, such a system will behave like a reconfigurable design during simulation. Still this is not a completely satisfying solution, since for every channel type a new switch has to be built. This leads to the concept of a reconfigurable interface (*sc_rec_interface*), which provides a framework for building these switches.

An *sc_rec_interface* is a special switch, designed to connect a static channel with a port of a reconfigurable module, as shown in Figure 7. During the simulation, accesses to the corresponding port from within the reconfigurable module are forwarded to the static channel. Additionally, any required events, the reconfigurable module is listening to (via sensitivity or dynamic wait() statements), is also forwarded from the static channel to the module. Multiple reconfigurable modules can be bound to a single *sc_rec_interface*.

During the simulation, the actual reconfiguration operations change the data-flow through the interface depending on the reconfiguration state of the connected modules. If all ports of a reconfigu-

Figure 7. A transaction-level based reconfigurable design

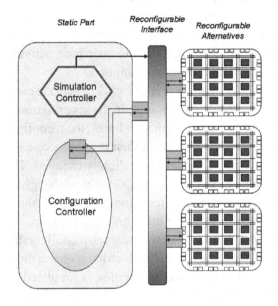

rable module are equipped with *sc_rec_interface*, no port can be triggered from the outside when the module is inactive (not configured). Therefore, no outbound traffic activity occurs since the module's processes are no longer triggered. Nevertheless, technically it is possible that a module keeps on triggering itself (for instance by a member of type *sc_clock*). In this case outbound traffic is suppressed and a warning is reported to the designer.

Network Analysis

Employing the modelling explained in section 5.2 and 5.3, a network with 16 nodes is modeled (shown in Figure 8). The aim of these simulations is to investigate the effect of dynamic reconfiguration of network parameters on the proposed drNoC. Network delay and data throughput is taken for different simulation scenarios to evaluate the proposed drNoC. Traditional NoC is taken as the one with fixed network parameters i.e. switching is packet switching, routing is adaptive, packet size is fixed. Clock frequency in these simulations is taken to be 100MHz.

Network delay is taken as the round-trip delay for a data packet within the network. Network delay comprises the sum of transmission delays and queuing delays experienced by a packet travelling through the collection of routers. Delay has huge impact on the processing capabilities of the concerned processing element waiting on data from

Figure 8. Block diagram of SystemC network model

Router.h
Packet_Reader ();
Packet_Interpreter ();
Reconfiguration_Controller();
Virtual_Channels_Controller();
Routing_Table_ Controller();
Routing_Decision_Module () ;
Switching_Decision_Module ();
Contention_detector ();
Handshaking_Module ();
TCP_STACK_Packetisation_Module ();
TCP_STACK_Depacketisation_Module();
TCP_STACK_Handshaking_Module ();

Processing_Element.h
Handskaing_module ();
Packetisation_module ();
Depacketisationn_module ();
Packet_generator ();
Packet_Receiver ();

Main ();	
Router.h	**Processing_Element.h**
Router_inst_1 ();	Processing_Element_Inst_1 ();
Router_inst_2 ();	Processing_Element_Inst_2 ();
-	-
-	-
-	-
Router_inst_16 ();	Processing_Element_Inst_16 ();

Figure 9. Simulated 4x4 2D mesh network

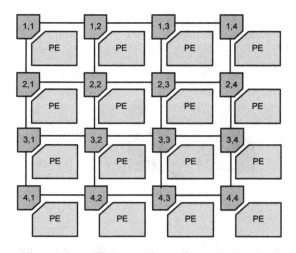

the other processing core. Data throughput on the other hand is taken as the amount of data transferred from one processing core to another through the network over a particular period of time.

In the simulations the terms normal, medium and high data throughput refers to the data threshold defined in section 4.1. Thus, in order to mimic this traffic pattern, normal data threshold traffic will be simulated as a small transaction of data injected to the network at a rate of 60 packets/sec or 9.6 Kbits/sec. In the case of medium data threshold packets are injected at a rate of 5000 packets/sec or 800 Kbits/sec. In case of high data threshold, packets are injected at a rate of 65000 packets/sec or 10.4 Mbits/sec. Another difference between high data threshold and medium data threshold is the way data is injected. Medium data threshold follows a bursty traffic burst where data will be injected for only a small period of time, while, high data threshold refers to continuous data getting injected in the network. Data packet size is taken as 160 bits. The 4x4 network used for simulation is shown in Figure 9. A 16 core size for NoC is a realistic simulation for near future applications. Different master-slave pairs considered are mentioned in different simulation scenarios.

• **Scenario 1: Network delay - Traditional NoC vs. drNoC - no contention in the network**

The aim of this simulation is to monitor the network delay when there is no contention in the network i.e. routers/channels are only utilised for one master-pair communication. Table 3 lists the mater-slave pair scenario considered. 6 cores (3 master-slave pairs) are involved in communication. Link 1 and Link 2 are medium data threshold links i.e. the packet injection will be bursty. Link 3 is high data threshold link and will be injecting packets continuously, hence the reason, the total number of packets is not specified. The simulation is run for 20 seconds.

Figure 10 shows the communicating cores and the simulation results. Under the medium data threshold, the links perform the same way. However, under high data threshold, network delay is decreased by 35.7% when drNoC is used. As mentioned above, network delay is the sum of transmission delay and queuing delay. In the case of Link 1 and link 2, the difference in network delay is due to the number of intermediate nodes involved.

Table 3. Scenario 1: Resource configuration

	Source	Destination	Traffic Type	No. of Packets
Link 1	2,3	3,4	Medium	80000
Link 2	1,3	3,2	Medium	50000
Link 3	1,1	4,2	High	Continuous

Table 4. Scenario 2: Resource configuration

	Source	Destination	Traffic Type	No. of Packets
Link 1	1,3	2,4	Medium	80000
Link 2	1,2	3,2	High	Continuous
Link 3	1,1	3,4	High	Continuous

Figure 10. Scenario 1: Network delay results

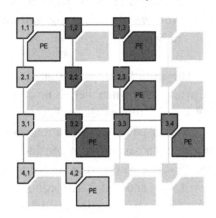 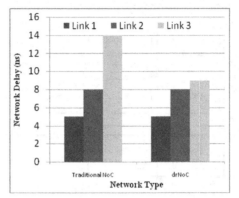

- **Scenario 2: Network delay - Traditional NoC vs. drNoC - in network with contention**

The aim of this simulation is to monitor the network delay in network with resource contention. Table 4 lists the master-slave pair scenario considered. 6 cores (3 master-slave pair) are involved in communication. Link 1 is medium data threshold links i.e. the packet injection will be bursty. Link 2 and Link 3 are high data threshold link and will be injecting packets continuously, hence the reason, total number of packets are not specified. The simulation is run for 30 seconds. Contention is taken as the sharing of router (3,2) by the data packets of Link 2 and Link 3.

Figure 11 shows the formation of two circuit switched paths in the network. In the case of tra-

ditional NoC, Link 3 takes the shortest possible route with the least resource contention. Link 2 takes the shortest possible route. Both these routes share a common router (3,2). On the establishment of circuit switched network by Link 2, the router (3,2) becomes invisible to the other network and hence Link 3 takes a two hop penalty forming a new route. As Link 3 carries high threshold data, a circuit switched path is formed between (1,1) and (3,4).

Figure 12 shows the simulation results of scenario 2. The network delay is decreased by 22% in Link 3, and 37% in Link 2. As per the last scenario, the difference in delay decrease ratio is due to the number of intermediate nodes involved.

- **Scenario 3: Network delay - Traditional NoC vs. drNoC – Fully simulated network**

Figure 11. Scenario 2 - Path formation in drNoC

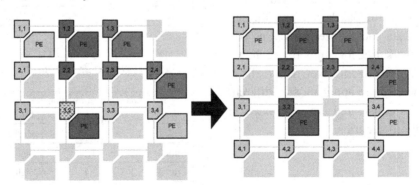

In order to analyse network delay in a complicated fully simulated network, eight master-slave pairs are made to communicate with different type of data thresholds. Table 5 lists the master-slave pairs. This network was simulated for 60 seconds.

The aim of this simulation is to analyse network delay in an environment where the proposed drNoC is not able to perform effectively due to its limitations. Figure 13 shows the paths that were established in the considered master-slave scenario. Due to the locality of master-slaves pairs, it became impossible for drNoC to form circuit switching paths in the high data threshold links. There was one circuit switched path formed in Link 5, however, it was formed after the communication ended in Link 8. The other two high data

threshold links; Link 3 and Link 4, are forced to continue communication as packet switched due to not being able to establish a path that is not used for any other communication. Even after the medium data threshold links ended communication, due to Link 4 having no alternative path to route its data, Link 3 was forced to communicate as packet switched network.

Figure 14 displays the network delay results of the simulation. Unlike the previous scenarios, disappointingly, network delay of Link 1, Link 2, Link 3 and Link 6 increased. This is due to the additional burden on the routers by re-routing of Link 1 packets due to formation of a circuit switched path by Link 5. drNoC did decrease the network delay by 17.7% for Link5, but overall,

Figure 12. Scenario 2 - Network delay results

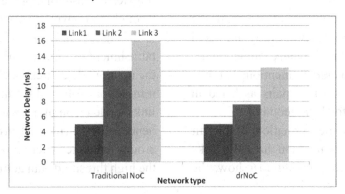

Figure 13. Scenario 3 - Path formation in drNoC

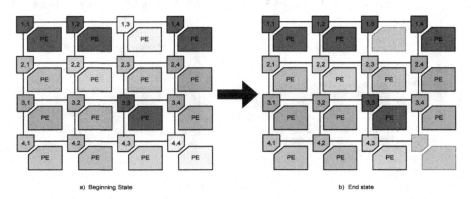

a) Beginning State

b) End state

Table 5. Scenario 3: Resource configuration

	Source	Destination	Traffic Type	No. of Packets
Link 1	1,1	3,3	Medium	50000
Link 2	2,1	4,2	Medium	100000
Link 3	3,1	2,4	High	Continuous
Link 4	4,1	2,3	High	Continuous
Link 5	1,2	1,4	High	Continuous
Link 6	2,2	4,3	Medium	150000
Link 7	3,2	3,4	Medium	200000
Link 8	1,3	4,4	Normal	200

there is a 2.7% increase in network delay. This situation can be avoided by communication centric placement of cores in the network.

- **Scenario 4: Throughput comparison - Traditional NoC vs. drNoC**

The aim of this simulation is to compare the throughput and network delay of a fully simulated drNoC with that of traditionally implemented NoC with fixed parameters as mentioned above. A network with master-slave pairs as listed in Table 6 is simulated for 120 seconds.

Figure 15 shows the formation of circuit switching paths once the medium threshold links have ended communication. Figure 16 shows the

percentage increase in throughput and percentage decrease in average network delay. It can be seen that over a period of 100 seconds, throughput is increased by 32%, and network delay is decreased by 37%.

Data throughput is taken as the amount of data transferred from one processing core to another through the network in a particular time. Increase in throughout in the simulation is due to the establishment of three circuit switched paths once the medium data threshold links have completed transactions. Most of the medium data threshold links ended communication after 60 seconds, hence achieving the peak in data throughput curve around that time. After the paths were established, the high threshold data getting communicated by

Figure 14. Scenario 3 - Network delay results

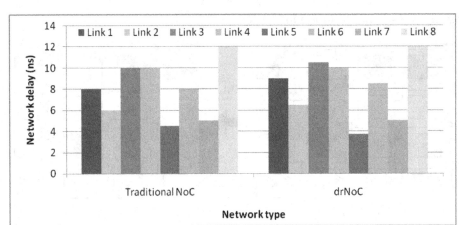

Table 6. Scenario 4: Resource configuration

	Source	Destination	Traffic Type	No. of Packets
Link 1	1,1	1,4	High	Continuous
Link 2	2,1	4,2	Medium	100000
Link 3	3,1	2,4	High	Continuous
Link 4	4,1	4,4	High	Continuous
Link 5	1,2	3,2	Medium	500000
Link 6	2,2	4,3	Medium	200000
Link 7	3,3	3,4	Medium	300000
Link 8	1,3	2,3	Normal	2000

drNoC increased as the traditional was still using packet switching. The same applies in case of network delay.

As mentioned in scenario 3, when changing one part of the drNoC, the load from that part is shifted to the packet switched part of the network. Figure 17 shows the effect on data throughput of the links by load shifted to normal data threshold and medium data threshold links.

It is noticed that the links can accommodate a 39% increase of shifted load without any degradation to data throughput in case of normal data threshold links and 23% in case of medium data threshold links. An interesting thing to note is the increase in data throughput of the links when the network load is shifted. The decrease in data throughput of the links is compensated for by the increase in delivery time for that certain link.

Figure 18 shows the increase in time required to accommodate the decrease in throughput due to network load shifted. In the case of a normal data throughput link, the delivery time increase by 2% in case of 60% increase in network load causing a 30% decrease in data throughput for that link. In the case of medium data threshold links the delivery time is increased by 3.3% to accommodate the 60% increase in network load that caused a 33% decrease in data throughput for the medium data threshold link. Increase in delivery time affects the QoS requirement

Figure 15. Scenario 4- Path formation in drNoC

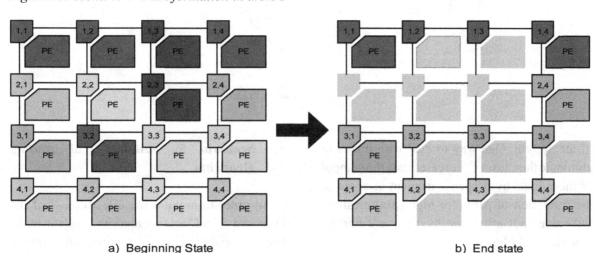

a) Beginning State

b) End state

Figure 16. Scenario 4 – Percentage Increase in data throughput and decrease in average network delay

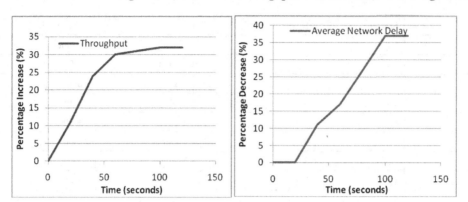

Figure 17. Effect of drNoC path formation on rest of the network

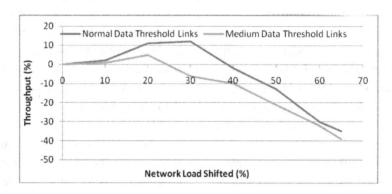

Figure 18. Effect of network load shifting on time

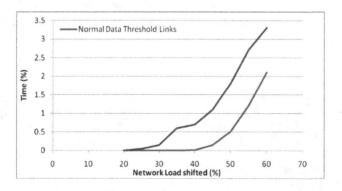

of the system. One way to deal with this is to eliminate the cause of decreased data throughput of these links by proper placement of cores so the formation of paths for high data threshold links does not affect the medium and normal data threshold links.

• **Scenario 5: Suitability of routing algorithm**

In order to monitor the effectiveness of the proposed drNoC routing algorithm versus traditional wormhole and XY routing algorithms, a

Figure 19. Scenario 5- comparison of routing algorithms on drNoC

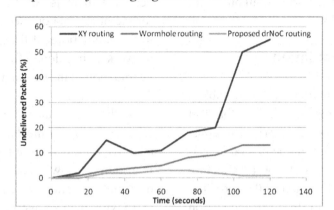

simulation was carried out under mixed traffic pattern and different master-slave pairs communicating at different times. Table 7 lists the resource configuration for this simulation. Unlike the scenarios above, the continuous nodes were made to stop and start after some time in order to check the adaptability of the routing algorithm.

Figure 19 shows the comparison of routing algorithms. It can be seen that by employing XY routing, which is deterministic in nature, the reliability of network, in terms of delivery of packets has decreased. However, wormhole routing increased in delivered packets (decrease of non-delieverd packets in the Figure 19). On the other hand, the proposed drNoC routing algorithm which is based on wormhole routing but with virtual channels, the percentage of delivered packets is higher than that of womhole routing. This is due to the highly adaptive nature and the use of virtual channels as a guard against livelocks and deadlocks. On the change of a link to circuit switching due to a high data threshold link, it appears invisible to the rest of the network. Thus, routing that is deterministic in nature cannot cope with the traffic. Comparing the graph with the resource configuration table, whenever a circuit switched link is formed, the percentage of undelivered packets increases. Leading up to a

Table 7. Scenario 5: Resource configuration

	Source	Destination	Traffic Type	No. of Packets	Start Time (second)
Link 1	1,1	1,4	Medium	30000	0
Link 2	2,1	4,2	Medium	100000	0
Link 3	3,1	2,4	High	Continuous	100
Link 4	4,1	4,4	High	Continuous	30-45
Link 5	1,2	3,2	Medium	50000	0
Link 6	2,2	4,3	Medium	20000	0
Link 7	3,3	3,4	Medium	30000	45
Link 8	1,3	2,3	Normal	2000	60
Link 9	2,2	3,4	Medium	100000	60
Link 10	2,1	2,3	Medium	80000	0
Link 11	3,3	4,2	High	Continuous	100

maximum of 50% when network is left only with high data threshold links.

Explanation of Simulations

The aim of these simulations was to establish the effectvieness of proposed drNoC for data intensive applications. For comparison reasons, a traditional NoC was first implemented in systemC based on the concepts of NoC architecture proposed by (Benini and Micheli, 2002) and (Sun, Kumar and Jantsch, 2002). In these application specific architectures, the architectural parameters, switching, routing and packet size is fixed at design time. This approach is effective for SoC architectures of a particular application domain, but for future SoC architctures with the existence of multi-domain traffic on the communication network, this will result in a waste of resources and valuable bandwidth.

In the proposed drNoC, the switching, routing and packet size in the network changes depending on the traffic getting transmitted, thus, allowing effective utilisation of available bandwidth. As seen in the simulation results, the proposed drNoC outperforms the traditional approach of fixed parameters. With the formation of a circuit switched path, the network load is shifted to the packet switched links. The advantage of this load shifting is the optimal utilisation of link bandwidth, but it also comes at a cost of affecting the performance of packet switched links if the load increases over 23% in the case of medium data threshold links (data throughput requirement between 100Kbits/sec to 10Mbits/sec) and over 39% in normal data threshold links (data throughput requirement under 100Kbits/sec).

Analysing the network delays in a fully simulated network reveals the importance of placement of cores on the network. Due to the formation of circuit switched links, it appears invisible to the remaining routers i.e. these links cannot be used for switching data packets of other communicating cores. Formation of circuit switching paths only occurs if there is an alternative path available for the other data on the network. Thus, as seen in the simulations, if the placement of cores on network is not done effectively, this can cause degradation of the overall performance (scenario 3 in section 5.2).

With the abundance of wiring resources in SoC, the control signal mechanism implemented for adaptive routing does not require a big overhead in resources. However, the formation of different data paths (circuit switching/packet switching) carry with them an overhead in terms of latency, which again, is negligible in high data intensive links. The only significant overhead is in terms of area of the router, especially with support for OCP socket standard. However, the advantages of drNoC exceed its disadvantages.

SUMMARY

In this chapter a dynamically reconfigurable NoC was proposed. In a traditional NoC design, the architectural parameters, like, switching, routing and packet size are fixed. With communication in future heterogeneous SoC architectures, especially with reconfigurable cores, this will prove highly in-efficient due to the network resources not getting utilised effectively, causing wastage of bandwidth. The proposed drNoC endeavours to solve this problem, by dynamically reconfiguring its nodes to match the required bandwidth. On-chip traffic from high data throughput links is shifted to links with low data throughput and a circuit switched path is formed for the high data throughput links. Simulation results have shown the effectiveness of this approach in cases where cores are placed according to their throughput requirement.

The router in drNoC, that also acts as an interface point for integration of cores, is given an OCP socket based interface, allowing cores to be integrated without having the need for wrappers. This makes drNoC a very attractive communication centric platform for future high data throughput applications. The increase in area is compensated by the increase in bandwidth provided by drNoC.

REFERENCES

Ahmad, B., Ahmadinia, A., & Arslan, T. (2005, September 18-21). Dynamically Reconfigurable NoC for Reconfigurable MPSoC. In *Proceedings of IEEE 2005 Custom Integrated Circuits* (pp. 277-280). San Jose, CA.

Ahmad, B., Ahmadinia, A., & Arslan, T. (2008, June 22-25). Dynamically Reconfigurable NoC with Bus Based Interface for Ease of Integration and Reduced Design Time. In *Proceedings of IEEE NASA/ESA Conference on Adaptive Hardware and Systems* (AHS 2008), (pp. 309-314). Noordwijk, Netherlands: European Space Agency

Ahmad, B., & Erdogan, A. T. A., & Khawam, S. (2006, June 15-18). Architecture of a Dynamically Reconfigurable NoC for Adaptive Reconfigurable MPSoC. In *Proceedings of IEEE NASA/ESA Conference on Adaptive Hardware and Systems* (AHS 2006), (pp. 405-411). Istanbul, Turkey.

Atienza, D., Angiolini, F., Muralia, S., Pullinid, A., Benini, l., & Micheli, G. D. (2007). Network-on-Chip design and synthesis outlook. *The VLSI journal, 41*, 340-359.

Benini, L., & Micheli, G. D. (2001). Powering Networks on Chips: Energy-Efficient and Reliable Interconnect Design for SoCs. In *Proceedings of the 14th international symposium on Systems synthesis* (ISSS '01), (pp.33-38)

Benini, L., & Micheli, G. D. (2002). Networks on chip: a new paradigm for systems on chip design. In *Proceedings of Conference on Design, Automation and Test in Europe*.

Bijlsma, B. (2005, September). *Asynchronous Network-on-Chip Architecture Performance Analysis* (Master's thesis). The Netherlands: Department of Electrical Engineering, Faculty of Electrical Engineering, Mathematics and Computer Science, Delft University of Technology.

Bjerregaard, T., & Mahadevan, S. (2006). A survey of research and practices network-on-chip. *ACM Computing Surveys, 38*(1), 1–54. doi:10.1145/1132952.1132953

Dally, W., & Towles, B. (2001, June 18-22). Route packets, not wires: On-chip interconnection networks. In *Proceedings of DAC 2001*. Las Vegas, NV.

Dally, W. J. (1992). Virtual channel flow control. *IEEE Transactions on Parallel and Distributed Systems, 3*(2), 194–205. doi:10.1109/71.127260

Dally, W. J., & Aoki, H. (1993). Deadlock -free adaptive routing in multicomputer networks using virtual channels. In *Proceedings of IEEE Trans. on Parallel and Distributed Systems*.

Duato, J. (1994). A necessary and sufficient condition for deadlock free adaptive routing in wormhole networks. In *Proceedings of the International Conference on Parallel Processing*.

Guerrier, P., & Greiner, A. (2000). A generic architecture for on-chip packet switched interconnections. In *Proceedings of the Design, Automation and Test in Europe Conference*, (pp. 250–256).

Hemani, A., Jantsch, A., Kumar, S., Postula, A., Öberg, J., Millberg, M., & Lindqvist, D. (2000). Network on a Chip: An architecture for billion transistor era. In *Proceeding of the IEEE NorChip Conference*.

Hu, J., & Marculescu, R. (2004, June). DyAD - Smart Routing for Networks-on-Chip. In *Proceedings of DAC 2004*. San Diego, CA.

Kim, D., Kim, M., & Sobelman, G. E. (2005, October 20-21). Design of a High-Performance Scalable CDMA Router for On-Chip Switched Networks. In *Proceedings of the International SoC Design Conference*, Korea.

Phi-Hung, P., Kumar, Y., & Chulwoo, K. (2006). High Performance and Area-Efficient Circuit-Switched Network on Chip Design. In *Proceedings of The sixth IEEE International Conference on Computer and Information Technology, 2006* (pp. 243 -243)

Shashi, K., et al. (2002). A Network on Chip Architecture and Design Methodology. In *Proceedings of the IEEE Computer Society Annual Symposium on VLSI*. Pittsburgh, PA.

Shin, K. G., & Daniel, S. W. (1996). Analysis and implementation of hybrid switching. In *Proceedings of the IEEE Tran. on Computers.*

Su, C. C., & Shin, K. G. (1996). Adaptive Fault tolerant dead lock free routing in meshes and hypercubes. *IEEE Transactions on Computers, 45*(6), 666–683. doi:10.1109/12.506423

Sun, Y., Kumar, S., & Jantsch, A. (2002). Simulation and Evaluation for a Network on Chip Architecture Using Ns-2. In Proceeding of NorChip. Copenhagen, Denmark.

Wu, J. (2002). A deterministic fault-tolerant and deadlock-free routing protocol in 2-D meshes based on odd-even turn model. In *Proceedings of the 16th international conference on Supercomputing.*

Zhong, M. (2005, June). Evaluation of deflection-routed on-chip networks (Master's thesis), In *Proceedings of KTH*. Stockholm.

Chapter 11
Reliability Aware Performance and Power Optimization in DVFS-Based On-Chip Networks

Aditya Yanamandra
The Pennsylvania State University, USA

Soumya Eachempati
The Pennsylvania State University, USA

Vijaykrishnan Narayanan
The Pennsylvania State University, USA

Mary Jane Irwin
The Pennsylvania State University, USA

ABSTRACT

Recently, chip multi-processors (CMP) have emerged to fully utilize the increased transistor count within stringent power budgets. Transistor scaling has lead to more error-prone and defective components. Static and run-time induced variations in the circuit lead to reduced yield and reliability. Providing reliability at low overheads specifically in terms of power is a challenging task that requires innovative solutions for building future integrated chips. Static variations have been studied previously. In this proposal, we study the impact of run-time variations on reliability. On-chip interconnection network that forms the communication fabric in the CMP has a crucial role in determining the performance, power consumption and reliability of the system. We manage protecting the data in a network on chip from transient errors induced by voltage fluctuations. Variations in operating conditions result in a significant variation in the reliability of the system, motivating the need to provide tunable levels of data protection. For example, the use of Dynamic Voltage and Frequency Scaling (DVFS) technique used in most CMPs today results in voltage variation across the chip, giving rise to variable error rates across the chip. We investigated the design of a dynamically reconfigurable error protection scheme in a NoC to achieve a desired level of reliability. We protect data at the desired reliability while minimizing the power and performance overhead incurred. We obtain a maximum of 55% savings in the power expended for error protection in the network with our proposed reconfigurable ECC while maintaining constant reliability. Further,

DOI: 10.4018/978-1-61520-807-4.ch011

35% reduction in the average message latency in the network is observed, making a case for providing tunability in error protection in the on-chip network fabric.

INTRODUCTION

Advancements in semiconductor technology have lead to diminutive feature sizes for a transistor. This has lead to a dramatic increase in the overall number of transistors available on a modern chip. To take advantage of the ever increasing transistor budget, there has been a paradigm shift towards having multiple processors on a chip [Nayfeh, 1999]. Chip multiprocessors (CMPs) have found a niche in embedded markets, the mainstream laptop and desktop computers as well as high-end servers. Network on chip (NoC) has been suggested as a solution to the exacerbating global wire delay problem in newer technology generations and for a scalable number of cores. Currently, commercially designed NoC topologies include ring (Intel's Larrabee), mesh (Tilera) and clustered networks. The on-chip interconnect has become an important focus of research as the CMPs and system-on-chip scale to hundreds of cores. The network fabric on chip plays a vital role in the performance, and power consumption in such a system. The significance of the underlying communication architecture is well understood in designing multiprocessors over the years. In the case of the CMPs, however, due to the small transistor size and high density of transistors, power consumption is becoming a first order design metric as opposed to the pin-bandwidth which is the major limiting factor in off-chip networks. It is predicted that NoC power can be a significant part of the chip power and can account for up to 40 to 60 watts [Borkar, 2007] with technology scaling for a mesh based network with 128 nodes. A few commercial designs also support this trend, where up to 28% of the entire chip power is devoted to the interconnect [Hoskote, Vangal, Singh, Borkar, 2007]. Thus, on-chip interconnects that can opti-

mize both performance and power pose intriguing research challenges. This is evident from the large body of literature covering multiple facets of NoC design [Kim, Davis, Oskin, & Austin, 2008; Kim, Dally, Scott & Abts, 2008 ;Muralimanohar & Balasubramonian, 2007].

Reliability has been identified as one of the key limiters to future transistor scaling. Process variation, wear-out and transient errors are the major contributors to faults in chips. Process variations are posing a big challenge for the semi-conductor industry in terms of the yield. In addition, dynamic variations due to non-uniformity of workload activity across the chip can also cause erroneous scenarios. The on-chip network which is the basic medium of communication among the components is a distributed resource and thus will play a major role in determining the overall system reliability. The impact of process and temperature variations on the on-chip interconnect were studied in [Bin, Peh & Patra, 2008; Nicopoulos, Yanamandra, Srinivasan, Narayanan & Irwin, 2007].

Recent work done by Kim et al. [Kim, W., Gupta, M. S., Wei, G.-Y. & Brooks, D., 2008] demonstrates the use of on-chip voltage regulators for a per-core Dynamic Voltage Frequency Scaling (DVFS) in a CMP. With technologies such as this, DVFS can become an integral part of the on-chip network as well. DVFS can also be applied to routers for saving power as well as for controlling congestion i.e. increasing the throughput of the system. Under congestion, techniques such as throttling the upstream routers and/or increasing the frequency of the congested router can help controlling the congestion. For using such a variable frequency system, micro-architectural changes such as using a dual clock I/O buffer for the input buffers would be required.

In this chapter, we handle transient errors induced due to voltage fluctuations for an on-chip interconnection network. In addition, power and performance optimizations such as per-router DVFS that are reliability-agnostic will result in significant variation in the error rate. We also take into account the voltage variations due to employing such a per-router DVFS. The transmitted data in the NoC can be protected using error correcting codes (ECC) such as Single Error Correcting (SEC), double error correcting (DEC) codes and Triple error correction (TEC). The ECC schemes differ in their protection capabilities, power consumption and the speeds at which they operate. Better coverage schemes have higher power consumption and have a higher latency penalty and thus these must be used wisely. Prior work does similar analysis in the context of buses which is a centralized communication media [Li, Narayanan, Kandemir & Irwin, 2003]. On-chip interconnects pose different challenges as it is a distributed system. This work identifies the variation of error rate on different regions of a chip due to non-uniformity in voltage profile that result from using tweaks such as DVFS. It provides a comprehensive evaluation of the overheads involved in supporting a dynamically reconfigurable hybrid ECC in the presence of per-router DVFS.

In the following sections, we motivate the necessity to accommodate differing levels of error correction capabilities in the NoC to tackle the voltage variations. In order to facilitate tunable levels of reliability, we need to accommodate for multiple types of encoder/decoder combinations that increases the area overhead of ECC from 1.8% to 3.2% of the total network area. We monitor the instantaneous vulnerability to transient errors at the granularity of an individual router and choose the ECC scheme capable of meeting the reliability target. With a small area penalty, we obtain up to 55% of savings in the power expended for error protection for synthetic traffic and up to 8% savings for applications. Further, we also decrease the

latency penalty of reliability mechanisms. These latency improvements along with those obtained due to per-router DVFS, result in up to 35% in average flit latency.

NETWORK ARCHITECTURE

Base Router Architecture

In this section, we outline the details of the underlying network architecture. Every network node is a processing core along with its private L1 and shared L2 bank. Every node has a router which handles the communication for it and the routers are connected through inter-router links. A simple mesh is used as the network topology. Thus, the router has five input/output ports; four of them are used to communicate with its neighbors in the four cardinal directions and one port for the processing element (PE). One or more nodes in the network have an extra injection/ejection port attached to the memory controller.

A flit is the smallest fraction of data that can be transmitted through the network. Every packet consists of a header flit and/or multiple body and tail flits. The base router is a pipelined structure with 5 stages namely Routing Computation (RC), Virtual channel allocation (va), Switch Allocation (SA), Switch Traversal (ST) and Link Traversal (LT) as shown in Figure 1. Only the header flits go through the RC and VA stages. SA, ST and LT stages are common to all flit types. The main router components are the RC unit, input buffer, arbiters for VA and SA stage and the crossbar switch as shown in the

Figure 1. Router pipeline

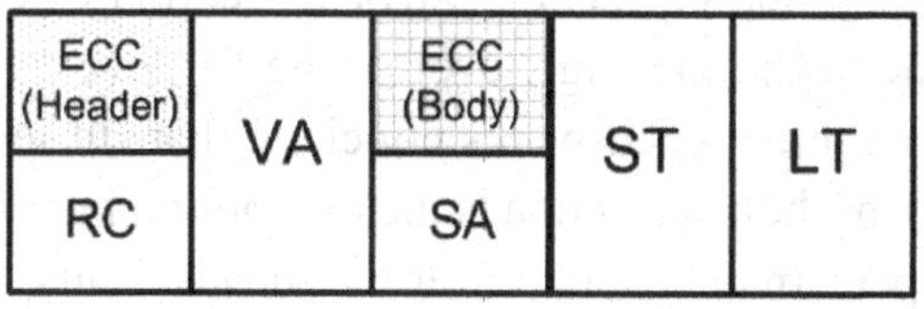

ECC (Header)	VA	ECC (Body)	ST	LT
RC		SA		

Figure 2. Router microarchitecture

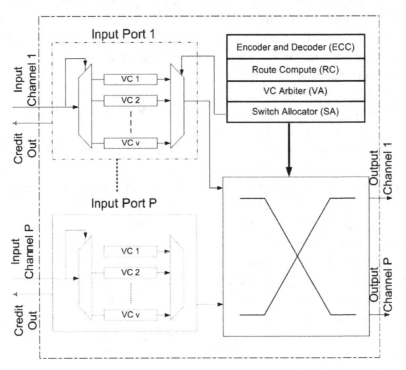

Figure 2. The RC unit computes the output port based on the packet destination. The input buffer is composed of multiple virtual channels. A virtual channel is allocated to a packet at a time. The VA unit arbitrates for a virtual channel in the input buffer of the next router. The winners of VA compete for the switch in the SA stage. Finally, the winners of the SA stage traverse the crossbar switch and the corresponding link to reach the next router.

Support for Error Correction

Protection from data errors in a NoC has been studied previously in literature. Srinivasan [2005] and Park [2006] investigate the various options and compare end-to-end correction or hop-by-hop correction. Error correction can be done at the message level or the flit level. The above mentioned works conclude that flit level hop-by-hop correction is the best option in terms of performance and power overheads. Thus, we use hop-by-hop error correction as the basis for our error correction scheme.

Consequently, for providing error correction every router is augmented with an encoder and a decoder. Every flit is encoded at its source and decoded at every hop to make sure the data remains correct. As part of the inter-router link traversal, the flit is written into the next router's buffer. The decoding of the flit is done in parallel with the RC stage (if it is a header flit) and checked for errors and corrected if any. If it is not a header flit then the decoding is done in parallel with the SA stage. If the router is running at its base frequency then our synthesis analysis shows that the decoding delay is less than a cycle. More details are discussed in section 4.1. The extra state information calculated during RC and VA stages etc., for the header flit are protected by triple modular redundancy as this is a very small amount of data that is store locally in the router. The architecture for the reconfigurable ECC is discussed in Section 5.

IMPLEMENTATION OF DVFS ON NOCS

We build a CMP framework in which every router performs DVFS independently in order to gain network performance. We increase the frequency of congested routers which reduces the amount of time spent by packets in the network.

In order to achieve per-router DVFS, each router should be supplied with different voltages and frequencies from which one can be selected. Instead of distributing multiple supply voltages across the network, we can use an on-chip voltage regulator for every router. We use the 2-step voltage regulator configuration as proposed by Kim's work. An off-chip regulator performs the initial step down from 3.7V to 1.8V followed by multiple on-chip voltage regulators where each of them steps it down from 1.8V to 1.2V. This approach amortizes the degradation in conversion efficiency of employing only off chip regulators. A multi-phase buck converter that can provide three voltages and a programmable voltage controlled ring oscillator provides the three frequency levels [Mosley, Hausman, Gaudenzi & Tempest, 1990]. The on-chip regulators operate at *125MHz* switching frequency and provide voltage transitions from *1V* to *0.75V*. Though we can support more levels of voltage and frequency, we use three levels one for each ECC scheme as it suffices to demonstrate our goal and the benefits of reconfigurable ECC schemes. The voltage settling time for every *100mV* change is *13ns*. The power consumption (at activity factor of *0.5*) of our regulator with conversion efficiencies similar to that in (Kim, Gupta, Wei & Brooks, 2008) is given in Table 1.

Table 1. On chip regulator power consumption

Output Voltage (V)	Power (mW)
0.75	34.1
0.90	41.5
1.00	52.3

The area overhead in a 6x6 mesh for *36* on-chip regulators is *2.25mm²* which is around 25% of the area of all routers in the network. This is based on our router area of *0.245mm²* and power of *0.2W* at 65nm technology and *1.3GHz* based on our synthesized design.

In our DVFS scheme, we use buffer utilization as the indicator for changing the voltage. The thresholds for transition were chosen to accommodate for the voltage settling time. We perform our DVFS at a coarse window of time as we expect that the buffer utilization takes some time to stabilize and to prevent unnecessary overhead due to transitions. A time window of *500* cycles is chosen for this work.

RELIABILITY CONSIDERATIONS UNDER DYNAMIC VARIATIONS OF VOLTAGE

In order to evaluate our reconfigurable scheme, we consider use of three error correction schemes of different strengths based on the BCH (acronym comprises the initials of the authors Bose, Ray and Chaudhuri) codes [Irving, 1960].

- **Single error correction (SEC)** is the most popular error coding scheme for single errors. For a code length of 127 bits, (127, 120) Hamming code is used.
- **Double Error Correction (DEC)** detects and corrects up to two errors. If there are more than two errors then it can erroneously indicate two or less errors. We use (7, 2) BCH code (n = 127 and data = 113) for this.
- **Triple Error Correction (TEC)** detects and corrects up to three errors. BCH code of (7, 3) is used which produces a codelength of 127 bits for 106 data bits. It cannot detect more than three errors.

In the following subsections, we establish the motivation for our reconfigurable ECC schemes. We present three main reasons in support of this additional flexibility. The first reason is that the error rates increase exponentially with decrease in voltage. Consequently, at higher voltages and higher speeds, lower protection schemes can be adopted. The second reason is that the amount of power expended to protect the network at all times with the highest protection capability scheme is significant and thus wasteful if we can do better than that. Finally, in the presence of DVFS when the routers are being operated at a higher frequency, the more powerful ECC scheme will require multiple cycles to execute and consequently may lead to performance degradation.

Error Rate Variation

Though DVFS is attractive for optimizing performance per watt, it results in a non-uniform profile of voltage and frequency on the chip. These variations in voltage lead to a change in the reliability of the router.

To understand this further, we model the errors due to noise in voltage that was presented in [Hedge & Shanbhag, 2000]. In addition to the noise in supply voltage, we also vary the supply voltage. Let ε be the probability that an error occurs either when a transfer occurs across a wire or a data bit is stored in the buffer. Let V_{sw} be the supply voltage. Let the noise voltage (V_N) have a variance of σ_N^2. Then we get,

$$\varepsilon = Q\left(\frac{V_{sw}}{2\sigma_N}\right) \tag{1}$$

where $Q(x)$ is the Gaussian noise pulse

$$Q(x) = \int_x^\infty \frac{1}{\sqrt{2\pi}} \, e^{-\frac{y^2}{2}} \, dy$$

From Equation 1, we conclude that decreasing the voltage of a router in the network will increase the probability of an error in the data passing through the router. It is to be noted that not all bit errors result in program errors. Every application has an inherent tolerance to errors. Thus, it is sufficient to keep the error rate below the tolerance limit of the application. Figure 3a and 3b show that as the supply voltage is increased, in order to maintain a constant residual error rate, a less powerful ECC scheme suffices. We find that for a supply noise of *0.2* or more the residual word error rate is extremely high (of the order of *0.9*) and no protection scheme will help. Thus, circuit designers will need to do design tricks to keep the noise at tangible levels. We assume a noise margin of *0.15* for this work and we believe that

Figure 3. Residual word error rate at varying supply voltage and error protection

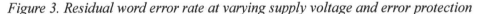

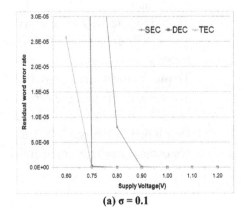

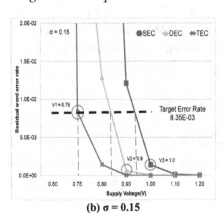

(a) σ = 0.1   (b) σ = 0.15

it is in the ballpark for state-of-art circuits. The tolerance limit is a characteristic of the application and varies for different applications. We statically chose the target residual error rate to be at *0.0085* for this work. This can be a user input depending on how much reliability is required. Tightening this tolerance limit reduces the opportunities for optimizations and relaxing this limit will give rise to more scope for improvements.

Figure 3b shows that between voltages V_1 and V_2, TEC is essential to keep the error rate below the target residual error rate. For voltages between V_2 and V_3, TEC is not required as it overprotects and DEC is sufficient to meet the target residual error rate. Beyond V_3 voltage, SEC can keep the error rate at the desired level. Thus, we chose voltages (V_1, V_2 and V_3) for our current experiments.

Error Correction Power

Table 2 shows the latency, power and area of each type of encoder and decoder. These were obtained from RTL synthesis of BCH encoder and decoders using Synopsys design compiler at 90nm technology and scaled down to 65nm. The corresponding router frequencies at these voltages chosen were *1.3GHz, 1.56GHz,* and *1.73GHz* respectively. The power numbers shown in Table 1 are scaled to these frequencies and voltages chosen in the previous section. The last row shows the power of each component normalized to the SEC encoder. From the table, it is clear that the most powerful TEC scheme is also the most expensive.

In order to maintain a target residual error rate in the network with no adaptability to dynamic variations, we would need to use the most powerful protection scheme all the time. The total power expended to provide TEC protection at all times in every router in the network can be as high as 15% of the total network power and is larger for bigger network sizes. Figure 4a shows the percentage of power expended by ECC to the total network power for various injection rates. This motivates us to look into the ECC power as a potential opportunity to save power. In addition in the presence of DVFS, if a router's frequency is boosted up then the TEC scheme takes multiple cycles to execute and can eat into the performance benefits obtained due to the higher frequency.

For the highest frequency in the system (1.73GHz), from Table 2 we can see that both DEC and TEC will require multiple cycles for decoding. Consequently, providing TEC protection at all times is not the best option.

In summary, we identify that the reliability at a given instant of time is varying across the NoC due to non-uniform voltages. We propose to have a more power and performance optimal approach by having differing levels of correction capabilities and switching between them dynamically along with the voltage domains.

EXPERIMENTAL SETUP

We used an in-house cycle-accurate on-chip network simulator that models the 5-cycle ECC router

Table 2. ECC scheme power, delay and area

	SEC @ 1V		DEC @ 0.9V		TEC @ 0.75V	
	Encoder	Decoder	Encoder	Decoder	Encoder	Decoder
Power(mW)	0.25	2.76	1.517	3.65	2.036	5.83
Delay(ns)	0.25	0.27	0.25	0.6	0.26	0.65
Area(μm²)	164.1	1158	271.7	1813.6	343.1	4204.9
Normalized Power	1	10.84	5.96	14.36	8	22.93

Figure 4. (a)Percentage of power expended for ECC to the total network power and (b) 16-way CMP layout where MC is the memory controller

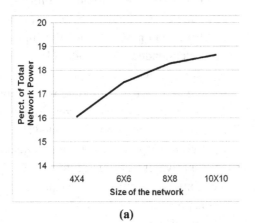

(a)

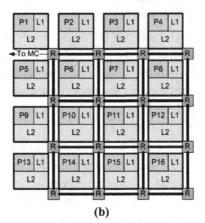

(b)

shown in Figure 1. The network used wormhole switching, virtual channel flow control and deterministic routing algorithm. We experiment with several synthetic traffic patterns namely uniform random (UR), transpose (TP), bit complement (BC) and tornado (TN) [Dally & Towles, 2004]. The number of nodes was also varied for node configurations: 4X4 (16), 6X6 (36), 8X8 (64) and 10X10 (100).

Application results of benchmarks chosen from SPLASH2 and SPECOMP suites are also presented. Traces for network traffic, with Simics [Magnusson, Christensson, Eskilson, Forsgren, Hallberg, Hogberg, Larsson, Moestedt, Werner 2002] as front-end, were collected for a 16-node CMP as shown in the Figure 4(b). Simics was made aware of the NoC layout with a 5 cycle router hop. Traces obtained were fed into our NoC simulator to observe the characteristics of applications.

We assume the number of VCs as two per PC with a buffer depth of four. The injected packet size is *512* data bits. The flit size (the channel width) is *129* bits with *127* code bits (*106* data and *21* maximum ECC bits) and *2* bits to indicate the type of ECC. The number of ECC bits used is *21* bits for TEC, *14* bits for DEC and *7* bits for SEC. Even though, more data can be sent with DEC and SEC, we cannot dynamically repartition the data in the flits. Wormhole switching prevents us from

waiting to form the entire packet in a router. Thus, we pad the rest of the data bits in DEC and SEC. At the destination node, when we eject the flit to the PE, depending on the ECC type indicated by the *2* bits in each flit, we drop the extra pad and restore the original data.

Reconfigurable ECC Support

In the case of the reconfigurable scheme, when a new flit arriving and the current router have different levels of ECC, then in addition to decoding and making sure the data received is correct, we also need to encode this data with the current router ECC scheme. Thus, the critical path for ECC now includes decoder delay of the arrived flit and the encoder delay of the new scheme of the current router. As the encoder delay is very small, this does not incur additional cycle penalty. If the ECC schemes of current flit and arriving flit are the same then no encoding is required. Thus, we incur an extra encoder power during transitions. This extra cost during transitions is small compared to the benefits obtained by saving decoder power. The total area overhead of supporting the three schemes increases from 1.85% to 3.24% of the network area.

DVFS was performed at three voltage and frequency levels of *1.3GHz* (*0.75V*), 1.56GHz

(*0.90V*) and 1.73GHz (*1.0V*). In the reconfigurable ECC scenario, TEC maps to *0.75V* domain, DEC to *0.90V* domain and SEC to *1V* domain. In the non-reconfigurable case however, the TEC needs to operate at frequencies of *1.56GHz*, and *1.73GHz* for higher voltage domains and thus will always require two cycles to finish its operation.

EXPERIMENTAL RESULTS

As explained earlier, the reconfigurable reliability fabric provides us with opportunities to obtain both performance and power benefits in the reliability component. The performance benefits are obtained because the operation of TEC and DEC requires multiple cycles at higher frequencies. In this section, we capture the performance benefits of switching to a lower level of protection which is faster while still satisfying the reliability specifications.

Performance Benefits

For a non-reconfigurable fabric, the highest level of ECC protection (TEC) should be enabled at all voltage domains. However, when the router is working at the high voltage (and thus high frequency) domain, the EC stage will require two cycles. Thus, whenever the router is operating at these voltages due to the DVFS technique, the router pipeline increases from 5 stages to 6 stages. We model this extra bubble in our simulator at the higher voltage domains. This effects the overall performance improvement attained by the DVFS technique. Figure 5 shows the drop in performance for UR and TP traffic for a various network sizes at an injection rate of *0.20* flits/node/cycle.

From the figure, we also see a consistent drop in performance improvement for the non-reconfigurable scenario for both UR and TP traffic. Thus, having reconfigurability is critical to realize the full potential of performance benefits that can be achieved by DVFS technique in CMPs. The performance improvement decreases with increasing network size. The reconfigurable fabric provides about 91% improvement over the non-reconfigurable fabric for UR traffic at *16* nodes and about 44% for TP pattern.

Power Benefits

The power saved by switching to lower level of ECC varies with several factors including threshold levels, network size, load rate and, traffic pattern. We explore each of these in the following

Figure 5. Average latency (per flit) improvement with DVFS

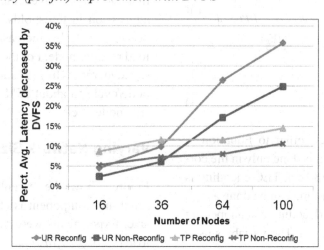

sections. The base case in all these experiments is the network using TEC at all voltage levels. The power consumed by the TEC scheme at a given voltage is given by Equation 2. Thus, the total power consumed in the base case is the power consumed at each of the voltages as given by Equation 3.

$$Power_{base-case}(V) = Power_{TEC-enc}(V) * N_{enc}(V)$$
$$+Power_{TEC-dec}(V) * N_{dec}(V)$$

(2)

where $Power_{base-case}(V)$ is the power in the base case at a voltage of 'V', $Power_{TEC-encoder}(V)$ and $Power_{TEC-decoder}(V)$ are the encoder and decoder power of TEC at voltage 'V', $N_{enc}(V)$ and $N_{dec}(V)$ are the number of encodes and decodes that take place at voltage 'V'.

$$Power_{base-case} = Power_{base-case}(V_1)$$
$$+Power_{base-case}(V_2) + Power_{base-case}(V_3)$$

(3)

where V_1, V_2 and V_3 are the three voltage domains.

$$Power_{reconfig} =$$
$$Power_{TEC-enc}(V_1) * N_{TEC-enc}(V_1)$$
$$+\left(\sum_{V=V1,V2,V3} Power_{TEC-dec}(V) * N_{TEC-dec}(V) \right)$$
$$+Power_{DEC-enc}(V_2) * N_{DEC-enc}(V_2)$$
$$+\left(\sum_{V=V1,V2,V3} Power_{DEC-dec}(V) * N_{DEC-dec}(V) \right)$$
$$+Power_{SEC-enc}(V_3) * N_{SEC-enc}(V_3)$$
$$+\left(\sum_{V=V1,V2,V3} Power_{SEC-dec}(V) * N_{SEC-dec}(V) \right)$$

(4)

where the variables are similar to Equation 2. Note that while encoding is done only in the corresponding voltage levels (e.g. TEC encoding is always done in V_1), decoding can be done at any voltage level (e.g. TEC decoding can happening in a router with voltage V_3). The variables used in these equations are collected from simulation to estimate the power consumed in each case.

Experiments with Threshold Levels

In our framework, the routers transition into higher voltage and frequency domains based on certain threshold buffer utilization levels. These pre-determined levels are critical for determining the network performance improvement. In this section, we estimate the power reduction caused in the reliability component due to the presence of reconfigurable EC component. Figure 6 shows the power reduction for three different threshold domains.

- **T1** Thresh-050-075: Switch to 0.9V at buffer utilization of 50% and 1V at buffer utilization of 75%
- **T2** Thresh-025-050: Switch to 0.9V at buffer utilization of 25% and 1V at buffer utilization of 50%
- **T3** Thresh-025-030: Switch to 0.9V at buffer utilization of 25% and 1V at buffer utilization of 30%

These threshold levels are ordered in the increasing order of aggressive scaling. Thus, T3 is more aggressive scaling compared to T2, which is turn more aggressive than T1. Figure 6 show that the power saved increases with increased aggression in thresholds. For each domain, the power saved also increases with an increase in load rate per node. For the T3 domain, the power saved increases from *2%* at a load rate of *0.10* flits/node/cycle to about 55% at a load rate of *0.35* flits/node/cycle.

Varying Network Size

Figure 7 shows the percentage power saved in the reliability component as a function of the network size. Experiments were performed at a constant load rate of *0.20 flits/node/cycle*. As shown earlier,

Figure 6. Savings in EC power for various threshold levels

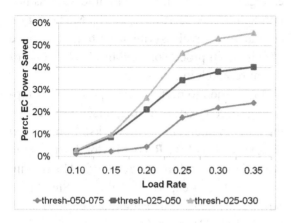

Figure 7. Impact of network size on savings in EC power

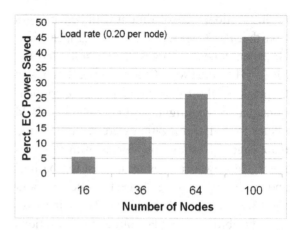

the power consumed by the reliability component increases with node size. Figure 7 shows that the best of the savings are also obtained with an increase in network size. Thus, we believe reconfigurability in the reliability component is critical for reliable CMP fabrics of the future.

Varying Traffic Patterns

Traffic patterns determine the congestion in the network and the buffer utilization across the NoC. Thus, the opportunity to perform DVFS varies with the injected traffic pattern. Figure 8 shows the percentage of power saved in the reliability component as a function of the injected traffic pattern. Transpose (TP) has the highest savings of 28% and Tornado (TN) has the least savings of 11%. A network size of *64* nodes with an injection rate of *0.2flits/node/cycle* was chosen for this experiment.

Varying Injection Rate

Injection rate is one of the key determining factors for buffer utilization in the NoC routers. Higher injection rates lead to larger buffer occupancy and thus will impact our DVFS technique. An increase in injection rate will cause the routers to transition to higher voltage levels and thus lead to improved

performance of the congested network. This gives us an opportunity to decrease the ECC scheme to be used and with it the power consumed by the reliability component.

Figure 9 shows the effect load rate on the power savings. The experiment was performed for a network size of *64* nodes with UR traffic. The power savings steadily increase with the injection rate. At an injection rate of *0.35flits/node/cycle*, the percentage savings of the reliability component are as high as 56%.

Figure 8. Impact of traffic patterns on power savings

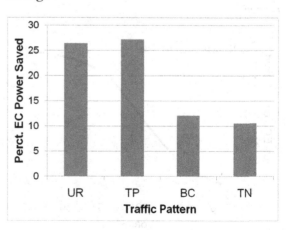

APPLICATION RESULTS

In this section, we experiment with benchmarks from the SPECOMP [Aslot, Domeika, Eigenmann, Gaertner, Jones & Parady, 2001] suite and the SPLASH2 suite [Woo, Ohara, Torrie, Singh, and Gupta (1995)] for power saved in those applications due to reconfigurable ECC. The threshold values for buffer utilization were set at *0.25* and *0.30*. Figure 10 shows the results of this experiment. The power savings are highest for `barnes' and `radix' from the SPLASH2 suite of about 8%. We anticipate these benefits to increase with the ability to simulate larger configurations of processor nodes (as observed from figure 7) and emerging applications with larger memory footprints and higher injection ratio.

RELATED WORK

Buses, a traditional solution to multi processor communication, are infeasible in the presence of large number of cores as they don't scale well. The main advantages of a bus are the ease of design, low power consumption and snoop base cache coherence. Kumar et al. [2005] presented a comprehensive analysis of interconnection mecha-

nisms for small scale CMPs. Network-on-Chip has emerged as the de-facto solution to address this problem [Dally & Towles, 2001]. Managing on chip data effectively is critical for the performance of chip multi processors and has lead to a lot of recent research in this area.

In addition to performance, power consumption has become one of the major factors in designing on-chip interconnect networks.. Commercial designs such as MIT RAW [Taylor, Kim, Miller, Wentzlaff, Ghodrat, Greenwald, Hoffmann, Johnson, Lee, Lee, Ma, Saraf, Ski, Shnidman, Strumpen, Frank, Amarasinghe & Agarwal, 2002] and ALPHA 21364 network [Mukherjee, Bannon, Lang, Spink & Webb, 2002)] show trends of 36% power consumption. Thus, reducing power consumption of NoC is one of the major challenges facing network design. Power-aware designs of the router micro-architecture such as segmented crossbar, write-through buffer have been investigated [Wang, Peh & Malik, 2003]. Also, throttling a flit traversal in a router [Shang, Peh & Jha, 2003] has been shown as a means to manage peak power constraints in off-chip networks. Prior works have proposed DVFS for links [Shang, Peh & Jha, 2003; Kim, Link, Yum, Narayanan, Kandemir & Irwin, 2003] to manage power in off-chip networks. Dynamic voltage and frequency scaling (DVFS) techniques are being

Figure 9. Impact of load rate on savings in EC power

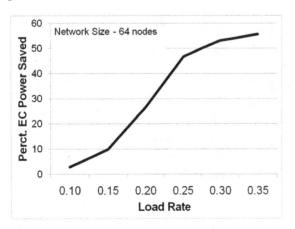

Figure 10. Power saved on multi-core benchmarks

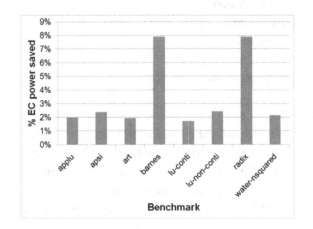

used in state-of-art chip-multi processors (CMPs) for saving power. Several works have proposed DVFS in processors [Brooks & Martonosi, 2001; Xie, Martonosi & Malik, 2003; Semeraro, Magklis, Balasubramonian, Dwarkadas & Albonesi, 2002] for power management. With increasing number of cores on the chip, the network connecting the components on-chip is expected to consume as much as 40W to 60W [Borkar, 2007].

Managing the reliability of the NoC was studied by [Murali, Theocharides, Narayanan, Irwin, Benini & Micheli, 2005]. They consider various methods by which on-chip reliability can be achieved and evaluate the tradeoffs. The tradeoff points include error correction/error detection and retransmission, end-to-end/hop-by-hop, message level/flit level protection. They conclude that hop-by-hop flit level error correction was best suited for reliability of the NoC. The reliability of data passing through the buses was investigated in [Li, Narayanan, Kandemir & Irwin, 2003]. Our work uses the same model of reliability based on supply voltage and variation in the voltage. They evaluate the variation in supply voltage (σ_V) only. Our study maintains a constant supply voltage variation and studies the effect of varying supply voltage on residual word error rate. Thus, our work differs in both the medium of communication and the cause for the data corruption.

This work investigates the power benefits that can be obtained from the reliability component under an iso-reliability scenario where the residual error rate changes due to changing voltage. The voltage in our work is changed to improve the performance of the system. In [Benini & Micheli, 2006], the authors investigate achieving iso-reliability by changing the voltage domain to appropriate levels whenever the residual error rate changes. Thus, they propose changing voltages as a solution to changing error rate. In our work however, voltage change is the cause for changing error rates. Our solution is to provide a reconfigurable fabric of ECC to provide the required reliability at minimum energy consumption.

CONCLUSION

In this work, we propose a dynamically reconfigurable reliability module which changes its coverage based on the operating conditions. This work models multiple errors and various multi-error correction codes. We provide an adaptive DVFS framework for a Network-on-chip that optimizes performance and power under a given reliability budget with a reduced power overhead. The savings in power expended in reliability can be as high as 55% for synthetic traffic and upto 8% for applications. In addition the adaptive scheme also reduces latency overhead due to reliability. We incur an additional area overhead penalty of 3.2% (increases from 1.8%) to accommodate for multiple encoders and decoders. We conclude that adaptive error correction schemes can be used to achieve balance in reliability, power and performance tradeoffs.

ACKNOWLEDGMENT

This research is supported in part by NSF CAREER grants 0702617, 0811687, 0720645, 0720749, 0702519, 0444345, 0454123 and a grant from GSRC.

REFERENCES

Aslot, V., Domeika, M. J., Eigenmann, R., Gaertner, G., Jones, W. B., & Parady, B. (2001). SPEComp: A New Benchmark Suite for Measuring Parallel Computer Performance. In *Proceedings of the International Workshop on OpenMP Applications and Tools* (WOMPAT '01), (pp. 1-10). London: Springer-Verlag.

Benini, L., & Micheli, G. D. (2006). *Networks on Chips: Techonology and Tools*. San Francisco: Morgan Kaufmann.

Bertozzi, D., Benini, L., & de Micheli, G. (2002). Low Power Error Resilient Encoding for On-Chip Data Buses, In *Proceedings of the conference on Design, automation and test in Europe* (DATE '02)

Borkar, S. (2007). Networks for multi-core chips: A contrarian view. In *Proceedings of the Special Session at ISLPED 2007.*

Brooks, D., & Martonosi, M. (2001). Dynamic thermal management for high performance microprocessors. In *Proceedings of the 'Int. Symp. High Performance Computer Architecture.*

Dally, W. J., & Towles, B. (2001). Route packets, not wires: On-chip interconnection networks. In *Proceedings of the 38th conference on Design automation* (DAC '01), (pp. 684-689). New York: ACM

Dally, W. J., & Towles, B. (2004). *Principles and Practices of Interconnection Networks.* San Francisco: Morgan Kaufmann.

Hausman, G. Gudenzi, J. M., & Tempest, S. (1990). *Programmable voltage controlled ring oscillator.* US Patent 4978927.

Hegde, R., & Shanbhag, N. R. (2000). Toward achieving energy efficiency in presence of deep submicron noise. *IEEE Transactions on Very Large Scale Integration Systems, 8*(4), 379–391. doi:10.1109/92.863617

Hoskote, Y., Vangal, S., Singh, A., Borkar, N., & Borkar, S. (2007, September- October). *A 5-ghz mesh interconnect for a teraflops processor* (volume 27, pp. 51–61).

Kim, E.-J., Link, G., Yum, K. H., Narayanan, V., Kandemir, M., Irwin, M. J., & Das, C. (2005). A Holistic Approach to Designing Energy-Efficient Cluster Interconnects. In *Proceedings of the IEEE Trans. on Computers* (pp. 660-671).

Kim, J., Dally, W. J., Scott, S., & Abts, D. (2008). Technology-Driven, Highly-Scalable Dragonfly Topology. In *35th International Symposium on Computer Architecture (ISCA)*, 2008.

Kim, M. M., Davis, J. D., Oskin, M., & Austin, T. (2008). Polymorphic on-chip networks. In *Proc. of the 35th International Symposium on Computer Architecture,* (ISCA-2008).

Kim, W., Gupta, M. S., Wei, G.-Y., & Brooks, D. (2008). System level analysis of fast, per-core DVFS using on-chip switching regulators. In *Proceedings of the 14th International Symposium on High-Performance Computer Architecture* (HPCA).

Kumar, R., Zyuban, V., & Tullsen, D. M. (2005). Interconnections in multi-core architectures: Understanding mechanisms, overheads and scaling. *SIGARCH Comput. Archit. News, 33*(2), 408–419. doi:10.1145/1080695.1070004

Li, B., Peh, L.-S., & Patra, P. (2008). Impact of Process and Temperature Variations on Network-on-Chip Design Exploration. In *Proceedings of NOCS* (pp. 117-126).

Li, L., Vijaykrishnan, N., Kandemir, M., & Irwin, M. J. (2003). Adapative Error Protection for Energy Efficiency. In *Proceedings of the 2003 IEEE/ACM international conference on Computer-aided design* (ICCAD '03), (pp. 2) Washington, DC: IEEE Computer Society.

Magnusson, P. S., Christensson, M., Eskilson, J., Forsgren, D., Hallberg, G., & Hogberg, J. (2002). Simics: A full system simulation platform. *Computer, 35*(2), 50–58. doi:10.1109/2.982916

Mukherjee, S., Bannon, P., Lang, S., Spink, A., & Webb, D. (2002, January-February). The Alpha 21364 network architecture. In *Proceedings of IEEE Micro* (pp. 26–35).

Murali, S., Theocharides, T., Vijaykrishnan, N., Irwin, M. J., Benini, L., & Ovanni De Micheli, G. (2005). Analysis of Error Recovery Schemes for Networks on Chips. In *Proceedings of the IEEE Des. Test.*

Muralimanohar, N., & Balasubramonian, R. (2007). Interconnect design considerations for large nuca caches. In *Proceedings of the 34th Annual International Symposium on Computer Architecture*, (ISCA '07), (pp. 369–380).

Nayfeh, B. A. (1999). *The case for a single-chip multiprocessor* (PhD Thesis). Stanford, CA.

Nicopoulos, C., Yanamandra, A., Srinivasan, S., Narayanan, V., & Irwin, M. J. (2007). Variation-Aware Low-Power Buffer Design. In Proceedings *of The Asilomar Conference on Signals, Systems, and Computers.*

Park, D., Nicopoulos, C., Kim, J., Narayanan, V., & Das, C. R. (2006). Exploring Fault-Tolerant Network-on-Chip Architectures. In *Proceedings of DSN 2006.*

Peh, L.-S., & Dally, W. J. (2001). A Delay Model and Speculative Architecture for Pipelined Routers. In *Proceedings of the 7th International Symposium on High-Performance Computer Architecture* (HPCA '01).

Reed, I. S. (1960). *Error-Control Coding for Data Networks.* Boston: Kluwer Academic Publishers.

Semeraro, G., Magklis, G., Balasubramonian, R. D. H., Albonesi, S. D., & Scott, M. L. (2002). Energy-efficient processor design using multiple clock domains with dynamic voltage and frequency scaling. In *Proceedings of the International Symposium on High-Performance Computer Architecture.*

Shang, L., Peh, L.-S., & Jha, N. K. (2003). PowerHerd: Dynamic Satisfaction of Peak Power Constraints in Interconnection Networks. In *Proceedings of the 17th Annual International Conference on Supercomputing* (ICS '03), (pp. 98-108).

Shang, L., Peh, L.-S., & Jha, N. K. (2003). Dynamic Voltage Scaling with Links for Power Optimization of Interconnection Networks. In *Proceedings of HPCA.*

Taylor, M. B., Kim, J., Miller, J., Wentzlaff, D., Ghodrat, F., & Greenwald, B. (2002). The Raw Microprocessor: A Computational Fabric for Software Circuits and General Purpose Programs. *IEEE Micro*, *22*(2), 25–35. doi:10.1109/MM.2002.997877

Wang, H., Peh, L.-S., & Malik, S. (2003). Power-driven Design of Router Microarchitectures in On-chip Networks. In *Proceedings of the 36th annual IEEE/ACM International Symposium on Microarchitecture* (MICRO 36), (pp. 105). Washington, DC: IEEE Computer Society.

Woo, S. C., Ohara, M., Torrie, E., Singh, J. P., & Gupta, A. (1995). The SPLASH-2 programs: Characterization and methodological considerations. In *Proceedings of the 22nd Annual International Symposium on Computer Architecture* (pages 24-37). New York: ACM Press.

Xie, F., Martonosi, M., & Malik, S. (2003). Compile-time Dynamic Voltage Scaling Settings: Opportunities and Limits. In Proceedings of PLDI: The Conference on Programming Language Design and Implementation.

KEY TERMS AND DEFINITIONS

BCH Codes: A specific type of error correction codes.

DVFS: Dynamic Voltage Frequency Scaling

ECC: Error Correction Code

Network-on-Chip: The on-chip communication fabric that connects the multiple cores on a CMP.

Reconfigurable: System whose configuration can be changed.

Reliability: The reliability of a system is defined as the conditional probability that the system performs correctly thoughout an interval of time given that it works correctly at the beginning of the interval how does that sound.

Chapter 12
SpaceWire Inspired Network-on-Chip Approach for Fault Tolerant System-on-Chip Designs

Björn Osterloh
Technical University of Braunschweig, Germany

Harald Michalik
Technical University of Braunschweig, Germany

Björn Fiethe
Technical University of Braunschweig, Germany

ABSTRACT

Today FPGAs with large gate counts provide a highly flexible platform to implement a complete System-on-Chip (SoC) in a single device. Specifically radiation tolerant space suitable SRAM-based FPGAs have significantly improved the flexibility of high reliable systems for space applications. Currently the reconfigurability of these devices is only used during development phase. A further enhancement would be using the reconfigurability of SRAM-FPGAs in space, either to statically update or dynamically reconfigure processing modules. This is a major improvement in terms of maintenance and performance, which is essential for scientific instruments in space. The requirement for this enhanced system is to guarantee the system qualification and retain the achieved high reliability. Therefore effects during the reconfiguration process and interference of updated modules on the system have to be prevented. Updated modules need to be isolated physically and logically by qualified communication architecture. In this chapter the advantage of a specialized Network-on-Chip architecture to achieve a high reliable SoC with dynamic reconfiguration capability is presented. The requirements for SoC based on SRAM-FPGA in high reliable applications are outlined. Additionally the influences of radiation induced particles are described and effects during the dynamic reconfiguration are discussed. A specialized Network-on-Chip architecture is then proposed and its advantages are presented.

DOI: 10.4018/978-1-61520-807-4.ch012

INTRODUCTION

In this chapter a specialized Network-on-Chip approach is presented, which logically isolates a host system from reconfigurable modules: System-on-Chip Wire (SoCWire). First the conditions for SRAM-based FPGAs in space environment are introduced. Then, the Virtex-4 hardware architecture is outlined to understand the fault behavior due to radiation or during the dynamic reconfiguration process. Furthermore dynamic partial reconfiguration in Virtex-4 is presented and limitations of a bus structure are discussed. Finally, a fault tolerant Network-on-Chip approach, *SoCWire* is presented and its performance is outlined.

Data Processing Units (DPUs) on-board of spacecrafts are used as interface between spacecraft and several instrument sensor electronics or heads, providing the operational control and specific data processing of scientific space instruments. These systems have to provide sufficient computer power at low volume, mass and power consumption. Furthermore these instruments have to be suitable for the harsh space environment conditions (e.g. temperature, radiation). They need to be robust and fault tolerant to achieve an adequate reliability at moderate unit costs. Different implementation approaches for DPUs exist in traditional designs (i) based on rad-hard discrete components, (ii) Application-Specific integrated Circuits (ASICs) of high quality and (iii) Commercial Off-The-Self (COTS) devices. The major disadvantage of these three approaches is reduced design flexibility. A change in a specific data processing algorithm results in a major hardware design change (Fiethe, Michalik, Dierker, Osterloh, & Zhou, 2007). The Configurable System-on-Chip (CSoC) approach is based on radiation tolerant SRAM-based FPGAs and provides the capability for both flexibility and reliability. The suitability of this approach was successfully demonstrated in many space missions (Osterloh, Michalik, Fiethe, & Kotarowski, 2008). But the reconfigurability of SRAM-based FPGA has only

been used in the development phase. A further enhancement of this approach would be to use the partial and dynamic reconfiguration capability of these devices in-flight. Processing modules could be updated, which is especially an improvement for DPUs in space because of the non-accessibility of maintenance points. Furthermore processing modules could be updated to improve functionality e.g. a generic data compression core can be replaced by a sophisticated core to calculate scientific parameters directly in-flight. Dynamic partial reconfiguration enables run-time adaptive functionality and is an improvement in terms of power, device utilization and device count. To achieve theses advanced design goals the system requirements for high reliable space applications have to be considered. DPUs are exposed to the harsh space environment. Radiation has a major impact on the system and can lead to functional interruption. Therefore mitigation techniques e.g. Triple Modular Redundancy (TMR) in combination with configuration scrubbing are required. Another important requirement for DPUs is the system qualification. The system is qualified on ground by intensive stress-tests. This qualification has to be guaranteed in space even after a module update or during the dynamic reconfiguration process. Therefore the modules need to be isolated from the host system logically and physically. Detailed knowledge of the hardware architecture is necessary (i) to indentify effects during the dynamic reconfiguration process and fault behavior caused by radiation and to (ii) apply appropriate mitigation techniques.

RADIATION ENVIRONMENT

The space radiation environment is composed of a variety of charged particles originating from the sun or as cosmic ray background with energies ranging between keV to GeV and beyond. These particles are either trapped by the Earth's magnetic field or passing through the solar system.

Fluxes of charged particles of electrons, protons and heavy ions have a major impact and can lead to temporary (soft error) and permanent (hard error) damage of electronic devices. The effects have to be distinguished in accumulative effects through impacts of many particles and effects that are caused by a single high-energetic particle. Radiation damage to on-board electronics may be separated into two categories: Total Ionizing Dose (TID) and Single Event Effects (SEE). TID is an accumulative non-reversible effect (hard error) with long-term degradation of the device when exposed to ionizing radiation. It depends not only on the radiation environment, but also on the duration of a specific mission. The result of TID is a shifted threshold, which causes a depletion-mode transistor to be turned into an enhancement-mode transistor and thus a permanent malfunction of the device. SEEs are individual events which occur when a single incident ionizing particle deposits enough energy to cause an effect in a device. This effect produces different failure modes, depending on the incident particle and specific device.

From their appearance two categories are important, (i) Single Event Latch-up (SEL) and (ii) Single Event Upset (SEU). A SEL leads to ignition of a parasitic thyristor in the silicon structure. The result is a current increase which may destroy the device if the current is not limited. A SEU is caused by charged particles losing energy by ionizing the medium through which they pass, leaving behind a wake of electron-hole pairs. If this happens in a flip-flip or memory cell, the particle could deposit enough charge to cause the flip-flop or memory cell to change state (bit-flip), corrupting the data being stored. These effects require mitigation techniques e.g. Triple Modular Redundancy (TMR) within the devices (radiation hardened devices) or have to be implemented in the design to achieve an adequate reliability. If a particle deposits charge in a controller function within a device, a major malfunction may be the result, because a logic state of a state machine may be falsified. These SEUs are called Single Event Functional Interrupts (SEFIs). The effects of these SEFIs may be much more severe than just a bit flip, but typically the SEFIs are very rare. Although SEU effects and related mitigation techniques are well established in space applications it becomes also an issue for microelectronics on Earth. Particles of the cosmic ray interact with the Earth's atmosphere and spallation produces high energy neutrons. Some of these high energy neutrons can interact with semiconductor devices and deposit a huge amount of energy to cause a SEU. The SEU rate increases with increased logic density (Ziegler, 1996), which is the trend in the continuous technology evolution and scaling down process of transistor dimensions (Moore, 1975). The probability of a SEU depends on the number of neutrons passing an area in a time unit (flux). At high altitudes, e.g. airplane altitude a higher neutron flux is presented than on sea level. At sea level the neutron flux is low and therefore also the probably for an SEU, but with increasing logic density the probability for an SEU will become an issue in near future and requires mitigation techniques also on Earth bounded applications.

VIRTEX-4 ARCHITECTURE

The FPGA technology today provides attractive device solution for custom designs with up to several millions of equivalent gate counts per device. The following sections describe explicitly Xilinx FPGAs but should also apply for other SRAM-based FPGA vendors. Currently the radiation tolerant Virtex-4 QPro-V family from Xilinx has the largest gate count and is available for space applications in high reliable ceramic package. Radiation tolerant denotes in this case TID and SEL immunity up to a high threshold. The Xilinx Virtex-4 family is SRAM-based and customized by loading configuration data into the internal memory cells. These memory cells (logic, routing, RAM) are sensitive to SEU. This is a major issue for the system reliability and requires detailed knowledge of the hardware architecture to indentify sources of errors and to introduce mitigation techniques. The advantage of SRAM-based FPGAs is their capability of re-configuring the complete or partial system even during operation. Effects during the dynamic reconfiguration process can have impact on the system and, again requires deep knowledge of hardware architecture.

Xilinx Virtex FPGAs are user programmable gate arrays with different configurable elements. The internal configurable logic is organized in a regular array and includes four major elements, (i) Configurable Logic Blocks (CLB), (ii) Block SelectRAM memory (BRAM), (iii) multiplier blocks and (iv) Digital Clock Manager (DCM). The Active Interconnects Technology (AIT) interconnects all of these elements. Each element is tied to a switch matrix and permits multiple connections to an array of routing switches, the General Routing Matrix (GRA). The overall programmable interconnect is hierarchical organized with interconnections of different length, i.e. Long Lines, Direct Lines and Fast Connect. Most of the interconnections within a design are locally limited; they connect adjacent CLBs or close elements (spatial location) with short interconnections. Therefore they are available with large quantity in the routing resources. Long Lines span the full height/width of the device. They are rarely used and limited in resources. Clock distribution is provided through global clock lines. Each CLB comprises 4 Slices and a Switch Matrix to the GRA. Slices provide the functional elements for combinatorial and synchronous logic. They comprise 2 function generators, configurable as Look-Up Table (LUT), as 16 Bit SelectRam (RAM16) or 16 Bit shift register. Additional carry logic (CY), arithmetic logic gates, multiplexer (MUXFx) and 2 register/latch elements are available (Xilinx, 2007), as depicted in Figure 1.

All programmable elements and routing resources are controlled by configuration data stored in static memory cells. This configuration data is loaded into the memory cells during configuration and can be reloaded to change the functions of programmable elements even during system operation. The configuration memory is

Figure 1. Virtex FPGA with GRM, BRAM and detail view of CLB and Slice (Xilinx, 2007)

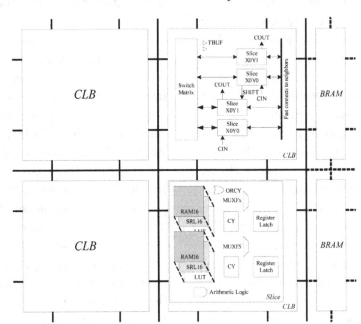

arranged in frames, each comprises 16 CLBs, which are tiled about the device. These frames are the smallest addressable segment of the configuration memory space (Xilinx, 2008). The configuration and reconfiguration process does not have an explicit activation. New frames become active as they are written. If bits are identical to the current value, the bits will not momentarily glitch to some other value. If the bits change its logic state, those bits could glitch when the frame write is processed. Some selections, e.g. the input multiplexers on CLBs, have their control bits split over multiple frames and thus do not change atomically. The glitch effect has a major impact on the system reliability. The interface logic of dynamic reconfigurable modules could have an unpredictable effect on the communication architecture during the reconfiguration process. The worst case scenario would be a malfunction of the communication architecture and thus operational interruption of the system.

RADIATION EFFECTS IN VIRTEX FPGAS

Each configurable element in the Xilinx Virtex device is susceptible to SEUs. A SEU in the switch matrix and routing resources can cause an unintended open circuit, e.g. in a feedback path of a finite state machine, or a short to ground (will not damage the device). The functional element within a Slice comprises the combinational and sequential logic. A SEU in a LUT modifies the implemented combinational logic or causes a change of state in the register. A transient upset can cause a pulse on the clock or data net. Figure 2 shows SEU effects in Virtex FPGAs.

These upsets are permanent errors and can only be corrected by "refreshing" the configuration memory, i.e. reload of the initial configuration bitstream. But before the refreshing of configuration memory is completed, the errors in the combinational logic are latched and forwarded to the next logic stage. These errors require fault tolerant techniques e.g. Triple Modular Redundancy (TMR). The concept of TMR is to triplicate the combinational and latch logic, and pass the outputs to a majority voter. If an upset occurs e.g. in one latch, the voter still delivers the correct output. This scheme is ineffective since the voter is implemented in the configuration memory cells and thus is susceptible to SEU, too. Therefore a modified TMR scheme with triplicate voters is used. Additional transients on the clock net have to be considered and therefore the clock nets are also triplicated. Figure 3 shows a finite state machine TMR design with feedback voters. Upsets in the data path are corrected by redundant voters on the next clock cycle.

This TMR fault tolerance technique requires by factor 3 higher area utilization. The triplicate voters add an additional delay and performance of the design decrease. Power consumption increases by a factor of 3 (Rezgui, et al., 2005). In most cases for DPUs TMR would not be fully applied since they are typically non mission critical and have strong resource limitations. Therefore partial TMR

Figure 2. SEU effects in Virtex FPGAs

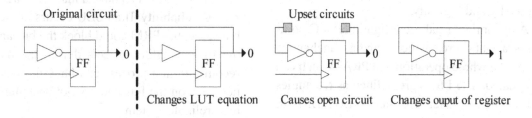

Figure 3. Finite state machine TMR design

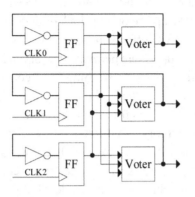

on critical modules and paths have to be carefully chosen dependent on the space environment and mission requirements.

DYNAMIC PARTIAL RECONFIGURATION

Partial reconfiguration permits to reconfigure pre-defined Partial Reconfigurable Areas (PRAs) in the configuration memory space. Dynamic partial reconfiguration enhances the system with at run-time adaptive functionality and in-flight re-configurability is a maintenance and performance improvement. The cost for this are increased development efforts and intensive failure mode investigations. To update a partial reconfigurable module (PRM) within a PRA the qualification has to be guaranteed in a high reliable system. Under all circumstances the dynamic reconfiguration process or an updated module must not interfere with the system because this could lead to an operation interruption of the system. Therefore PRMs have to be isolated from the host system physically and logically.

A dynamic partial reconfigurable SoC comprises a static area which remains unchanged during the whole operation and PRAs which can be updated. The PRAs are defined as rectangles and encompass all logic resources (LUT, BRAM, multiplier, IOB) within this area. The static area is distributed over the device by the place & route tool and encompasses the PRAs. Communication between the static area and PRAs requires a locking of routing to provide an unchanged interface during reconfiguration. For this, Xilinx provides pre-placed and pre-routed hard bus macros. The bus macros are implemented in the top level of the design and manually placed between the static area and PRAs. These macros provide a unidirectional communication from left-to-right, right-to-left, top-to-bottom and bottom-to-top in the Virtex-4 device. The PRAs can be physically isolated from the static area by area constraints. Since the static area remains unchanged one approach is to sub-divide the system. The static area stores all critical interfaces, e.g. processor, interfaces to spacecraft and is qualified on ground. The PRAs can be updated or dynamically reconfigured. This offers the advantage that only the updated module has to be qualified in a delta qualification step.

DYNAMIC RECONFIGURABLE SOC BASED ON BUS STRUCTURE

The SoC communication architecture is usually based on a bus structure. The Xilinx bus macros are suitable for handshaking techniques on busses like AMBA or Wishbone. The most resource efficient approach to implement a bus structure in Virtex FPGA would be to use wires that distribute signals across the whole device. Theses signals (long wires) are limited in resources in the Virtex-4 device and no dedicated bus macros are provided to connect to these signals, which leads to intensive manual routing. The susceptibility of PRMs to SEU can have a major impact on the system reliability. If an SEU occurs in the PRM bus logic, the PRM could block the bus and stop the system. Glitch effects during the dynamic reconfiguration process can have unpredictable behavior on the bus and are not acceptable in a high reliable system.

DPUs in space applications are typically structured as macro-pipeline with pre- and post-processing steps and require high data rate point-to-point communication. In each stage, processed data is forwarded to subsequent stage; macro-pipeline here describes a pipeline in consecutive hardware processing stages. To realize this architecture in a bus structure, multi master capability and bus arbitration are needed which may lead to unpredictable performance (Jantsch & Tenhunen, 2003). With these limitations a fault tolerant bus structure with hot-plug ability is necessary to guarantee data integrity. A bus structure based dynamic reconfigurable SoC would encounter the following disadvantages:

- No dedicated bus-macros are provided by Xilinx to access long lines which lead to manual time consuming routing.
- Dynamic reconfiguration process or an SEU in the PRM bus logic could block the bus and stop the system.
- Multi master bus structure may have unpredictable behavior.
- Failure tolerant bus structure (high efforts) with hot-plug ability is necessary to guarantee data integrity.

NETWORK-ON-CHIP APPROACH FOR DYNAMIC RECONFIGURABLE SOC

The PRMs need to be physically and logically isolated from the host system to guarantee system qualification. Physically the PRMs are isolated by the constraint of the sub-divided system of static area and PRAs. To isolate the PRMs logically, hot-plug ability is required with a communication architecture buildup a link automatically between two nodes. To overcome initial glitch

effects, the communication architecture should provide an exchange of characters between two nodes until a secure connection is established. To support the adaptive macro-pipeline structure, a reconfigurable point-to-point communication is necessary via crossbar for parallel data transfer between PRMs, as depicted in Figure 4.

Therefore, instead of the bus structure with its limitations addressed before we consider a communication architecture with a Network-on-Chip (NoC) approach providing:

- Reconfigurable point-to-point communication
- Support of adaptive macro-pipeline
- High speed data rate
- Hot-plug ability to support dynamic reconfigurable modules
- Easy implementation with standard Xilinx bus-macros

In order to achieve these requirements we have developed our own NoC architecture: System-on-Chip Wire (*SoCWire*).

Figure 4. Macro-pipeline network

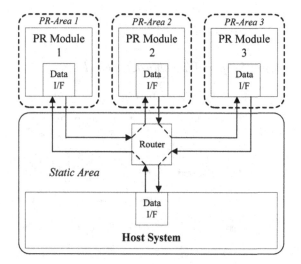

SPACECRAFT COMMUNICATION ARCHITECTURES

The development of new spacecraft communication standards requires a long term qualification process with intensive tests. Therefore our approach was to analyze existing spacecraft communication standards on their applicability for a NoC. The constraint for the NoC is a point-to-point and switched topology to support the adaptive macro-pipeline and the feasibility to extend the network with redundant schemes. Furthermore the NoC implementation shall have small resource utilization, error detection mechanisms and a straightforward interface. Communication architectures on-board of spacecrafts are mostly based on busses, e.g. MIL-STD-1553, CAN, ESA/CCSDS OBDH. In contrast, the ESA SpaceWire interface is the only currently available switched topology in space applications. The interface is a well established standard, providing a layered protocol (physical, signal, character, exchange packet, network) and proven interface for space applications. It is an asynchronous communication, serial link, bi-directional (full-duplex) interface including flow control, error detection and recovery in hardware, hot-plug ability and automatic reconnection after a link disconnection.

SPACEWIRE

SpaceWire is based on IEEE 1355-1995 with additional improvements (ESA-ESTEC, 2003). The IEEE 1355-1995 needs significant small resource utilization, i.e. 1/3 of a UART implementation in an FPGA (Walker, 1999). SpaceWire uses Data Strobe (DS) encoding. DS consists of two signals: Data and Strobe. Data follows the data bit stream whereas Strobe changes state whenever the Data does not change from one bit to the next. The clock can therefore be recovered by a simple XOR function. The performance of the interface depends on skew, jitter and the implemented technology. Off-chip data rates up to 400Mb/s can be achieved.

Character Level

The SpaceWire character level describes data and control characters used to manage the flow of data across the link. It is based on the IEEE 1355-1995 with additional Time-Code distribution. A data character (10 bit length) is formed by 1 parity bit, 1 data-control flag and 8 data bits to be transmitted. The data-control flag indicates if the current character is a data (0) or control character (1). Control characters (4 bit length) are used for flow control: Flow Control Token (FCT), End of Packet (EOP), Error End of Packet (EEP) and an Escape Character (ESC) which is used to form higher level control codes (8-14bit length),

Figure 5. SpaceWire character level

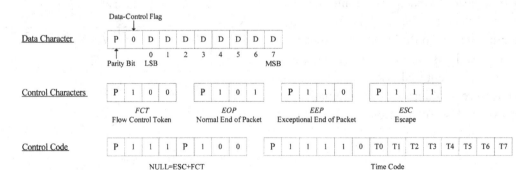

e.g. NULL (ESC+FCT) and Time-Code (ESC + Data Character). Odd parity is assigned to data and control characters to support the detection of transmission errors. The parity bit covers the previous eight bits of a data character or two bits of a control character. SpaceWire does not provide a global time base; nodes are synchronized through a system time distribution with Time-Codes, which have a high priority in the network. Figure 5 shows the different SpaceWire characters.

Exchange Level

The exchange level manages the connection and flow across the link. The exchange level is separated into two types: Link-Characters (L-Char) and Normal-Characters (N-Char). N-Char comprises data character, EOP and EEP and are passed to the network level. L-Char are used in the exchange level and are not passed to the network level. They comprise FCT and ESC characters and are responsible for link connection and flow control. Link connection is performed automatically and initialized with an exchange of silence. Both nodes are in reset state and wait for a certain time period. The receivers are then enabled and again both nodes wait for a certain time period. These time periods makes sure that both ends of the link are ready to receive data before either end begins transmission. Then, the transmitters are enabled and NULL characters are transmitted over the link. When the receiver detects NULL character

the link is connected, if no NULL characters are received a time out procedure re-initializes the link and restarts with sending NULL characters. Figure 6 shows the SpaceWire link connection procedure.

To avoid buffer overflows and therefore data loss, a credit-based flow control is required. After link connection is established, FCTs are transmitted over the link. FCT signify that one end of the link is ready to receive 8 N-Chars.

Figure 7 shows the SpaceWire credit management. The receiver consists of a buffer for 7*8 N-Char. If a slot for 8 N-Chars is free (right hand side), the receiver sends a FCT to transmitter to signify free space for 8 more N-Chars. The transmitter receives the FCT and increases its credit counter by 8. Additionally the credit counter authorizes the transmitter to send data. Each transmitted N-Char vice versa decrease the credit counter by 1 (left hand side). If the credit counter reaches 0, the transmitter stops to send N-Chars. This credit-flow provides an efficient mechanism to prevent buffer overflow.

Packet Level

The packet level describes the format to support routing of packets over a SpaceWire network. A SpaceWire packet comprises Destination Address plus Cargo plus EOP/EEP. The Destination Address is required to send packets over a SpaceWire network to a certain target. Depended

Figure 6. SpaceWire link connection

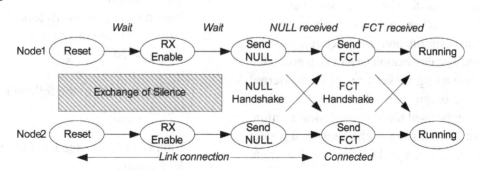

Figure 7. SpaceWire credit management

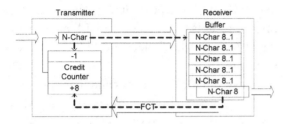

on the network topology multiple destination addresses can follow, before the Cargo begins. The Cargo contains the user data. Regular packets are completed with an EOP marker. For point-to-point communication the destination address is not required. EEP marker is exclusively send by the user to indicate an erroneous packet. To this the target can react accordingly and reject the packet. In summary, the SpaceWire packet level is highly flexible and permits to implement a wide range of protocols.

Network Level

SpaceWire network nodes can be either connected by links or connected by routing switches. The network operates exclusively with the objects of the packet level. All lower levels are completely masked from the network. The network topology can be configured as e.g. tree, cloud or cube. Space-Wire router support wormhole routing; packets arriving at one port are routed immediately to the output port, if the port is free, which reduces buffer space and latency. The SpaceWire network level supports the simple and effective header deletion technique to transfer packets across an arbitrary sized network. When a packet is received at a routing switch the destination port is determined from the header. The destination header is then deleted and the remaining packet content is transferred through the output port. If a second identifier exists it can be used for any subsequent routing. Therefore at each stage of the network a packet can be regarded as a packet comprising a single

destination identifier header, cargo and end of packet. The SpaceWire network level comprises several different packet-addressing schemes e.g. path addressing, logical addressing, regional addressing. With path addressing a sequence of destination identifier within a packet is used to guide the packet across the network. In logical addressing each physical port is associated with a configurable logical address. In contrast to path addressing the packet comprises only one address and is routed across the network. To support this routing scheme each routing switch is provided with a routing table and a configuration port to initialize the routing table. SpaceWire networks provide the feasibility to choose the right balance between performance, fault-tolerance and power consumption.

SpaceWire Error Recovery Schemes

The SpaceWire error recovery scheme covers the exchange and network level. In the exchange level the following errors can be detected:

- Disconnect error
- Parity error
- Escape error
- Character sequence error (invalid token at invalid time)
- Credit error

The response to any of these errors is:

1. Detect error
2. Disconnect link
3. Report error to network level
4. Attempt to reconnect the link if the link is still enabled

In the network level the following errors can be detected:

- Link error (exchange level error)
- EEP received

- Invalid destination address

If a link error is detected the network level response as follows:

1. Error is received by the network level
2. Current received packet is terminated with EEP
3. If the error occurred in destination or source node, the error shall reported to the host system
4. If the error occurred in a routing switch, the error can be reported by a status, either through the interface or a status register

The SpaceWire standard covers all requirements for a fault tolerant and robust network. It comprises (i) a link connection flow, which is suitable to overcome glitch effects during the dynamic partial reconfiguration process, (ii) recovery from link disconnection, which could occur during the dynamic partial reconfiguration process or an SEU induces transient error and (iii) error recovery, which makes it suitable for a dynamic reconfigurable system in high reliable space environment. Different network topologies can be implemented for fault tolerant schemes. Additionally the interface implementation uses significantly small resource utilization and comprises a straightforward interface.

SOCWIRE CODEC

As mentioned before SpaceWire is a serial link interface and performance of the interface depends on skew, jitter and the implemented technology. For our NoC approach we are in a complete on-chip environment with up to 31 reconfigurable modules, which can be operated by one switch. The maximum character length the in the SpaceWire standard without Time-Code is 10bit (data character). Time-Code characters are not necessarily required because a global time base distribution

can be easily implemented in a complete on-chip environment through dedicated signals, which saves resources. Therefore we have modified the SpaceWire interface to a 10bit parallel data interface. The advantage of this approach is that significantly higher data rates can be achieved as compared to the SpaceWire standard. Additionally, we have implemented a scalable data word width (8-128 bit) to support medium to very high data rates. On the other hand we keep in our implementation the advantageous features of the SpaceWire standard including flow control and hot-plug ability. Also error detection and link re-initialization is still fully supported making it suitable even for a SEE sensitive environment. For parallel data transfer the FCTs need to be included in the transfer. After initialization phase, every eight data characters are followed by one FCT. The maximum data rate for a bi-directional (full-duplex) transfer can therefore be calculated to:

$$DRate_{Bi}\left[\frac{\text{Mb}}{\text{s}}\right] = f_{Core(MHz)} \times DWord\ Width \times \frac{7}{8}$$

For a unidirectional data transfer the FCTs are processed in parallel and the maximum data rate can be calculated to:

$$DRate_{Uni}\left[\frac{\text{Mb}}{\text{s}}\right] = f_{Core(MHz)} \times DWord\ Width$$

Figure 8 shows data rates for different data word width, unidirectional and bi-directional (full-duplex) data transfer at a core clock frequency of 200 MHz. The SoCWire CODEC has been implemented and tested in Xilinx Virtex-4 LX60-10. Figure 9 shows the occupied area, absolute values and maximum clock frequency. Both figures show that very high data rates are achieved with small resource utilization occupied by the SoCWire CODEC.

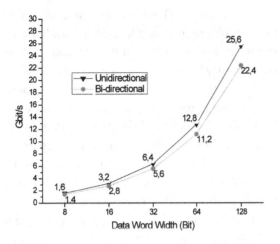

Figure 8. SoCWire CODEC data rates at core clock frequency 200 MHz

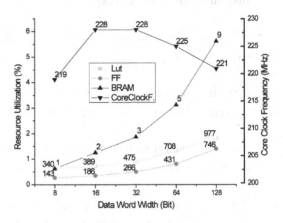

Figure 9. SoCWire CODEC synthesis report

SOCWIRE SWITCH

The SoCWire Switch enables the transfer of packets arriving at one link interface to another link interface on the switch. The SoCWire Switch provides a configurable number of ports, realized by internal SoCWire CODECs. In contrast to the SpaceWire standard the SoCWire Switch waives the configuration port. Therefore logical addressing is not supported because the macro-pipeline star network topology (Figure 4) does not require this routing scheme. This saves resources of the routing table and control logic. The SoCWire Switch provides up to 32 ports. Incoming packets are analyzed by an entrance module. The module verifies the validation of the header, processes header deletion and passes the packet to the matrix. The matrix (crossbar) manages the cargo transfer to the destination port by cell modules. Each cell module represents an inter-connection between two ports and therefore a switch with 4 ports comprises 16 cells or with 32 ports 1024 cells. Figure 10 shows the configuration of the SoCWire Switch. Entrance module and the matrix support the parallel data transfer of packets through the network.

The SoCWire Switch is a fully scalable design supporting many data word widths (8-128bit) and 2 to 32 ports. It is a totally symmetrical input and output interface with direct port addressing including header deletion. The SoCWire Switch has been implemented and tested in a Xilinx Virtex-4 LX60-10. Figure 11 shows the occupied area and maximum clock frequency for an 8 bit data word width switch dependent on the number of ports.

DYNAMIC RECONFIGURABLE SOC

A dynamic reconfigurable macro-pipeline system with four SoCWire CODECs, one in the host system, three in the PRMs and one SoCWire Switch

Figure 10. 4 Ports SoCWire switch

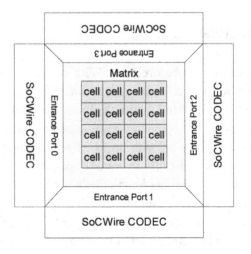

Figure 11. SoCWire Switch 8 bit data word width synthesis report

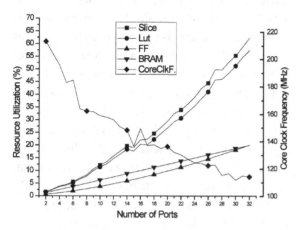

Figure 12. Implemented SoCWire macro-pipeline system

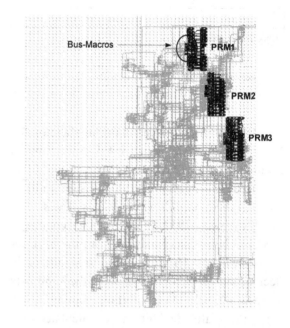

has been implemented and tested, as depicted in Figure 4. The host system and SoCWire switch were placed in the static area and the PRMs in the PRAs. All SoCWire CODECs were configured with 8 bit data word width. The system including reconfigurable areas can be easily implemented with the standard unidirectional Xilinx bus macros. Figure 12 shows a cut out of the placed and routed SoCWire macro-pipeline system: PRMs (PRM1, PRM2 and PRM3) and bus macros in a Virtex-4 LX60. The static area is distributed over the FPGA.

The PRMs where configured as packet forwarding modules. Different configuration of packet forwarding, e.g. between modules and through the whole macro-pipeline system under the condition of parallel communication between nodes, were tested. The system runs at 100 MHz and the maximum data rate of the simulation was validated to be 800 Mbps according to the selected 8 bit data word width. Additionally one PRM was dynamically reconfigured in the system. During the reconfiguration process the communication between the PRM and SoCWire Switch was interrupted, the other PRMs connections were still established. After the reconfiguration process was completed, the communication between the two nodes was built up automatically without

any further external action (e.g. reset of node or switch). The dynamic reconfiguration forces a link disconnection. In the SpaceWire standard, this leads to a link error recovery with unlimited attempts. If the link re-initialization is successful, SoCWire signals an active state through its interface.

There is not much known about the details of the reconfiguration process of the Virtex FPGAs. In the frame structure the CLK lines are activated in the middle of the frame. Our experience of reconfiguring the whole device with complex design, e.g. processor system, a reset is not required and the system starts without external action. However, for a safe dynamic reconfiguration we propose to keep the PRM in reset state and release this state after reconfiguration process has finished.

An error scenario was simulated; the link was disconnected and reconnected to measure the error detection and recovery time. Table 1 shows the results.

The Partial Reconfiguration Time (PRT) depends on the bitstream size of PRM and the

Table 1. SoCWire initialization/detection/recovery time

	Clock cycles
Link initialization	7
Error detection	8
Error recovery	6

reconfiguration interface. In this application the SelectMap interface with 8 bit data word width running at 50 MHz with a data rate of 400 Mb/s was used. The area for one PRM was set to utilize 0.6% of the overall logic resources. The size of one PRM was 37912 Bytes (64 bytes command + 37848 bytes data) and therefore the PRT resulted in 758µs. This time could be significantly decreased by a factor of 8 in using the 32bit data word width SelectMap interface running at 100 MHz. Generally, the PRT can be calculated to:

$$PRT[s] = \frac{PRM[Bytes]}{CCLK[Hz] \times SelectMap_{DWordWidth}[Bytes]}$$

SoCWire meets all requirements for a high reliable dynamic reconfigurable architecture. Error detection and link re-initialization overcome glitch effects. SEUs in the PRM SoCWire interface would only block one port of the SoCWire Switch but does not affect the host system. The system qualification can be guaranteed with additional PRM self-testing features (timing constraints and test vectors). The main difference between Space-Wire and SoCWire is the parallel data transfer and the non-distribution of time codes through the SoCWire network. Apart from these points, SoCWire is fully compliant with the SpaceWire standard.

FUTURE RESEARCH

The SoCWire architecture is a suitable approach for dynamic reconfigurable systems in high reliable environments. Currently it supports the macro-pipeline architecture with star topology. The star topology is suitable for a macro-pipeline system because data is processed and transferred from module to module in a pipeline. Therefore, scheduling mechanisms are not required. To extend the network and to implement other network topologies with redundant path schemes further research is necessary. Additional switches, network arbitration and packet scheduling mechanism e.g. FlexRay are required to prevent packet deadlocks. A global time base distribution can be easily implemented in the SoC design to support Time Division Multiplexing (TDM) schemes

CONCLUSION

In this chapter a specialized NoC architecture for dynamic reconfigurable SoC to guarantee system qualification has been presented. SRAM-based FPGAs provide a highly flexible platform to implement a complete SoC in a single device. These devices are even suitable for space environment applications requiring high reliability with additional mitigation techniques and have improved system flexibility for scientific instruments. Currently the reconfiguration ability of these devices has only been used in the development phase. A further enhancement for DPUs in terms of system maintenance, performance improvement and run-time adaptability is in-flight update of hardware functions and dynamic partial reconfiguration. To achieve these enhancements the system qualification has to be guaranteed to retain the achieved reliability. Therefore PRM need to be isolated physically and logically from the host system. To meet these requirements detailed knowledge of the hardware architecture and failures induced by radiation are necessary. Glitch effects during the dynamic reconfiguration process have to be considered and SEU errors need appropriate mitigation techniques. A bus structure based approach has the limitation of high effort to guarantee system reliability. Therefore we have developed the NoC approach *SoCWire*. It meets all requirements for a high speed reconfigurable architecture. High data rates are achieved with significantly small implementation efforts. Hardware error detection, hot-plug ability and support of the adaptive macro-pipeline are provided. *SoCWire* can be easily implemented with standard Xilinx bus macros. The reference design has shown the suitability for dynamic reconfigurable systems and allows the designer to adapt the SoCWire system quickly to any possible basis architecture. Validation and verification are simplified by low resource use. Robustness and error mitigation are guaranteed by a clear and straight RTL design. The network can be extended for failure tolerant redundant network topology schemes with additional switches and the requirement for a global time base distribution to prevent packet deadlocks. Although this NoC approach has been developed for space applications it is also appropriate for highly reliable industrial applications. Furthermore with the increasing logic density and scaling down process of transistor dimensions the SEU sensitivity and appropriate mitigation techniques need also be considered for microelectronics on Earth in near future.

REFERENCES

ESA-ESTEC. (2003). *Space Engineering: SpaceWire–Links, nodes, routers, and networks*. Norrdwijk Netherlands: ESA-ESTEC.

Fiethe, B., Michalik, H., Dierker, C., Osterloh, B., & Zhou, G. (2007). Reconfigurable System-on-Chip Data Processing Units for Miniaturized Space Imaging Instruments. In *Proceedings of the conference on Design automation and test in Europe* (DATE), (pp. 977-982). New York: ACM.

Jantsch, A., & Tenhunen, H. (2003). *Networks on Chip*. Boston: Kluwer Academic Publishers.

Moore, G. E. (1975). Progress in Digital Integrated Electronics. In *Proceedings of Digest of the 1975 International Electron Device Meeting* (p. 1113). New York

Osterloh, B., Michalik, H., Fiethe, B., & Kotarowski, K. (2008). SoCWire: A Network-on-Chip Approach for Reconfigurable System-on-Chip Designs in Space Applications. In *Proceedings of the NASA/ESA Conference on Adaptive Hardware and Systems* (AHS-2008), (pp. 51-56). Noordwijk, The Netherlands.

Rezgui, S., George, J., Swift, G., Somervill, K., Carmichael, C., & Allen, G. (2005). *SEU Mitigation of a Soft Embedded Processor in the Virtex-II FPGAs*. San Joese, CA: North American Xilinx Test Consortium, Xilinx Inc.

Walker, P. (1999). *IEEE 1355 Why yet another high-speed serial interface standard? IEE Electronics and Communications Open Forum*. Washington, DC: IEEE.

Xilinx. (2007). *Virtex-II Platform FPGAs: Complete Data Sheet* (3.5). San Jose, CA: Xilinx Inc.

Xilinx. (2008). *Virtex-4 FPGA Configuration User Guide* (1.1). San Jose, CA: Xilinx Inc.

Ziegler, J. F. (1996). IBM Experiments in Soft Fails in Computer Electronics. *IBM Journal of Research and Development, 40*(3).

KEY TERMS AND DEFINITIONS

Data Processing Unit (DPU): Scientific space instrument on-board of spacecrafts which is used as interface between spacecraft and several instrument sensor electronics or heads, providing the operational control and specific data.

Macro-Pipeline: DPUs in space applications are typically structured as macro-pipeline with pre- and post-processing steps and require high data rate point-to-point communication. In each stage, processed data is forwarded to subsequent stage. Macro-pipeline here describes a pipeline in consecutive hardware processing stages.

Partial Reconfiguration: Permits to reconfigure pre-defined Partial Reconfigurable Areas (PRAs) in the configuration memory space within an FPGA.

Single Event Effects (SEEs): Are individual events which occur when a single incident ionizing particle deposits enough energy to cause an effect in a device. This effect produces different failure modes and can lead to temporary (soft error) and permanent (hard error) damage of electronic devices.

Single Event Latch-Up (SEL): Is an event caused by a charged particle passing the device and ignites a parasitic thyristor in the silicon structure. The result is a current increase which may destroy the device if the current is not limited.

Single Event Upset (SEU): Is caused by charged particles losing energy by ionizing the medium through which they pass, leaving behind a wake of electron-hole pairs. If this happens in a flip-flip or memory cell, the particle could deposit enough charge to cause the flip-flop or memory cell to change state (bit-flip), corrupting the data being stored.

Spacewire: Is a well established serial interface standard, providing a layered protocol (physical, signal, character, exchange packet, network) and proven interface for space applications. It is an asynchronous communication, serial link, bi-directional (full-duplex) interface including flow control, error detection and recovery in hardware, hot-plug ability and automatic reconnection after a link disconnection

System-on-Chip Wire (SoCWire): Is a Network-on-Chip (NoC) approach based on the ESA SpaceWire interface standard to support dynamic reconfigurable System-on-Chip (SoC). SoCWire has been developed to provide a robust communication architecture for the harsh space environment and to support dynamic partial reconfiguration in future space applications.

Total Ionizing Dose (TID): Is an accumulative non-reversible effect (hard error) with long-term degradation of the device when exposed to ionizing radiation. The result of TID is a shifted threshold, which causes a depletion-mode transistor to be turned into an enhancement-mode transistor and thus a permanent malfunction of the device.

Triple Modular Redundancy (TMR): The concept of TMR is to triplicate the combinational and latch logic, and pass the outputs to a majority voter to achieve fault tolerance against errors in a single sub-module.

Chapter 13
A High–Performance and Low–Power On–Chip Network with Reconfigurable Topology

Mehdi Modarressi
Sharif University of Technology, Iran

Hamid Sarbazi-Azad
Sharif University of Technology, Iran

ABSTRACT

In this chapter, we present a reconfigurable architecture for network-on-chips (NoC) on which arbitrary application-specific topologies can be implemented. The proposed NoC can dynamically tailor its topology to the traffic pattern of different applications, aiming to address one of the main drawbacks of existing application-specific NoC optimization methods, i.e. optimizing NoCs based on the traffic pattern of a single application. Supporting multiple applications is a critical feature of an NoC as several different applications are integrated into the modern and complex multi-core system-on-chips and chip multiprocessors and an NoC that is designed to run exactly one application does not necessarily meet the design constraints of other applications. The proposed NoC supports multiple applications by configuring as a topology which matches the traffic pattern of the currently running application in the best way. In this chapter, we first introduce the proposed reconfigurable topology and then address the two problems of core to network mapping and topology exploration. Experimental results show that this architecture effectively improves the performance of NoCs and reduces power consumption.

INTRODUCTION

With the advance of the semiconductor technology in recent years, current multi-core system-on-chips (SoCs) have grown in size and complexity and future SoCs will consist of complex integrated components communicating with each other at very

DOI: 10.4018/978-1-61520-807-4.ch013

high-speed rates. Moreover, the microprocessor industry is moving from single-core to multi-core and eventually to many-core architectures, containing tens to hundreds of identical cores arranged as chip multiprocessors (CMPs) [Owens et al. 2007]. The lack of scalability in bus-based systems and large area overhead and unpredictability of electrical parameters of point-to-point dedicated links have motivated researchers to propose packet-switched Network-on-Chip (NoC) architectures to overcome complex on-chip communication problems [Benini and De Micheli 2002; Guerrier and Greiner 2000; Dally and Towles 2001; Jantsch and Tenhunen 2003].

However, it has been shown that, in future technologies (especially 22 nm), the power consumption of the current NoCs, when providing the required performance of typical CMP and multi-core SoC applications, is about 10 times greater than the power budget that can be devoted to them [Owens et al. 2007]. Application-specific optimization is one of the most effective approaches to bridge this exiting gap between the current and the ideal NoC power/performance metrics. This class of optimization methods tries to customize the architecture of an NoC for a target application, where the application and its traffic characteristics are known at design time. Most state-of-the-art NoC architectures and their design flows provide design-time NoC optimization (including topology selection and mapping) for a single application [Bjerregaard and Mahadevan 2006]. In other words, they try to generate an optimized NoC based on the traffic pattern of a single application and then, synthesize the chip. However, today's (often programmable) multi-core SoCs are highly complex and cost-effective chips, and as technology advances, it becomes more cost-effective to integrate several different applications onto a single SoC chip. As a result, NoC architectures should closely match the traffic characteristics and performance requirements of different applications. Since different applications have different functionalities, the inter-core communication characteristics can be very different across the applications. Consequently, an NoC that is designed to run exactly one application does not necessarily meet the design constraints of other applications. Prior work [Kim et al. 2008] shows, by conducting simulation over 1500 different NoC configurations (topology, buffer size, and bit-width), that no single NoC provides optimal performance across a range of applications.

In this chapter, we tackle this problem for two effective application-specific optimization methods, i.e. topology generation and core to network mapping, by introducing a reconfigurable NoC architecture designed based on a regular mesh-based NoC. It enables the network topology to dynamically match the communication pattern of the currently running application. Afterwards, with a set of different applications as input, we develop a two-phase algorithm which first maps all of the IP-cores used by the applications onto the reconfigurable NoC nodes and then implements a suitable topology for each input application by appropriately configuring the NoC.

Optimizing the network topology and core to network mapping are two application-specific NoC customization methods which dramatically affect the network performance-related characteristics such as average inter-core distance, total wire length, and communication flow distributions. These characteristics, in turn, determine power consumption and average network latency of NoC architectures. Topology determine the connectivity of the NoC nodes while mapping determines on which node each processing core should be physically placed. Mapping algorithms generally try to place the processing cores communicating more frequently near each other; when the number of routers between two cores reduces, the power consumption and delay of their communications decreases linearly.

The reconfiguration of the proposed architecture is achieved by inserting several simple switches in the network which allow the network to dynamically change the inter-node connections and implement the topology that best matches the running application demands. More precisely, we try to reduce the network hops (or number of routers) between the source and destination nodes of high volume communication traces (or ideally connect them directly) by bypassing one or more intermediate routers. This can lead to considerable performance improvement since the power/latency of the router pipeline stages has a significant contribution to the total NoC power/latency. For example, in Intel's 80-core TeraFlops processor, more than 80% of the NoC power is consumed by routers [Hoskote et al. 2007].

The topologies proposed for on-chip networks vary from regular tiled-based [Murali and De Micheli 2004a; Hu and Marculescu 2005] to fully customized [Ogras and Marculescu 2005b; Chan and Parameswaran 2008] structures. Since fully customized NoCs are designed and optimized for a specific application, they give the best performance and power results for that application. On the other hand, regular NoC architectures provide standard structured interconnects which ensure well-controlled electrical parameters. In these topologies, designers can solve usual physical design issues like crosstalk tolerance, timing closure, and wire routing for the regular NoC topology and reuse it in several designs. This reusability alleviates the predictability problem in deep sub-micron technologies [Angiolini et al. 2006] and can effectively reduce the NoC design effort.

Our proposed NoC architecture can be placed between these two extreme points in NoC design and benefit from both worlds. While this NoC architecture is designed and optimized like regular NoCs, it can be dynamically configured to a topology that best matches the traffic pattern of the currently running application. In other words, this architecture realizes application-specific topologies over structured components.

The rest of the chapter is organized as follows. The next section discusses previous work. Afterwards, the proposed reconfigurable NoC architecture is presented and evaluated in terms of its implementation cost. We then describe the mapping and topology selection algorithms developed for our reconfigurable NoC followed by some experimental results. The last sections address some future work in this line and conclude the chapter.

BACKGROUND

The need for scalable on-chip communication architectures is pointed out in [Benini and De Micheli 2002; Guerrier and Greiner 2000; Dally and Towles 2001]. Many NoC architectures, implementations, and design flows have therefore been proposed in the literature [Jalabert et al. 2004; Bjerregaard and Sparsø 2005; Goossens et al. 2005; Milberg et al. 2004].

Substantial prior work has proposed different optimization methods for on-chip networks to reduce the power consumption and message latency of NoCs. Some of these methods focus on reducing the power consumption and latency within a network hop by optimizing the microarchitecutre of NoC routers and network switching methods [Mullins and Moore 2004; Peh and Dally 2001; Kim et al. 2005; Abed et al. 2007]. To reduce the message latency, most of these (often general-purpose) optimization methods try to cut down the router critical path delay by parallelizing multiple pipeline stages [Mullins and Moore 2004; Peh and Dally 2001; Kim et al. 2005]. Similarly, the power reduction is mostly achieved by reducing the router activity and the total capacitance switched per cycle [Kim, Nicopoulos, et al. 2006; Meloni et al. 2006]. These methods are all orthogonal to our reconfigurable topology, which targets the hop-count rather than per hop energy and latency.

Mapping, routing, and topology selection mechanisms, on the other hand, try to decrease the average number of hops messages travel by exploiting the information about the traffic pattern of an application. The problem of topology selection and mapping cores onto NoC nodes have been explored by many researchers [Hu and Marculescu 2003 1a; Hu and Marculescu 2003b; Murali and De Micheli 2004b; Bertozzi et al. 2005]. The methods in [Murali and De Micheli 2004a] and [Hu and Marculescu 2005] consider regular mesh-based topologies, while [Ogras and Marculescu 2005b] and [Chan and Parameswaran 2008] develop irregular topologies customized to a target application. In [Murali and De Micheli 2004b], for example, the authors evaluate different standard topologies in terms of power, performance, and area, and select the one giving the best results. In [Hu and Marculescu 2005], a branch-and-bound algorithm for mapping the cores of a task-graph into a mesh-based NoC is developed. Exploring efficient application-specific topologies using evolutionary algorithms is addressed in [Ascia et al. 2005], [Lei and Kumar 2003], and [Srinivasan and Chatha 2005]. Optimizing the power and performance of mesh-based NoCs by inserting some physical and virtual connections to connect non-adjacent nodes is addressed in [Ogras and Marculescu 2005a] and [Kumar et al. 2007], respectively. However, all of these works are limited to optimize an NoC based on a set of communication constraints, obtained from a single application.

Only few work has been performed to optimize an NoC for multiple applications [Kim et al. 2008] [Stensgaard and Sparsø 2008][Vassiliadis and Sourdis 2006][Murali et al. 2006a][Murali et al. 2006b]. In [Murali et al. 2006a], for example, a worst-case NoC design flow based on the delay and throughput constraints of a set of input applications has been proposed. By applying DVS and DFS at run-time, the power consumption of the NoC is optimized while the performance constraints of the currently running tasks are met. Authors in [Kim et al. 2008] introduce a polymorphic NoC, a configurable set of buffers, crossbars and links, on which an arbitrary network can be constructed. The network can be configured to offer the same performance as a fixed function network, while incurs an average of 40% area overhead. However, the authors have not analyzed the power consumption of their NoC. In [Stensgaard and Sparsø 2008], reconfiguration is achieved by wrapping the NoC routers by a logic called topology switch. In this architecture, topology can be dynamically reconfigured by bypassing some routers through these switches. The authors report a 56% reduction in NoC power consumption and

10% area overhead when running a multimedia application. As will be shown in the next section, our proposed architecture differs from all of these works in terms of the flexibility it offers and as the results show, it can give a more appropriate trade-off between flexibility and area overhead.

NoC Topology

Network topology determines the physical arrangement of channels and nodes in an NoC. Selecting a suitable topology is one of the most important steps in designing NoCs, as it has a significant effect on the cost and power, and performance-related metrics of an NoC.

The network cost depends heavily on the number and complexity of the routers and wiring resources required to realize the network. Topology has the most important contribution in determining these two cost factors. Topology specifies the number of required routers and the input/output ports of each router. On the other hand, the topology largely affects the routing scheme and flow control method of the network, which in turn determines the buffering requirements and routing/switching logic complexity of the routers.

In addition to router complexity, topology specifies the number of channels of a network, as well. It also affects the length and placement of channels. The length and VLSI layout of the channels, in turn, influence the design time/effort and area of the chip.

In addition to cost, the power consumption and performance of an NoC greatly depend on the topology. Average message latency is the most important performance metrics of an NoC. The average message latency of the network is defined as the average time required for packets to traverse the network [Dally and Towles 2004]. Although the network latency also depends on the routing complexity and switching schemes of the NoC, the topology has the most significant contribution to the latency, as it determines the network diameter and average message hop count.

The average message hop count is the average number of network hops a message traverses to travel from the source to destination node. The latency and power consumption of an NoC linearly relates to the average number of hops the packets traverse, hence to the network topology.

Most NoCs implement regular topologies that can be laid out on a 2-dimensional plane. 60% of reported NoC architectures use the mesh and torus, which are commonly known as regular tile-based topologies [Salminen et al. 2008]. Despite the advantages of meshes for on-chip implementation, some packets may suffer from long latencies due to lack of short paths between remotely located nodes. A number of previous works, consequently, try to tackle this shortcoming by applying some network topologies of traditional interconnection networks with lower diameter in on-chip networks. In our previous work, for example, we have developed some novel topologies for on-chip networks which are inspired from the topologies of traditional interconnection networks. The topologies include 2D-SEM [Sabbaghi et al. 2008a], developed based on traditional shuffle-exchange topology, and 2D-DBM [Sabbaghi et al. 2008b], which is developed based on the traditional De-Bruijn topology. Both of these topologies offer a lower diameter than an equivalent mesh-based NoC with rather the same cost.

Unlike traditional interconnection networks, in addition to regular topologies, on-chip networks may implement irregular application-specific topologies, as well. This is due to the fact that most multi-core SoCs are typically composed of heterogeneous cores with non-uniform sizes where regular topologies result in poor performance and power, with large area overhead [Atienza et al. 2008]. Moreover, the communication traffic characteristics of multi-core SoCs are non-uniform and can usually be obtained statically. As a result, an application-specific NoC with a custom topology, which satisfies the design objectives and constraints of the target application, is more appropriate.

Similarly, the algorithms developed for homogeneous CMPs should be created to suit the multiprocessor topology in order to exploit all the physical network resources and maximize the performance. In [Vassiliadis and Sourdis 2006], it is shown that the performance of some sorting algorithms can be 1.5 to 2 times higher when running on the topology which best suits them.

In the previous section, we discussed the advantages and disadvantages of regular and irregular topologies. As mentioned there, our proposed NoC provides a reconfigurable architecture on which a vast range of regular and application-specific topologies can be constructed. The next section introduces the proposed architecture and describes its different architecture- and system-level aspects.

THE RECONFIGURABLE NOC ARCHITECTURE

In this section, we first present the proposed reconfigurable network-on-chip architecture and then evaluate its cost in terms of area overhead and reconfiguration procedure latency. Afterwards, we present a mapping and topology generation scheme which exploits the reconfigurability of our NoC to adapt its topology to each target application. The last part of this section presents the experimental results which are obtained by implementing some multi-core SoC designs over the proposed reconfigurable architecture.

The NoC Architecture

The system under consideration is composed of $n \times n$ nodes arranged as a 2D mesh network. In the proposed NoC architecture, the routers are not connected directly to each other, but rather are connected through a simple switch box, called configuration switch (Figure 1). Each square in Figure 1 represents a network which is composed of a processing element and a router while each circle represents a configuration switch. This

Figure 1. (a) The reconfigurable NoC architecture and, (b) three possible switch configurations

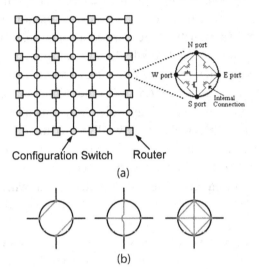

(a)

(b)

Figure 2. The reconfigurable NoC configured as a (a) mesh (b) binary tree. The black node is the root of the binary tree

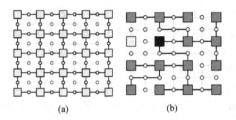

(a)　　　　　(b)

architecture has been inspired from the reconfigurable mesh interconnection networks presented in [Hwang & Briggs 1984]. The topology of these NoCs is determined by the configuration of the configuration switches which allows the connectivity between the network nodes to be dynamically changed.

Figure 1.a shows the internal structure of a configuration switch. It consists of some simple transistor switches which can establish connections between incoming and outgoing links. Figure 1.b displays three possible switch configurations.

In the current implementation, the configuration switches consider the incoming and outgoing sub-links of a bidirectional link as a single link and route them together. For example, by connecting the N port of a configuration switch to its S port, the incoming sub-link of the N port is connected to outgoing sub-link of the S port and the incoming sub-link of the S port is connected to outgoing sub-link of the N port. Independent routing of the sub-links can be considered as a future work. This NoC can be configured as arbitrary topologies, including some regular topologies, by properly setting the configuration switches. For example,

figures 2.a and 2.b display the network configured as a mesh and a binary-tree, respectively.

In general, dynamic hardware reconfiguration can only be implemented on dynamically reconfigurable devices; hence, most of the reconfigurable architectures are implemented using FPGAs [Vaidyanathan and Trahan 2004]. However, since the reconfigurable part of the proposed NoC is limited to some simple configuration switches, not only this NoC can be implemented using FPGAs, but also can be realized on ASIC platforms.

An important consideration in the proposed topology is the long links that may be generated by merging a number of wire segments. Such long wires in a traditional NoC may decrease the NoC clock frequency, if flits cannot pass through them during a single NoC cycle. Even with the repeater insertion, the delay in the wires can exceed one clock period. This problem is addressed by segmenting the long links into fixed length links connected by registers (1-flit buffers) and sending the data over them in a pipelined fashion. Since the connection between two adjacent nodes (on which flits travel in a single NoC cycle) consists of two wire segments, the registers should be distributed among the switches in such a way that each flit is latched after passing through two wire segments.

Figure 3 shows the switches in which the flits are latched. Similar approaches have been used in [Ogras and Marculescu 2005a] and [Carloni et al. 2001] where it has been shown that by sending the flits of a message over a long link in a

Figure 3. The switch boxes on which the flits are buffered are shown as solid circles

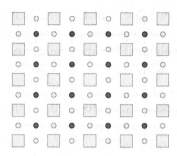

Figure 4. Reconfigurable NoC architectures with more configurable connections and more complex configuration switches (a) and wider configurable connections (b)

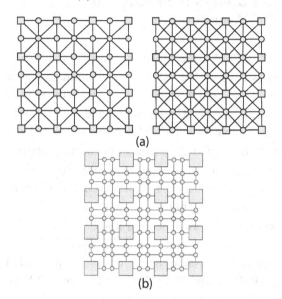

(a)

(b)

pipelined fashion, latency-insensitive operation is guaranteed.

Figure 4 depicts some other implementations of a reconfigurable NoC. These architectures offer more connections between a router and adjacent configuration switches, hence providing more flexibility to reduce the number of hops traveled by messages. This flexibility comes at the price of larger network area. Most of the multi-core SoC designs can be effectively implemented on the simpler reconfigurable NoC shown in Figure 1. As will be shown later, the structure in Figure 1 have a modest implementation cost, which makes it a suitable choice for embedded systems. Nonetheless, evaluating the more flexible reconfigurable NoC structures of Figure 4 for implementing high-performance CMP applications (with more complex traffic) can be considered as a future work in this line.

Although the proposed NoC is not restricted to specific switching and routing schemes, the NoC routers, in this study, implements a wormhole switching mechanism, since this switching method is best suits the limited buffering resources and low-latency communication requirements of on-chip networks. Like most application-specific optimized NoCs, this NoC applies a table-based routing scheme. This allows the NoC to support any static routing algorithm and is suitable to the irregular nature of application-specific topologies. It also allows the designer to exploit its understanding of the

application traffic characteristics and avoid congestion by appropriately allocating paths to traffic flows. In [Hu and Marculescu 2003b], the authors compared the advantages and drawbacks of the dynamic and static routing schemes, and concluded that static routing is more suitable for the NoC based systems due to its low cost, ease of implementation, and power efficiency.

In this work, we assume that the NoC reconfiguration process is initiated by a configuration manager (which may be implemented in the application layer) as a result of switching between one application to another. Switching between network configurations is done in parallel with application switching. In most SoC designs, the application switching time, is of the order of few milliseconds which consists of the time needed to load the data and code of a new application into the SoC, sending control signals to different parts of the SoC, and shutting down the old application [Murali et al. 2006a]. Since the NoC configuration-switching time is by far smaller than the time needed to switch between applications, it do not impose any additional delay to the application switching process.

The Cost of the Proposed NoC Architecture

Like FPGAs, the proposed reconfigurable network pays for the flexibility with additional area overhead. In this section, we evaluate the area of the proposed NoC architecture and compare it to a traditional packet-switched NoC. The traditional NoC is considered as a mesh-connected network with 5 input and 5 output input-buffered routers. The same structure is considered for a reconfigurable NoC, but it includes some additional configuration switches and links.

The area estimation is done based on the hybrid synthesis-analytical area models presented in [Balfour and Dally 2006], [Kim et al. 2006], and [Kim et al. 2008]. In these papers, the area of the router building blocks (e.g. the area of a memory cell) is calculated in 90nm standard cell ASIC technology and then analytically combined to estimate the router total area. Table 1 outlines the architecture-level and circuit-level parameters of our reconfigurable NoC and a

traditional NoC. The analytical area models for a traditional NoC and its components are displayed in Table 2. In Table 3, the models are extended for a reconfigurable NoC. The area of a router is estimated based on the area of the input buffers, network interface queues, and crossbar switch, since the router area is dominated by these components. Although the table assumes an $n \times n$ NoC, the models can be easily extended to an $m \times n$ network.

Each configuration switch is composed of a 4×4 crossbar to implement internal connections. As mentioned before, some switches have four additional 1-flit buffers, each of which latches incoming flits from one of the switch ports.

The area of a configuration switch is by far less than the area of a router since it has smaller buffers, no controller logic (virtual channel and switch allocator, routing logic), and no network adaptor. However, there are n^2 routers and $(2n-1)^2-n^2$ configuration switches in an $n \times n$ mesh-based reconfigurable NoC, respectively. $(n-1)^2$ (out of $(2n-1)^2-n^2$) configuration switches contain one crossbar switch and four 1-flit buffers, while the remaining $2n(n-1)$ switches are only made by one crossbar switch.

Table 1. The architecture-level and circuit-level parameters of a router in reconfigurable NoCs

Parameter	Symbol
Flit Size	F
Buffer Depth	B
No. of Virtual channels	V
Buffer area (0.00002 mm^2/bit [Kim et al. 2008])	B_{area}
Wire pitch (0.00024 mm [Kim et al. 2008])	W_{pitch}
No. of Ports	P
Network Size	N (= n×n)
Packetization queue capacity	PQ
Depacketization queue capacity	DQ
Channel Area (0.00099 mm²/bit/mm [Kim et al. 2006])	W_{area}
Channel Length (1mm [Kumar, Kundu, et al. 2007])	L
No. of Channels (Traditional NoC)	$N_{channel,trad}$
No. of Channels (Reconfigurable NoC)	$N_{channel,reconfig}$

Table 2. The analytical area models for traditional NoCs [Kim et al. 2008]

Component or Unit	Symbol	Model
Crossbar	RCX_{area}	$W^2_{pitch} \times P \times P \times F^2$
Buffer (per port)	RBF_{area}	$B_{area} \times F \times V \times B$
Router	R_{area}	$RCX_{area} + P \times RBF_{area}$
Network Adaptor	NA_{area}	$PQ \times B_{area} + DQ \times B_{area}$
Channel	TCH_{area}	$2 \times F \times W_{area} \times L \times N_{channel,trad}$
Traditional NoC	$TNoC_{area}$	$n^2 \times (R_{area} + NA_{area}) + TCH_{area}$

Table 3. The analytical area models for reconfigurable NoCs

Component or Unit	Symbol	Model
Configuration Switch Crossbar	SCX_{area}	$W^2_{pitch} \times (P-1) \times (P-1) \times F^2$
Configuration Switch Buffer	SBF_{area}	$B_{area} \times F$
Configuration Switch (without buffer)	$S_{area,cx}$	SCX_{area}
Configuration Switch (with buffer)	$S_{area,cxb}$	$SCX_{area} + (P-1) \times SBF_{area}$
Channel	RCH_{area}	$2 \times F \times W_{area} \times L \times N_{channel,reconfig}$
Reconfigurable NoC	$RNoC_{area}$	$n^2 \times (R_{area} + NA_{area}) + (n-1)^2 \times S_{area,cxb} + (n-1)^2 + 2 \times n \times (n-1) \times S_{area,cx} + RCH_{area}$

The area overhead due to the additional inter-router wires is analyzed by calculating the number of channels in a traditional and a reconfigurable mesh-based NoC. A traditional $n \times m$ mesh has $n(m-1)+m(n-1)$ channels while the proposed reconfigurable NoC has $(2m-1) \times (n-1)+(2n-1) \times (m-1)$ channels. We consider two wire-segments in the reconfigurable NoC (which has the same size as a channel in a traditional NoC) as one channel. We assume square-shaped processing elements, measuring 1 mm along an edge, as NoC cores. Consequently, each NoC channel is 1mm long. Figure 5 plots the area overhead of a reconfigurable NoC as a function of router buffer depth for different network sizes in a 128-bit wide system. In Figure 5, each bar shows the amount of extra area (percentage) that must be added to a traditional NoC in order to make it reconfigurable.

As the figure shows, the area of a reconfigurable NoC with 8-flit deep buffers, for example, is increased by 29-40% (depending on the network size) over a traditional NoC. A performance analysis in [Owens et al. 2007] shows that a typical router in next generation high-performance CMPs (in 22nm) will need 64 flit buffers at each port,

Figure 5. The area overhead of the reconfigurable NoC over a traditional NoC for different network sizes and buffering spaces

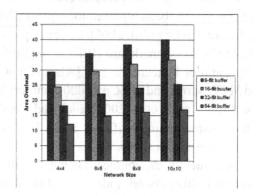

which results in a higher router/switch area and hence reduces the area overhead of the proposed reconfigurable NoC to 12-17%.

In this analysis, the length of packetization (depacketization) queue is assumed to be 32 flits. This is a modest length for the network interface queues (larger queues are used in some implementations, e.g. in [Goossens et al. 2005]). Obviously, by increasing the queue size, the network node/configuration switch area increases, leading to much reduction in the area overhead of the proposed architecture.

Another source of overhead is the data needed to configure the configuration switches and routing tables. Each configuration switch is configured by 6 bits, each of which controls one of the internal connections of the switch. As a result, the configuration data for ten or more applications can be kept in the switch with little storage requirements. Routing information has to be kept for every individual application, as well. Specifically, for the routed connections of each application, this information determines the values of the relevant routing table entries in the routers that are used by these connections.

However, the amount of routing information is small and can be stored in each router with reasonable area overhead. The reason for this small amount of routing information can be explained as follows. In most of the applications implemented as a multi-core SoC [Srinvasan et al. 2006; Murali and De Micheli 2004a; Srinivasan and Chatha 2006], each core communicates with a few (less than 3, on average) other cores. Similar behavior has been reported for typical CMP applications [Ding et al. 2005]. For example, in SPLASH-2 benchmark suit [Biler et al. 1999], NAS parallel benchmark suit [Afsahi 2000], and SPECweb99 and SPECjbb2000 commercial workloads [Jerger et al. 2007], each node tends to have a small number of favored destinations for the messages it sends.

Consequently, the number of communication flows among the network nodes is small, resulting in a small amount of routing information which has to be kept to route the corresponding packets.

Flow-Control Mechanism

We need to modify existing flow control mechanisms to adapt to the variable length of the proposed reconfigurable NoC links. In the current work, we have designed a simple hop-to-hop flow control scheme between the cores based on the traditional *on/off* flow control mechanism [Dally and Towles 2004]. An *off* signal is sent by the downstream node of a long link when the free buffers at its interface falls below the threshold T_{off}. If the number of free buffers rises above a threshold T_{on}, an *on* signal is sent. When the *off* signal is received at the long link upstream node, it stops transmission of data over the long link until an *on* signal is received.

These signals are sent along the path from the source to destination of a long link. This path is set up during the long link construction process. The signals propagate to the sender along this path one hop per cycle. When an *off* signal is transmitted to the upstream node, there must be still sufficient buffering at the receiver to store all of the data items in transit, as well as all of the data items that will be injected during the time it takes for the signal to propagate back to the sender. As a result T_{off} must be at least $2 \times d$, where d is the long link length (in terms of hop count).

Energy and Performance-Aware Mapping Algorithm

In this section, we address the mapping and routing problems in the proposed reconfigurable mesh. Simply stated, for the given set of input applications designed based on a specific set of IP cores, our objective is to (1) map the cores into different nodes of a reconfigurable NoC, (2) find a customized topology for each application, based on the mapping of the previous step and application traffic characteristics, and (3) find a route for the traffic flows of each application, based on the topology found for the application. We develop a two-step algorithm for this problem where core to network mapping is done at the first step and then topology and route generation are done concurrently (Figure 6) at the next step.

Each input application is described as a Communication Task Graph (CTG). CTG is a directed graph $G(V,E)$, where each $v_i \in V$ represents a task, and a directed edge $e_{i,j} \in E$ characterizes the communication from v_i to v_j. The communication volume (bits per second) corresponding to every edge is also provided and is denoted by $t(e_{i,j})$. It is assumed that each task is non-migratory and already mapped onto an IP-core; hence each task graph node represents its corresponding IP-core, as well.

Core to Network Mapping

At the first step, our objective is to figure out how to physically map the IP-cores required by input applications onto different tiles of a mesh network such that the distance between the communicating cores is minimized. We assign a weight to every task-graph based on its criticality, e.g. the percent-

age of time that the corresponding application is run on the NoC. Assigning weights enables the designer to bias the mapping for major or critical applications of the NoC.

This step is performed by constructing a synthetic task-graph (average task-graph) from the task-graphs of the given set of input applications. This average graph includes all the nodes of all task graphs of the input applications. For the edge between every pair of nodes, the weighted average of the corresponding edge volumes across all of the task-graphs is calculated and used in the average task-graph. If an edge does not exist in a task graph, its volume is considered to be 0 in the task graph. More formally, the weight of each edge is calculated as:

$$t_{avr}(e_{x,y}) = (\sum_{\forall applications} t_i(e_{x,y}) \times W_i) / n$$

where W_i represents the weight of the ith input task-graph and $t_i(e_{x,y})$ and $t_{avr}(e_{x,y})$ denote the volume of $e_{x,y}$, the edge between nodes v_x and v_y, in the ith input task-graph and the average task-graph, respectively. n is the number of input task-graphs.

The mapping problem can be formulated as follows. Given a synthetic average CTG constructed from the task graphs of the set of input applications and a reconfigurable NoC (satisfying $size(CTG)<size(NoC)$), find a mapping M from CTG to the NoC nodes as

$$Min\{\sum_{\forall e_{i,j}} t(e_{i,j}) \times dist(M(v_i), M(v_j))\}$$

such that

$$\forall v_i \neq v_j \in CTG, M(v_i) \neq M(v_j)$$

where $dist(a,b)$ is the Manhattan distance between the network nodes a and b and $M(v_i)$ is the network node to which CTG nodes v_i is mapped. $size(CTG)$ and $size(NoC)$ denote the number of

Figure 6. The reconfigurable NoC design flow

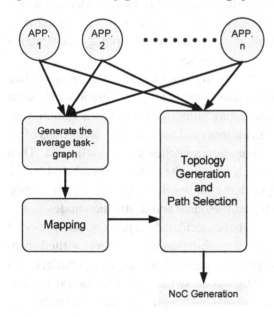

CTG and NoC nodes, respectively. Again, $t(e_{i,j})$ denote the volume of $e_{i,j}$, the edge between vertex v_i and vertex v_j. The constraint formulated by the second relation states that each IP-core should be mapped to exactly one NoC node and no node can host more than one IP-core.

Since core to network mapping is an NP-hard problem [Ascia et al. 2005], instead of searching for an optimal solution, it has been solved by heuristic algorithms in some prior work [Srinvasan et al. 2006; Ascia et al. 2004; Murali and De Micheli 2004a; Pinto et al. 2003]. In this chapter, we tackle the mapping problem for the average graph using NMAP, one of these existing heuristic methods presented in [Murali and De Micheli 2004a].

NMAP uses a heuristic algorithm for power-aware mapping of task graph nodes into a mesh-based network and generating a route for every task-graph edge. We use only the mapping algorithm of NMAP and then, in the next step, propose a topology and route selection algorithm based on the reconfigurable network links.

Initially in this algorithm, all cores are un-mapped. Then, the core mapping is accomplished in some steps, as follows.

Step 1. Map the core with the maximum communication demand onto one of the mesh nodes with maximum number of neighbors.

Step 2. Select the core that communicates most with the already mapped cores and examine all unallocated mesh nodes for placing it. Select the node which minimizes the communication cost between the core and already mapped cores. The communication cost of mapping vertex v_i of the CTG into node x of the NoC is given by:

$$\sum_{\forall j| \ e_{i,j} \in CTG} (t(e_{i,j}) \times dist(x, M(v_j)))$$

where $dist(x, M(v_j))$ is the Manhattan distance between x and the node to which CTG vertex v_j is mapped. We refer interested readers to [Murali and De Micheli 2004a] for the details of the NMAP algorithm.

The process is repeated until all cores are mapped.

Topology and Route Generation

Once the mapping is obtained from the average task-graph, a suitable topology will be constructed for each individual application. As mentioned before, the reconfigurability of the proposed NoC architecture can be exploited for different objectives, such as guaranteeing the QoS level (end-to-end delay, for example) required by an application. However, the algorithms presented in this section aims to reduce the average power consumption and message latency of the NoC for a specific application being processed at a given time.

To achieve this goal, we implement a topology for each application in which the number of hops between the source and destination nodes of high-volume communication flows is as minimum as possible.

The overall idea is to choose the heaviest communication flow that is not yet assigned a route and find a path with minimum possible hop counts. Finding this route may involve configuring the switches not yet configured to bypass some intermediate routers and make a shorter connection between the nodes (in terms of hop counts). As a result, route selection and topology construction is done in parallel.

Initially in the topology selection algorithm, all edges of an application task graph are stored in a decreasing order (in order of their communication volumes) and the internal connections of all configuration switches are unconfigured. Then for each edge in the order, a branch-and-bound algorithm chooses the path with the least cost between its source and destination nodes.

We have calculated the power consumption of a typical mesh-based NoC router (1 virtual channel per physical channel, 8-flit deep buffers) and a configuration switch for different traffic loads [Modarressi and Sarbazi-Azad 2007]. Traffic load denotes the probability of receiving a flit in a cycle. The power consumption of a configuration switch is 4 to 6 times less than a router with the same load. As the cost of a path is related on the number of routers/switches the path contains, we assign a cost of 1 to a configuration switch and a cost of 5 to a router.

The algorithm can configure the unconfigured internal connections of the configuration switches, but not the connections that have been configured at previous iterations of the algorithm (by edges with higher volumes).

The algorithm searches for the optimal path by alternating the following branch and bound steps.

Branch: starting from the source router of the selected edge, the algorithm makes a new branch by adding a router/configuration switch adjacent to the current node to the partial path. Current node is defined as the last node added to a partial path and through which the path will be extended. The added node must be located within the shortest-path area, i.e. along one of the shortest paths between the source and destination nodes of the edge. The shortest path area includes the nodes and configuration switches along one of the shortest paths between the source and destination nodes, as well as their adjacent configuration switches as shown in Figure 7.

If the current node is a router, the path is extended by including its neighboring configura-

tion switches along the shortest path towards the destination node. If the current node is a configuration switch, the path is extended by adding the neighboring routers or configuration switches along the shortest path. However, if the current configuration switch has been already configured (at previous steps of the algorithm) in such a way that the port through which the partial path is reached to the switch is connected to another port, the path can be only extended along the direction determined by the switch configuration, provided that the direction ends to a node along the shortest path towards the destination.

Bound: A branch (a path) is bounded (discarded) in some conditions. First, it is bounded if by adding the new node, the bandwidth constraints of the newly added link is violated. In addition, if the cost of a partial path that reaches to a node is larger than the minimum cost of the partial paths already reached to that node, the path is discarded. The minimum cost of the already completed paths is also kept by the algorithm and a partial path is bounded when its new cost exceeds this value. Finally, we perform a connectivity check to verify that there is at least one path between the source and destination nodes of all edges not yet mapped. If the current partial path configures the switches in such a way that all possible paths between the source and destination of at least one unmapped edge are blocked, the partial path is bounded.

After finding the path with minimum cost for an edge, it is established in the NoC by configuring all configuration switches accordingly within the path and then the algorithm continues with the next edge. The algorithm is repeated for all the edges of the application task-graph. Once all task-graph edges are mapped to a path in the NoC, the paths are analyzed for detecting potential deadlocks. To this end, all cyclic dependencies among paths are broken by adding a virtual channel in one of nodes of the cycle.

Figure 7. The shaded area is the shortest path area between S and D

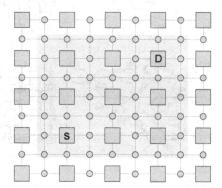

Power and Performance Evaluation

To evaluate the performance of the proposed NoC architecture and mapping/topology generation methodology, we perform simulations on some benchmark application task-graphs. The benchmark set includes two random graphs: random graph 1 with 25 nodes and 30 edges, and random graph 2 with 25 nodes and 20 edges, together with some existing SoC designs which have been widely used in the literature: MPEG4 decoder [Srinivasan and Chatha 2006], Multi-Window Display (MWD) [Srinivasan and Chatha 2006], Video Object Plane Decoder (VOPD) [Murali and De Micheli 2004a], and H263 decoder+mp3 decoder [Srinvasan et al. 2006]. Although, these SoCs have a single task graph, we generate two additional task graphs for each of them by modifying the base task graphs and integrate the three graphs into a single NoC. We also assign a weight of 0.5 to the base task-graph and 0.3 and 0.2 to two other modified graphs. For example, the task-graph of the object plane decoder is depicted in Figure 8.a and the two additional graphs we generated are shown in Figures 8.b and 8.c. The edge tags represent the communication volume between the source and destination nodes of the edge in mega bits per second. The physical mapping is accomplished based on the average graph and then, a network topology is generated for each task-graph. Figures 8.d, 8.e, and 8.f display the topology generated for the graphs of Figure 8.a, 8.b, and 8.c, respectively.

We have implemented the NoC architecture using Xmulator, a fully parameterized simulator for interconnection networks [Xmulator 2008]. The simulator is augmented with the Orion 1.0 power library [Wang et al. 2002] to calculate the power consumption of the networks.

Simulation experiments are performed for a 32-bit wide system with conventional 5-stage pipelined wormhole routers [Dally and Towles 2004] and 8-flit buffers. Moreover, the power results reported by Orion are based on an NoC

with 100nm process feature size and 250 MHz working frequency. The results are compared with a traditional NoC with the same parameters as the reconfigurable NoC. The core-to-NoC mapping in the traditional NoC is exactly the same as the reconfigurable one (i.e., is done by NMAP and based on the average graph), but its topology is fixed for all benchmarks.

In the simulation, packets are generated with exponential distribution and the communication rates between any two nodes are set to be proportional to the communication volume between them in the task-graph. This task-graph-based traffic generation approach is introduced and used in [Hu and Marculescu 2004].

Figure 8. (a) The task-graph for the Video Object Plane Decoder (VOPD), (b) and (c) two task-graphs based on the task-graph in 'a', (d) the network topology for task-graph 'a', (e) the network topology for task-graph 'b', (f) the network topology for task-graph 'c'

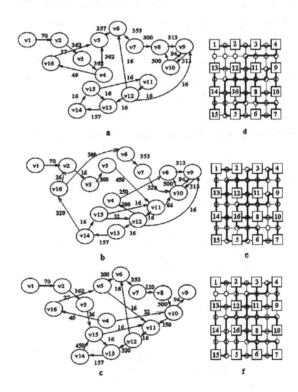

Table 4 displays the power consumption of the NoC. The results show considerable power and performance improvements over a traditional NoC. As the figure indicates, the reconfiguration can adapt the topology to the application and effectively reduce the power consumption of the NoC by 24%, on average. It also improves the average message latency by 19%, on average.

As the difference among the traffic patterns of the applications integrated into a NoC increases, the proposed reconfiguration method reduces the power consumption and message latency more effectively. For example, the two additional graphs generated for the MPEG benchmark are resulted from a slight modification to the original task-graph and thus the impact of reconfiguration on its power consumption is lower than the other benchmarks whose additional modified graphs are generated with more changes. Moreover, two random task-graphs show that the impact of reconfiguration increases with an increase in the number of graph edges. When the number of edges increases, putting the source and destination of all connections near each other becomes difficult which can be alleviated by reconfiguration.

FUTURE RESEARCH DIRECTIONS

In this work, we assume that the input set of applications are known at the design time and are described by a task-graph. In some NoC-based CMPs and programmable SoCs, however, the running applications may not be known at the design time. The proposed NoC can be adapted to these systems by developing a dynamic topology customization mechanism. This goal can be achieved by means of a light-weight setup network we have already proposed for dynamic point-to-point connection construction in hybrid point-to-point/packet-switched NoCs [Modarressi et al. 2009]. This setup-network monitors the on-chip traffic and configures the configuration switches to adapt the topology to the current on-chip traffic characteristics.

Most traditional SoCs have a bus-based shared-medium architecture, where all communication devices share the transmission medium and only one device can drive the bus at a time. Bus splitting, in which a traditional monolithic bus is split into multiple segments, is a mechanism for improving the power and performance of a traditional bus by allowing multiple transactions to proceed in parallel over different segments [Raghunathan et al. 2003]. A large number of bus-based communication protocols and cache coherency mechanisms can be found in the literature [Hennessy and Patterson 2006; Culler et al. 1999].

Supporting bus-based communication in an NoC platform would be highly beneficial for applications with asymmetric communication patterns in which information flows from few

Table 4. The power consumption (in Watts) and message latency (cycles for 8-flit packet) in a traditional and a reconfigurable NoC

Benchmarks	NoC Size	Traditional NoC		Reconfigurable NoC		Improvement (%)	
		Packet Latency	Power	Packet Latency	Power	Packet Latency	Power
VOPD	4×4	22.55	0.073	19.26	0.059	14.59	19.18
MWD	4×3	23.07	0.050	19.54	0.041	15.30	18.00
MPEG	4×3	23.30	0.190	21.33	0.167	8.48	12.11
MP3	4×4	26.09	0.117	20.06	0.086	23.11	26.50
Random Graph-1	5×5	36.34	0.337	26.38	0.210	27.41	37.69
Random Graph-2	5×5	33.05	0.235	25.18	0.152	23.81	35.32

transmitters to many receivers. It is also suitable for broadcast-based coherency protocols.

An interesting property of the proposed reconfigurable NoC architecture is its ability to configure as a global bus or multiple bus segments. A bus is constructed by appropriately configuring the switches in order to connect all nodes together, as shown in Figure 9. The cores, in this configuration, bypass the network adaptor, as well as the router, and directly connect to the constructed bus, so the communication among the cores is done according to their interface protocol, e.g. Open Core Protocol (OCP) [OCPIP 2003].

In addition, the bus configuration can be used along with other topologies to improve the performance of the NoC. For example, a bus can be constructed on demand for one or more cycles, when a processing element has some data to broadcast. In addition to global bus, this architecture also supports multiple concurrent multicasts by constructing multiple independent bus segments among some nodes (e.g. nodes having shared data). This technique is general and can be applied to a variety of bus standards. However, more work needs to be done in order to provide some features required by a bus-based communication system, such as bus arbitration to determine which processor should be granted access to the bus and for how many cycles when several processors attempt to use the bus simultaneously.

CONCLUSION

In this chapter, we presented a reconfigurable architecture for network-on-chips (NoC) on which arbitrary application-specific topologies can be implemented. Since entirely different applications may execute on an SoC at different times, the on-chip traffic characteristics can vary significantly across the applications. However, almost all of the existing NoC design flows and the corresponding application-specific optimization methods customize NoCs based on the traffic characteristics of a single application. The reconfigurability of the proposed NoC architecture allows it to dynamically tailor its topology to the traffic pattern of different applications. In this chapter, we first introduced the proposed reconfigurable topology and evaluated its implementation cost in terms of area overhead over a traditional NoC. The results showed that the proposed reconfigurability increases the NoC area by 12-40% (based on different NoC sizes and buffering spaces) over a traditional NoC. We then addressed the two problems of core to network mapping and topology exploration in which the cores of a given set of input applications are physically mapped to the network and then a suitable topology is found for each individual application. Experimental results, using some multi-core SoC workloads, showed that this architecture effectively improves the performance of the system by 19% and reduces the power consumption by 25%, over one of the

Figure 9. The reconfigurable NoC configured as a (a) global bus (b) multiple bus-segments

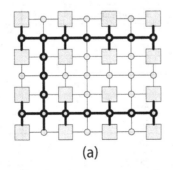
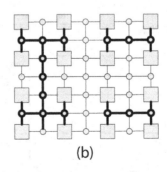

(a)   (b)

most efficient mapping algorithms for traditional NoCs. Regarding the imposed area overhead and power/performance gains, the proposed NoC introduces a more appropriate trade-off between the cost and flexibility, compared to previous related work reported in the literature.

REFERENCES

Abad, P., Puente, V., Gregorio, J. A., & Prieto, P. (2007). Rotary router: an efficient architecture for CMP interconnection networks. In *Proceedings of Symposium on Computer Architecture* (ISCA), (116-125).

Afsahi, A. (2000). *Design and Evaluation of Communication Latency Hiding/Reduction Techniques for Message-Passing Environments* (Unpublished doctoral dissertation). British Columbia, Canada: University of Victoria.

Angiolini, F., Benini, L., Meloni, P., Raffo, L., & Carta, S. (2006). Contrasting a NoC and a traditional interconnect fabric with layout awareness. In *Proceedings of the Design Automation and Test in Europe (DATE)*.

Ascia, G., Catania, M., & Palesi, M. (2005). An evolutionary approach to network-on-chip mapping problem. In *Proceedings of the IEEE Congress on Evolutionary Computation,* (112-119).

Ascia, G., Catania, V., & Palesi, M. (2004). Multi-objective mapping for mesh-based NoC architectures. In *Proceedings of ISSS-CODES,* (182–187).

Atienza, D., Angiolini, F., Murali, S., Pullini, A., Benini, L., & De Micheli, G. (2008). Network on chip design and synthesis outlook. *Journal of Integration, 41*(3), 340–359.

Balfour, J., & Dally, W. J. (2006). Design tradeoffs for tiled CMP on-chip networks. In *Proceedings of the International Conference of Supercomputing.*

Benini, L., & De Micheli, G. (2002). Networks on chip: a new paradigm for systems on chip design. *IEEE Computer, 35*(1), 70–78.

Bertozzi, D., Jalabert, A., Murali, S., Tamahankar, R., Stergiou, S., Benini, L., & De Micheli, G. (2005). NoC synthesis flow for customized domain specific multiprocessor systems-on-chip. *IEEE Transactions on Parallel and Distributed Systems, 16*(2), 113–129. doi:10.1109/TPDS.2005.22

Bilir, E., Dickson, R., Hu, Y., Plakal, M., Sorin, D., Hill, M., & Wood, D. (1999). Multicast snooping: a new coherence method using a multicast address network. In *Proceedings of ISCA.*

Bjerregaard, T., & Mahadevan, S. (2006). A survey of research and practices of network-on-chip. *ACM Computing Surveys, 38*(1), 1–51. doi:10.1145/1132952.1132953

Bjerregaard, T., & Sparsø, J. (2005). A router architecture for connection-oriented service guarantees in the MANGO clockless network-on-chip. In *Proceedings of Design, Automation and Testing in Europe Conference* (DATE), (1226–1231).

Carloni, L. P., McMillan, K. L., & Sangiovanni-Vincentelli, K. L. (2001). Theory of latency-insensitive design. *IEEE Transaction on Computer-Aided Design, 20*(9), 1059–1076. doi:10.1109/43.945302

Chan, J., & Parameswaran, S. (2008). NoCOUT: NoC topology generation with mixed packet-switched and point-to-point networks. In *Proceedings of the Asia and South Pacific Design Automation Conference.*

Culler, D., Singh, J., & Gupta, A. (1999). *Parallel computer architecture: a hardware/software approach* (1st ed.). San Francisco: Morgan Kaufmann Publishers.

Dally, W. J., & Towles, B. (2001). Route packets, not wires: On-chip interconnection networks. In *Proceedings of Design Automation Conference (DAC)*, (684–689).

Dally, W. J., & Towles, B. (2004). *Principles and practices of interconnection networks*. San Francisco: Morgan Kaufmann Publishers.

Ding, Z., Hoare, R., Jones, A., Li, D., Shao, S., Tung, S., et al. (2005). Switch design to enable predictive multiplexed switching in multiprocessor networks. In *Proceedings of IEEE IPDPS*.

Goossens, K., Dielissen, J., & Radulescu, A. (2005). Æthereal network on chip: Concepts, architectures and implementations. *IEEE Design & Test of Computers*, *22*(5), 414–421. doi:10.1109/MDT.2005.99

Guerrier, P., & Greiner, A. (2000). A generic architecture for on-chip packet-switched interconnections. In *Proceedings of the Design Automation and Test in Europe* (DATE), (250–256).

Hennessy, J. L., & Patterson, D. A. (2006). *Computer architecture: a quantitative approach, 4*[th]. San Francisco, CA: Morgan Kaufmann Publishers.

Hoskote, Y., Vangal, S., Singh, A., Bokar, N., & Bokar, S. (2007). A 5-GHz mesh interconnect for a Teraflops processor. *IEEE Micro*, *27*(5), 51–61. doi:10.1109/MM.2007.4378783

Hu, J., & Marculescu, R. (2003a). Energy-aware mapping for tile-based NoC architectures under performance constraints. In *Proceedings of Asia and South Pacific Design Automation Conference*, (233-239).

Hu, J., & Marculescu, R. (2003b). Exploiting the routing flexibility for energy/performance aware mapping of regular NoC architectures. In *Proceedings of the Design Automation and Test in Europe* (DATE).

Hu, J., & Marculescu, R. (2004). Application specific buffer space allocation for networks on chip router design. In *Proceedings of the International Conference on Computer Aided Design*.

Hu, J., & Marculescu, R. (2005). Energy- and performance-aware mapping for regular NoC architectures. *IEEE Transactions on Computer-Aided Design of Integrated Circuits and Systems*, *24*(1), 551–562.

Hwang, K., & Briggs, F. A. (1984). *Computer architecture and parallel processing*. New York: McGraw-Hill.

Jalabert, A., Murali, S., Benini, L., & De Micheli, G. (2004). XpipesCompiler: A tool for instantiating application specific networks-on-chip. In *Proceedings of Design, Automation and Testing in Europe Conference* (DATE). (884-889).

Jantsch, A., & Tenhunen, H. (2003). *Networks on Chip*. Boston: Kluwer Academic Publishers.

Jerger, N. E., Lipasti, M., & Peh, L. S. (2007). Circuit-switched coherence. *IEEE Computer Architecture Letters*, *6*(1), 5–8. doi:10.1109/L-CA.2007.2

Kim, J., Nicopoulos, C., Park, D., Narayanan, V., Yousif, M. S., & Das, C. R. (2006). A gracefully degrading and energy-efficient modular router architecture for on-chip networks. In *Proceedings of International Symposium on Computer Architecture* (ISCA), (4–15).

Kim, K., Lee, S. J., Lee, K., & Yoo, H. J. (2005). An arbitration look-ahead scheme for reducing end-to-end latency in networks-on-chip. In *International Symposium on Circuits and Systems* (ISCAS), (2357-2360).

Kim, M., Davis, J., Oskin, M., & Austin, T. (2008). Polymorphic on-chip networks. In *Proceedings of ISCA*.

Kim, M., Kim, D., & Sobelman, E. (2006). NoC link analysis under power and performance constraints. In *Proceedings of ISCAS*.

Kumar, A., Kundu, P., Singh, A., Peh, L. S., & Jha, N. K. (2007). A 4.6Tbits/s 3.6GHz single-cycle NoC router with a novel switch allocator in 65nm CMOS. In *Proceedings of the International Conference on Computer Design*.

Kumar, A., Peh, L. S., Kundu, P., & Jha, N. K. (2007). Express virtual channels: towards the ideal interconnection fabric. In *Proceedings of the International Symposium on Computer Architecture*, (150-161).

Lei, T., & Kumar, S. (2003). A two-step genetic algorithm for mapping task graphs to a NoC architecture. In *Proceedings of EuroMicro Digital System Design Conference*.

Meloni, P., Murali, S., Carta, S., Camplani, M., Raffo, L., & De Micheli, G. (2006). Routing aware switch hardware customization for networks on chips. In Proceedings of NanoNet.

Millberg, M., Nilsson, E., Thid, R., & Jantsch, A. (2004). Guaranteed bandwidth using looped containers in temporally disjoint networks within the nostrum network-on-chip. In *Proceedings of Design, Automation and Testing in Europe Conference* (DATE), (890–895).

Modarressi, M., & Sarbazi-Azad, H. (2007). Power-aware mapping for reconfigurable NoC architectures. In *Proceedings of the International Conference on Computer Design*.

Modarressi, M., Sarbazi-Azad, H., & Tavakkol, A. (2009). Low-power and high-performance on-chip communication using virtual point-to-point connections. In *Proceedings of the IEEE/ACM International Symposium on Network-on-Chip* (NoCS'09).

Mullins, R., & Moore, W. S. (2004). Low-latency virtual-channel routers for on-chip networks. In *Proceedings of the International Symposium on Computer Architecture*. (188–197).

Murali, S., Coenen, M., Radulescu, R., Goossens, K., & De Micheli, G. (2006a). Mapping and configuration methods for multi-use-case networks on chips. In *Proceedings of Asia and South Pacific Design Automation Conference (ASP-DAC)*, (146-151).

Murali, S., Coenen, M., Radulescu, R., Goossens, K., & De Micheli, G. (2006b). A methodology for mapping multiple use-cases onto networks on chips. In *Proceedings of Design Automation and Test in Europe* (pp. 118–123). DATE.

Murali, S., & De Micheli, G. (2004a). Bandwidth-constrained mapping of cores onto NoC architectures. In *Proceedings of Design Automation and Test in Europe* (pp. 896–901). DATE.

Murali, S., & De Micheli, G. (2004b). SUNMAP: A tool for automatic topology selection and generation for NoCs. In *Proceedings of Design Automation Conference (DAC)*.

OCPIP. (2003). *Open Core Protocol (OCP) Specification, Release 2.0*. Retrieved from http://www.ocpip.org

Ogras, U., & Marculescu, R. (2005a). Application-specific network-on-chip architecture customization via long-range link insertion. In *Proceedings of the Design Automation Conference (DAC)*.

Ogras, U., & Marculescu, R. (2005b). Energy and performance-driven customized architecture synthesis using a decomposition approach. In *Proceedings of the Design Automation and Test in Europe Conference*, (352–357).

Owens, J., Dally, W. J., Ho, R., Jayasimha, D. N., Keckler, S. W., & Peh, L. S. (2007). Research challenges for on-chip interconnection networks. *IEEE Micro, 27*(5), 96–108. doi:10.1109/MM.2007.4378787

Peh, L. S., & Dally, W. J. (2001). A delay model for router microarchitectures. *IEEE Micro, 2*(1), 26–34.

Pinto, A., Carloni, L. P., & Sangiovanni-Vincentelli, A. L. (2003). Efficient synthesis of networks-on-chip. In *Proceedings of the International Conference on Computer Design,* 146–150.

Raghunathan, V., Srivastava, B., & Gupta, R. K. (2003). A survey of techniques for energy efficient on-chip communication. In *Proceedings of the Design Automation Conference* (DAC).

Sabbaghi, R., Modarressi, M., & Sarbazi-Azad, H. (2008-1). The 2d DBM: an attractive alternative to the simple 2d mesh topology for on-chip networks. In *Proceedings of the International Conference on Computer Design.*

Sabbaghi, R., Modarressi, M., & Sarbazi-Azad, H. (2008-2). A novel high-performance and low-power mesh-based NoC. In *Proceedings of the 7th. IPDPS Workshop on Performance Modeling, Evaluation, and Optimization of Ubiquitous Computing and Networked Systems* (PMEO).

Salminen, E., Kulmala, A., & Hamalainen, T. D. (2008). *Survey of network-on-chip proposals.* OCP-IP White Paper. Retrieved December 20, 2008, from http:// www.ocp-ip.org

Srinivasan, K., & Chatha, K. (2005). ISIS: A genetic algorithm based technique for synthesis of on-chip interconnection networks. In *Proceedings of VLSI Design Conference.*

Srinivasan, K., & Chatha, K. (2006). A low complexity heuristic for design of custom network-on-chip architectures. In *Proceedings of Design Automation and Test in Europe.* DATE.

Srinvasan, K., Chatha, K., & Konjevod, G. (2006). Linear programming-based techniques for synthesis of network-on-chip architectures. In IEEE Transaction on VLSI, 14(4), 407-420.

Stensgaard, M., & Sparsø, J. (2008). ReNoC: a network-on-chip architecture with reconfigurable topology. In *Proceedings of International Symposium on Networks-on-Chip* (NoCs), (55-64).

Vaidyanathan, R., & Trahan, J. (2004). *Dynamic Reconfiguration: Architectures and Algorithms.* New York: Springer.

Vassiliadis, S., & Sourdis, I. (2006). Flux networks: Interconnects on demand. In *Proceedings of International Conference on Embedded Computer Systems: Architectures, Modeling and Simulation* (IC-SAMOS), (160-167).

Wang, H., Zhu, X., Peh, L. S., & Malik, S. (2002). Orion: A power-performance simulator for interconnection networks. In *Proceedings of the 35th International Symposium on Microarchitecture* (MICRO).

Xmulator NoC Simulator. (2008). *Xmulator NoC Simulator.* Retrieved December 20, 2008, from http:// www.xmulator.org

ADDITIONAL READING

Balfour, J., & Dally, W. J. (2006). Design tradeoffs for tiled CMP on-chip networks. In *Proceedings of the International Conference of Supercomputing.*

Benini, L., & De Micheli, G. (2002). Networks on chip: a new paradigm for systems on chip design. *IEEE Computer, 35*(1), 70–78.

Bjerregaard, T., & Mahadevan, S. (2006). A survey of research and practices of network-on-chip. *ACM Computing Surveys, 38*(1), 1–51. doi:10.1145/1132952.1132953

Dally, W. J., & Towles, B. 2001. Route packets, not wires: On-chip interconnection networks. In *Proceedings of Design Automation Conference (DAC)*, (684–689).

Dally, W. J., & Towles, B. (2004). *Principles and practices of interconnection networks*. San Francisco: Morgan Kaufmann Publishers.

Hennessy, J. L., & Patterson, D. A. (2006). *Computer architecture: a quantitative approach* (4th ed.). San Francisco: Morgan Kaufmann Publishers.

Hoskote, Y., Vangal, S., Singh, A., Bokar, N., & Bokar, S. (2007). A 5-GHz mesh interconnect for a Teraflops processor. *IEEE Micro*, 27(5), 51–61. doi:10.1109/MM.2007.4378783

Jantsch, A., & Tenhunen, H. (2003). *Networks on Chip*. Boston: Kluwer Academic Publishers.

Kim, M., Davis, J., Oskin, M., & Austin, T. (2008). Polymorphic on-chip networks. In *Proceedings of ISCA*.

Modarressi, M., & Sarbazi-Azad, H. (2007). Power-aware mapping for reconfigurable NoC architectures. In *Proceedings of the International Conference on Computer Design*.

Modarressi, M., Sarbazi-Azad, H., & Tavakkol, A. (2009). Low-power and high-performance on-chip communication using virtual point-to-point connections. In *Proceedings of the IEEE/ACM International Symposium on Network-on-Chip (NoCS'09)*.

Murali, S., Coenen, M., Radulescu, R., Goossens, K., & De Micheli, G. (2006a). Mapping and configuration methods for multi-use-case networks on chips. In *Proceedings of Asia and South Pacific Design Automation Conference (ASP-DAC)*, (146-151).

Murali, S., & De Micheli, G. (2004a). Bandwidth-constrained mapping of cores onto NoC architectures. In *Proceedings of Design Automation and Test in Europe* (pp. 896–901). DATE.

Owens, J., Dally, W. J., Ho, R., Jayasimha, D. N., Keckler, S. W., & Peh, L. S. (2007). Research challenges for on-chip interconnection networks. *IEEE Micro*, 27(5), 96–108. doi:10.1109/MM.2007.4378787

Compilation of References

Abad, P., Puente, V., Gregorio, J. A., & Prieto., P. (2007). Rotary router: an efficient architecture for CMP interconnection networks. In *Proceedings of Symposium on Computer Architecture* (ISCA), (116-125).

Afsahi, A. (2000). *Design and Evaluation of Communication Latency Hiding/Reduction Techniques for Message-Passing Environments* (Unpublished doctoral dissertation). British Columbia, Canada: University of Victoria.

Ahmad, B., Ahmadinia, A., & Arslan, T. (2008, June 22-25). Dynamically Reconfigurable NoC with Bus Based Interface for Ease of Integration and Reduced Design Time. In *Proceedings of IEEE NASA/ESA Conference on Adaptive Hardware and Systems* (AHS 2008), (pp. 309-314). Noordwijk, Netherlands: European Space Agency

Ahmad, B., Erdogan, A. T., A., & Khawam, S. (2006, June 15-18). Architecture of a Dynamically Reconfigurable NoC for Adaptive Reconfigurable MPSoC. In *Proceedings of IEEE NASA/ESA Conference on Adaptive Hardware and Systems* (AHS 2006), (pp. 405-411). Istanbul, Turk

Ahonen, T., Sigüenza-Tortosa, D., Bin, H., & Nurmi, J. (2004). Topology Optimization for Application Specific Networks on Chip. In *Proceedings of the 2004 international workshop on System level interconnect prediction* (pp. 53-60).

Al Fanique, M., Ebi. T., & Henkel, H. (2007), Run-Time Adaptive On-chip Communication Scheme, In *proceedings of the International Conference on Computer Aided Design*, (pp. 26-31).

Ali, M., ; Welzl, M., & ; Hellebrand, S. (2005, November 21-22). , "A dynamic routing mechanism for network on chip. In *Proceedings of the," 23rd NORCHIP Conference, 2005,*. 23rd, (pp. 70-73)., 21-22 Nov. 2005

Amamiya, M., & and Taniguchi, R. (1990). Datarol: a massively parallel architecture for functional languages. In *Proceedings of the Second IEEE Symposium on Parallel and Distributed Processing, 1990. Proceedings of the Second IEEE Symposium on*, (pages 726–735).

Angiolini F., Meloni, P., Carta, S., Benini L., & Raffo, L. (2006, March 6-10). Contrasting a NoC and a Traditional Interconnect Fabric with Layout Awareness. In *Proceedings of Design, Automation and Test in Europe, 2006. (DATE '06)*, (vol.1, pp.1-6).

Angiolini, F., Benini, L., Meloni, P., Raffo, L., & Carta, S. (2006). Contrasting a NoC and a traditional interconnect fabric with layout awareness. In *Proceedings of the Design Automation and Test in Europe (DATE).*

Arteris. (2005). A comparison of network-on-chip and buses. *White paper.*

Ascia, G., Catania, M., & Palesi, M. (2005). An evolutionary approach to network-on-chip mapping problem. In *Proceedings of the IEEE Congress on Evolutionary Computation,* (112-119).

Ascia, G., Catania, V., & Palesi, M. (2004). Multi-objective mapping for mesh-based NoC architectures. In *Proceedings of ISSS-CODES,* (182–187).

Ascia, G., Catania, V., Palesi, M., & Patti, D. (2006). Neighbors-on-path: A new selection strategy for on-chip networks. In *Proceedings of the 4ᵗʰ IEEE Workshop on Embedded Systems for Real Time Multimedia*, (pp. 79-84).

Ashcroft, E. A. (1986). Dataflow and eduction: data-driven and demand-driven distributed computation. In J. W. de Bakker, W. P. de Roever, and G. Rozenberg, (Eds.), *Current Trends in Concurrency. Overviews and Tutorials.* New York: , J. W. de Bakker, W. P. de Roever, and G. Rozenberg, Eds. Springer Lecture Notes In Computer Science, vol. 224. Springer-Verlag. New York, New York, NY, 1-50.

Aslot, V., Domeika, M. J., Eigenmann, R., Gaertner, G., Jones, W. B., & Parady, B. (2001). SPEComp: A New Benchmark Suite for Measuring Parallel Computer Performance. In *Proceedings of the International Workshop on OpenMP Applications and Tools* (WOMPAT '01), (pp. 1-10). London: Springer-Verlag.

Atat, Y., & Zergainoh, N. (2007). Simulink-based MP-SoC Design: New Approach to Bridge the Gap between Algorithm and Architecture Design. In *Proceedings of ISVLSI'07.* (pp. 9–14).

Athanas, P., & Silverman, H. (1993). Processor reconfiguration through instruction-set metamorphosis. *Computer, 26*(3), 11-18.

Atienza, D., Angiolini, F., Murali, S., Pullini, A., Benini, L., & De Micheli, G. (2008). Network on chip design and synthesis outlook. *Journal of Integration, 41*(3), 340-359.

Atitallah, et al. (2007). Multilevel MPSoC simulation using an MDE approach. In *Proceedings of SoCC 2007.*

Atmel (2009). *Atmel Corporation, Inc.* Retrieved February 7, 2009, from http://www.atmel.com

Balfour, J., & Dally, W. J. (2006). Design tradeoffs for tiled CMP on-chip networks. In *Proceedings of the International Conference of Supercomputing.*

Barat, F., Lauwereins, R., & Deconinck, G. (2002). Reconfigurable instruction set processors from a hardware/software perspective. *IEEE Transactions on Software Engineering, 28*(9), 847-862.

Bartic, T. A., ; Mignolet, J.-Y., ; Nollet, V., ; Marescaux, T., ; Verkest, D., ; Vernalde, S., & ; Lauwereins, R.

(2005, July 8). , "Topology adaptive network-on-chip design and implementation.," *IEE Computers and Digital Techniques,* IEE Proceedings - , vol.*152*, no.(4), pp. 467-472., 8 July 2005

Bartic, T., et al. (2005). Topology Adaptive Network-on-Chip Design and Implementation. *IEE Computers and Digital Techniques, 152*(4), 467-472.

Becker, J., Hubner, M., Hettich, G., Constapel, R., Eisenmann, J., & Luka, J. (2007). Dynamic and Partial FPGA Exploitation. *IEEE, 95*(2), 438-452.

Becker, J., Piontek, T., & Glesner, M. (2000). DReAM: A dynamically reconfigurable architecture for future mobile communications applications. In *Proceedings of the 10ᵗʰ Conference on Field Programmable Logic and Application* (pp. 312-321). Villach, Austria.

Becker. J., Donlin, A., & Hübner, M. (2007). New tool support and architectures in adaptive reconfigurable computing. In *Proceedings of the IFIP International Conference on Very Large Scale Integration VLSI-SoC,* (pp. 134–139).

BEE2. (2008). *BEE2- Berkeley Emulation Engine 2.* Retrieved from http://bee2.eecs.berkeley.edu/

Benini, L. (2006). Application Specific NoC Design. In *Proceedings of the Conference on Design, Automation and Test in Europe.* (pp. 491-495).

Benini, L. (2006, March 6-10). Application Specific NoC Design. In *Proceedings of Design, Automation and Test in Europe, 2006.* (DATE '06), (vol.1, pp.1-5).

Benini, L., & ; De Micheli, G. (2002, January). , "Networks on chips: a new SoC paradigm. ," *Computer,* , vol.*35*, no.(1), pp.70-78., Jan 2002

Benini, L., & De Micheli, G. (2001). Powering networks on chips: **energy-efficient and reliable interconnect design for SoCs.** In *Proceedings of the 14th international symposium on Systems synthesis,* (pp. 33–38).

Benini, L., & Micheli, G. D. (2001). Powering Networks on Chips: Energy-Efficient and Reliable Interconnect Design for SoCs. In *Proceedings of the 14th international symposium on Systems synthesis* (ISSS '01), (pp.33-38),

Benini, L., & Micheli, G. D. (2002). Networks on chip: a new paradigm for systems on chip design. In *Proceedings of Conference on Design, Automation and Test in Europe*.

Benini, L., & Micheli, G. D. (2002, January). Networks on Chips: a new SoC paradigm. *IEEE Computer, 35*(1), 70–78.

Benini, L., & Micheli, G. D. (2006). *Networks on Chips: Techonology and Tools*. San Francisco: Morgan Kaufmann.

Berthelot et al. (2008). A Flexible system level design methodology targeting run-time reconfigurable FPGAs. *EURASIP Journal of Embedded Systems, 8*(3), 1–18.

Bertozzi, D., & ; Benini, L. (2004). , Xpipes: a network-on-chip architecture for gigascale systems-on-chip. *IEEE, Circuits and Systems Magazine, IEEE, 4*(2), :18–31., 2004

Bertozzi, D., Benini, L., & de Micheli, G. (2002). Low Power Error Resilient Encoding for On-Chip Data Buses, In *Proceedings of the conference on Design, automation and test in Europe* (DATE '02)

Bertozzi, D., Jalabert, A., Murali, S., Tamahankar, R., Stergiou, S., Benini, L., & De Micheli, G. (2005). NoC synthesis flow for customized domain specific multiprocessor systems-on-chip. *IEEE Transactions on Parallel and Distributed Systems, 16*(2), 113-129.

Bertozzi, D., Jalabert, A., Murali, S., Tamhankar, R., Stergiou, S., Benini, L., & De Micheli, G. (2005, February). NoC synthesis flow for customized domain specific multiprocessor systems-on-chip. In Proceedings of IEEE Transactions on Parallel and Distributed Systems (vol.16, no.2, pp. 113-129).

Bijlsma, B. (2005, September). *Asynchronous Network-on-Chip Architecture Performance Analysis* (Master's thesis). The Netherlands: Department of Electrical Engineering, Faculty of Electrical Engineering, Mathematics and Computer Science, Delft University of Technology.

Bilir, E., Dickson, R., Hu, Y., Plakal, M., Sorin, D., Hill, M., & Wood, D. (1999). Multicast snooping: a new coherence method using a multicast address network. In *Proceedings of ISCA*.

Bjerregaard, T., & Mahadevan, S. (2006). A Survey of Research and Practices of Network-on-Chip. *ACM Computing Surveys, 38*(1), 1-51.

Bjerregaard, T., & Mahadevan, S. (2006). A survey of research and practices of network-on-chip. *ACM Computing Surveys, 38*(1), 1-51.

Bjerregaard, T., & Sparsø, J. (2005). A router architecture for connection-oriented service guarantees in the MANGO clockless network-on-chip. In *Proceedings of Design, Automation and Testing in Europe Conference (DATE)*, (1226–1231).

Bobda, C., & ; Ahmadinia, A. (2005, September-October). , Dynamic interconnection of reconfigurable modules on reconfigurable devices. *IEEE Design & Test of Computers, IEEE, 22*(5), :443–451., Sept.-Oct. 2005

Bobda, C., Ahmadinia A., Majer, M., Teich J., Fekete, P. S., & van der Veen, J. (2005). Dynoc: A dynamic infrastructure for communication in dynamically reconfigurable devices. In Tero Rissa et al. (Ed.), *Field Programmable Logic and Application FPL, In Proceedings of the 15th International Conference on Field Programmable Logic and Application* (FPL), (pp. 153–158). Washington, DC: IEEE Press.

Bobda, C., Majer, M., Koch, D., Ahmadinia, A., & Teich, J. (2004). A dynamic NoC approach for communication in reconfigurable devices. In Becker et al. (Ed.), *Field Programmable Logic and Application FPL, 14th International Conference volume 3203 of Lecture Notes in Computer Science* (pp. 1032–1036). New York: Springer.

Boden et al. (2008). GePARD - a High-Level Generation Flow for Partially Reconfigurable Designs. In *Proceedings of ISVLSI 2008*.

Borkar, S. (2007). Networks for multi-core chips: A contrarian view. In *Proceedings of the Special Session at ISLPED 2007*.

Boulet, P. (2007). *Array-OL Revisited, Multidimensional Intensive Signal Processing Specification*. (Tech. rep.), INRIA. Retreived from http://hal.inria.fr/inria-00128840/en/

Braun, L., Hubner, M., Becker, J., Perschke, T., Schatz, V., & Bach, S. (2007, August). *Circuit switched run-time adaptive network-on-chip for image processing applications*. In *Proceedings of the Field Programmable Logic and Applications International Conference.*

Bray, T., & Paoli, J. (2008). , Sperberg-McQueen, C. M., Maler, E., and Yergeau, F. (2008). *Extensible Markup Language (XML) 1.0* (Fifth Edition). Cambridge, MA: MIT Press

Brooks, D., & Martonosi, M. (2001). Dynamic thermal management for high performance microprocessors. In *Proceedings of the 'Int. Symp. High Performance Computer Architecture.*

Butts, M. (2007). Synchronization through communication in a massively parallel processor array. *IEEE Micro, 27*(5), :32–40.

Carloni, L. P., McMillan, K. L., & Sangiovanni-Vincentelli, K. L. (2001). Theory of latency-insensitive design. *IEEE Transaction on Computer-Aided Design, 20*(9), 1059–1076.

Carvalho, E., Calazans, N., Moraes, F., & Mesquita, D. (2004). Reconfiguration control for dynamically reconfigurable systems. In *Proceedings of the 19th Conference on Design Circuits and Integrated Systems* (pp. 405-410). Bordeaux, France.

Carver et al. (2008). *Relocation and Automatic Floorplanning of FPGA Partial Configuration Bit-Streams.* (Tech. rep.), Redmond, WA: Microsoft Research.

Ces'ario, W., Baghdadi, A., Gauthier, L., Lyonnard, D., Nicolescu, G., Paviot, Y., Yoo, S., Jerraya, A. A., & Diaz-Nava, M. (2002, June). Component-based design approach for multicore SoCs. In *Proceedings of the 39th Design Automation Conference*, pages 789–794. ACM Press.

Cesario et al. (2002). Component-Based Design Approach for Multicore SoCs. In *Proceedings of Design Automatic Conference,* (DAC'2002), (pp. 789).

Chan, J., & Parameswaran, S. (2004). Nocgen: A template based reuse methodology for networks on chip architecture. In *Proceedings of the 17th International Conference on VLSI Design* (VLSID), (pp.717-7120).

Chan, J., & Parameswaran, S. (2008). NoCOUT : NoC topology generation with mixed packet-switched and point-to-point networks. In *Proceedings of the Asia and South Pacific Design Automation Conference.*

Chen, W., Wang, Y., Wang, X., & Peng, C. (2008, July 29-31). A New Placement Approach to Minimizing FPGA Reconfiguration Data. *In Proceedings of the International Conference on Embedded Software and Systems, 2008* (ICESS '08), (pp.169-174).

Chiu, G. (2000). The odd-even turn model for adaptive routing. *IEEE Transactions Parallel Distributed Systems, 11*(7), 729–728.

Ciordas, C., Hansson, A., Goossens, K., Basten, T. (2006). A Monitoring-Aware Network-on-Chip Design Flow. In *Proceedings of the 9th EUROMICRO Conference on Digital System Design: Architectures, Methods and Tools* (DSD 2006), (pp.97-106).

Claus, C., Zeppenfeld, J., Muller, F., & Stechele, W. (2007). Using Partial-Run-Time Reconfigurable Hardware to accelerate Video Processing in Driver Assistance System. In *Proceedings of Design, Automation and Test in Europe Conference and Exposition* (pp. 1-6). Nice, France.

Claus, C., Zhang, B., Huebner, M., Schmutzler, C., & Becker, J. (2007, May). An XDL-based busmacro generator for customizable communication interfaces for dynamically and partially reconfigurable systems. In *Proceedings of the Workshop on Reconfigurable Computing Education at ISVLSI 2007.*

Cocco, M., Dielissen, J., Heijligers, M., Hekstra, A., Huisken, J., Hive, S., & and Eindhoven, N. (2004). A scalable architecture for LDPC decoding. In *Proceedings of Design, Automation and Test in Europe Conference and Exhibition, 2004* (Vol. 3). *Proceedings,* volume 3.

Coppola, M., Curaba, S., Grammatikakis, M., Locatelli, R., Maruccia, G., & Papariello, F. (2004). Occn: a noc modeling framework for design exploration. *Journal of Systems Architecture, 50*(2-3), 129–163.

CoreConnect. (1999). *CoreConnect Bus Architecture.* Retrieved from http://www-01.ibm.com/chips/techlib/techlib.nsf/literature/CoreConnect_Bus_Architecture

Cormen, T. H., Leiserson, C. E., Rivest, R. L., & Stein, C. (2001). *Introduction to Algorithms, Second Edition.* Cambridge, MA: The MIT Press.

Cozzi, D., Farè, C., Meroni,, A., Rana V., Santambrogio, M. D., & Sciuto, D. (2009, May). Reconfigurable NoC design flow for multiple applications run-time mapping on FPGA devices. In *Proceedings of the 19th ACM/IEEE Great Lakes Symposium on VLSI,* (pp. 421-424).

Culler, D., Singh, J., & Gupta, A. (1999). *Parallel computer architecture: a hardware/software approach,* 1st. ed. San Francisco: Morgan Kaufmann Publishers.

Dally, et al. (2001). Not Wires: On-Chip Interconnection Networks. In *Proceedings of the IEEE Design Automation Conf.,* (pp. 684–689).

Dally, W. J. (1992). Virtual channel flow control. *IEEE Trans on parallel and Distributed Systems, 3*(2), 194-205.

Dally, W. J., & Aoki, H. (1993). Deadlock -free adaptive routing in multicomputer networks using virtual channels. In *Proceedings of IEEE Trans. on Parallel and Distributed Systems.*

Dally, W. J., & Seitz, C. L. (1987). Deadlock-free message routing in multiprocessor interconnection networks. *IEEE Transactions on Computing, 36*(5), 547–553.

Dally, W. J., & Towles, B. (2001). Route packets, not wires: On-chip inteconnection networks. In *Proceedings of the 38th conference on Design automation* (DAC '01), (pp. 684-689). New York: ACM

Dally, W. J., & Towles, B. (2001). Route packets, not wires: On-chip interconnection networks. In *Proceedings of Design Automation Conference* (DAC), (684–689).

Dally, W. J., & Towles, B. (2004). *Principles and practices of interconnection networks.* San Francisco: Morgan Kaufmann Publishers.

Dally, W., & Towles, B. (2001). Route packets, not wires: On-chip interconnection networks. In *Proceedings of the Design Automation Conference* (pp. 684-689). San Diego, CA.

Damasevicius, R., & Stuikys, V. (2004). Application of UML for hardware design based on design process model. In *Proceedings of ASP-DAC'04.*

DaRT team. (2009). *GASPARD SoC Framework.* Retrieved from http://www.lifl.fr/DaRT

De Micheli, G., & Benini, L. (2006). *Network on chips.* San Francisco: Morgan Kaufmann.

Deb K., Pratap, A., Agarwal, S., Meyarivan, T. (2002, April), A fast and elitist multi-objective genetic algorithm. *IEEE Transactions on NSGA-II, Evolutionary Computation, 6*(2), 182–197.

Dehyadgari, M., Nickray, M., Afzali-Kusha, A., & Navabi, Z. (2006). A new protocol stack model for network on chip. In *Proceedings of the Annual Symposium on VLSI, ISVLSI* (pp 440–44). Karlsruhe, Germany: IEEE Computer Society.

Dehyadgari, M., Nickray, M., Afzali-kusha, A., & Navabi, Z. (2005). Evaluation of pseudo adaptive XY routing using an object oriented model for NOC. In *Proceedings of the International Conference on Microelectronics.* (pp. 204-208).

Diguet, J.- Ph., Evain, S., Vaslin, R., Gogniat, G., & Juin, E. (2007, May). *Noc-centric security of reconfigurable soc.* In *Proceedings of the 1st ACM/IEEE International Symposium on Networks-on-Chips.* Princeton, NJ.

Ding, Z., Hoare, R., Jones, A., Li, D., Shao, S., Tung, S., Zheng, J., & Melhem, R. (2005). Switch design to enable predictive multiplexed switching in multiprocessor networks. In *Proceedings of IEEE IPDPS.*

Donato, A., Ferrandi, F., Redaelli, M., Santambrogio, M., & Sciuto, D. (2005). Caronte: A complete methodology for the implementation of partially dynamically self-reconfiguring systems on FPGA platforms. In *Proceedings of the IEEE Symposium on Field-Programmable Custom Computing Machines* (pp. 321-322). Napa, CA.

Dong, W., Al-Hashimi, B., & Schmitz, M. (2006). Improving Routing Efficiency for Network-on-Chip through

Contention-Aware Input Selection. In *the Proceedings of the Asia and South Pacific Design Automation Conference* (pp. 36-41).

Dorairaj, et al. (2005). PlanAhead Software as a Platform for Partial Reconfiguration. *Xcell Journal, 55,* 68–71.

Duato, J. (1993). A new theory of deadlock-free adaptive routing in wormhole networks. *IEEE Transactions Parallel Distributed Systems, 4*(12), 1320–1331.

Duato, J. (1994). A necessary and sufficient condition for deadlock free adaptive routing in wormhole networks. In *Proceedings of the International Conference on Parallel Processing.*

Duato, J., et al. (2007). *Interconnection Networks.* San Francisco: Morgan Kaufmann.

Duato, J., Yalamanchili, S., & Ni, L. M. (1997, January). *Interconnection Networks: an engineering approach.* San Francisco, Morgan Kaufman.

Dyer, M., & Wirz, M. (2002). *Reconfigurable Systems of FPGA* (Diploma Thesis). Swiss Federal Institute of Technology Zurich (82 p.). Zurich, Switzerland.

Dyer, M., Plessl, C., & Platzner, M. (2002). Partially Reconfigurable Cores for Xilinx Virtex. In *Proceedings of the 12th Conference on Field Programmable Logic and Application* (pp. 292-301). Montpellier, France.

Eclipse. (n.d.). *Eclipse Modeling Framework Technology.* Retrieved from http://www.eclipse.org/emft

Elmiligi, H., Morgan, A. A., El-Kharashi, M. W., & Gebali, F. (2008) Power-Aware topology optimization for networks-on-chips. In *Proceedings of the IEEE International Symposium on Circuits and Systems,* (pp. 360–363).

Embedded linux/microcontroller project (2008). *Embedded linux/microcontroller project.* Retrieved from http://www.uclinux.org/

ESA-ESTEC. (2003). *Space Engineering: SpaceWire–Links, nodes, routers, and networks.* Norrdwijk Netherlands: ESA-ESTEC.

Eto, E. Xilinx Inc. (2004, September). , *XAPP290 - Two flows for partial reconfiguration: Module based or Difference based..* San Jose, CA: Xilinx Inc., September 2004.

Evain, S., Dafali, R., Diguet, J.-Ph., & Juin, E. (2007). *μspider CAD tool: Case Study of NoC IP Generation for FPGA.* In *Proceedings of the Workshop on Design and Architectures for Signal and Image Processing.* Grenoble, France.

Evain, S., & Diguet, J.-Ph. (2007 March). *Efficient space-time NoC path allocation based on mutual exclusion and pre-reservation.* In *Proceedings of the 17th ACM Great Lakes Symposium on VLSI* (GLSVLSI). Italy.

Evain, S., Diguet, J.-P., Houzet, D. (2004, November 18-19). A generic CAD tool for efficient NoC design," , 2004.. In *Proceedings of 2004 International Symposium on Intelligent Signal Processing and Communication Systems* (ISPACS 2004), (728-733).

Faruque, M., Ebi, T., & Henkel, J. (2007). Run-time Adaptive on-chip Communication Scheme. In *Proceedings of the International Conference on Computer Aided Design,* (pp. 26-31).

Fekete, S. P., van der Veen, J. C., Ahmadinia, A., Gohringer, D., Majer, M., & Teich, J. (2008). Offline and Online Aspects of Defragmenting the Module Layout of a Partially Reconfigurable Device. *IEEE Transactions on Very Large Scale Integration Systems* (VLSI), *16*(9), 1210-1219.

Ferrandi, F., ; Santambrogio, M. D., & ; Sciuto, D. (2005, April 4-8). , "A design methodology for dynamic reconfiguration: the Caronte architecture. In *Proceedings of the 19th IEEE International,*" *Parallel and Distributed Processing Symposium, 2005. Proceedings. 19th IEEE International,* (pgp. 4)., 4-8 April 2005

Ferrandi, F., Morandi M., Novati M., Santambrogio M. D., Sciuto D. (2006, November). Dynamic Reconfiguration: Core Relocation via Partial Bitstreams Filtering with Minimal Overhead. In *Proceedings of the International Symposium on System-on-Chip* (SoC 06), (pp. 33-36).

Fiethe, B., Michalik, H., Dierker, C., Osterloh, B., & Zhou, G. (2007). Reconfigurable System-on-Chip Data

Processing Units for Miniaturized Space Imaging Instruments. In *Proceedings of the conference on Design automation and test in Europe* (DATE), (pp. 977-982). New York: ACM.

Francalanci, C., & Giacomazzi, P. (2006, January). *High-performance self-routing algorithm for multiprocessor systems with shuffle interconnections.* In Proceedings of IEEE Transactions on Parallel and Distributed Systems, (vol. 17).

G"otz, M., & Dittmann, F. (2006, September). Reconfigurable microkernel-based RTOS: mechanisms and methods for run-time reconfiguration. In *Proceedings of the IEEE International Conference on Reconfigurable Computing and FPGAs*, (pp. 1–8).

Gailliard, et al. (2007). Transaction level modelling of SCA compliant software defined radio waveforms and platforms PIM/PSM. In *Proceedings of Design, Automation & Test in Europe, DATE'07*.

Gamatié, et al. (2008b). A model driven design framework for high performance embedded systems. *Research Report RR-6614*, INRIA. Retrieved from http://hal.inria.fr/inria-00311115/en

Genko, N., Atienza, D., De Micheli, G., Benini, L., Mendias, J.M., Hermida, R., & Francky C. (2005). A novel approach for network on chip emulation. In *Proceedings of the IEEE International Symposium on Circuits and Systems,* (Vol. 3, 2365–2368).

Genko, N., Atienza, D., De Micheli,G. & Benini., L. (2007). Feature - noc emulation: a tool and design flow for mpsoc. In *Circuits and Systems Magazine, 7*(4), 42 – 51, IEEE.

Gericota, M., Alves, G., Silva, M., & Ferreira, J. (2003). Run-Time Management of Logic Resources on Reconfigurable Systems. In *Proceedings of the Design, Automation and Test in Europe Conference and Exposition* (pp. 974-979). Munich, Germany.

Ghomsheh, V. S., Khanehsar, M. A., Teshnehlab, M. (2007). Improving the non-dominate sorting genetic algorithm for multi-objective optimization. *In Proceedings of CISW 2007*, (pp. 89–92).

Giraud-Carrier, C. (1994). A reconfigurable dataflow machine for implementing functional programming languages. *ACM Sigplan Notices, 29*(9), :22–28.

Glass, C., & Ni, L. (1992). Maximally fully adaptive routing in 2D meshes. In *Proceedings of International Conference Parallel Processing*, (pp. 101–104).

Glass, C., & Ni, L. (1994). The turn model for adaptive routing. *Journal of ACM, 31*(5), 874–902.

Goldstein, S., Schmit, H., Moe, M., Budiu, M., Cadambi, S., Taylor, R., & Laufer, R. (1999). PipeRench: A coprocessor for streaming multimedia acceleration. In *Proceedings of the 26th International Symposium on Computer Architecture.* (pp. 28-39). Atlanta, GA.

Gonzalez, R. (2000). Xtensa: A configurable and extensible processor. *IEEE Micro, 20*(2), 60-70.

Goossens, K., ; Dielissen, J., & ; Radulescu, A. (2005, September-October). , "AEthereal network on chip: concepts, architectures, and implementations. ," *IEEE, Design & Test of Computers, IEEE* , vol.22, no.(5), pp. 414-421., Sept.-Oct. 2005

Goossens, K., Bennebroek, M., Young ,J. H., & Wahlah, M. A. (2008). Hardwired Networks on Chip in FPGAs to unify Data and Configuration Interconnects. In *Proceedings of the Second Symposium ACM/IEEE International Symposium on Networks on Chip NOCS* (pp.45-54).

Gough, J. (2001). *Compiling for the.Net Common Language Runtime* (CLR). Upper Saddle River, NJ: Pearson Education., 1st edition.

Grecu, G., Ivanov, A., Pande, P. P., Jantsch, A., Salminen, E., Ogras, U., & Marculescu, R. (2007). Towards open network-on-chip benchmarks. In *Proceedings of the NOCS*, (pp. 205). Washington, DC: IEEE Computer Society.

Griese, B., Vonnahme, E., Porrmann, M., & Rückert, U. (2004). Hardware support for dynamic reconfiguration in reconfigurable SoC architectures. In *Proceedings of the 14th Conference on Field Programmable Logic and Application* (pp. 842-846). Leuven, Belgium.

Guccione, S., Levi, D., & Sundararajan, P. (1999). JBits: A Java Based Interface for Reconfigurable Computing. In *Proceedings of the 2nd Annual Military and Aerospace Applications of Programmable Devices and Technologies Conference*. Laurel, MD.

Guerrier, P., & and Greiner, A. (2000). A generic architecture for on-chip packet-switched interconnections. In *Proceedings of DATE '00: Proceedings of The conference on Design, automation and test in Europe*, (pages 250–256). , New York: , NY, USA. ACM.

Guerrier, P., & Greiner, A. (2000) A generic architecture for on-chip packetswitched interconnections. In *Design, Automation and Test in Europe Conference and Exhibition DATE* (pp. 250–256).

Guerrier, P., and Greiner, A. (2000). A generic architecture for on-chip packet switched interconnections. In *Proceedings of the Design, Automation and Test in Europe Conference*, (pp. 250–256).

Guo, et al. (2005). Optimized generation of data-path from C codes for fpgas. In *Proceedings of Design, Automation & Test in Europe, DATE'05* (112–117).

Hansson, A., Coenen, M., & Goossens, K. (2007, April). *Undisrupted quality-of-service during reconfiguration of multiple applications in networks on chip*. In *Proceedings of Design, Automation, Test in Europe Conference and Exhibition*.

Hansson, A., Goossens, K., & Radulescu, A. (2005). A unified approach to constrained mapping and routing on network-on-chip architectures. In *Proceedings of the 3rd IEEE/ACM/IFIP International Conference on Hardware/Software Codesign and System Synthesis*, (pp. 75-80).

Hartenstein, R. (2001). A decade of reconfigurable computing: A visionary retrospective. In *Proceedings of Design, Automation and Test in Europe Conference and Exposition* (pp. 642-649). Munich, Germany.

Hartenstein, R., & Kress, R. (1995). A datapath synthesis system for the reconfigurable datapath architecture. In *Proceedings of the Conference on Asia Pacific Design Automation* (pp. 479-484). Makuhari, Massa, Chiba, Japan.

Hashimoto, A., & Stevens, J. (1988). Wire routing by optimizing channel assignments within large apertures. In *Proceedings of 25 years of DAC: Papers on Twenty-five years of electronic design automation*, (pp. 35–49).

Hauck, S., Fry, T., Hosler, M., & Kao, J. (1997). The chimaera reconfigurable functional unit. In *Proceedings of the 5th IEEE Symposium on Field-Programmable Custom Computing Machines* (pp. 87-96). Napa, CA.

Hauser, J., & Wawrzynek, J. (1997). Garp: A MIPS processor with a reconfigurable coprocessor. In *Proceedings of the IEEE Symposium on Field-Programmable Custom Computing Machines* (pp. 12-21). Napa, CA.

Hausman, G. Gudenzi, J. M., & Tempest, S. (1990). *Programmable voltage controlled ring oscillator*. US Patent 4978927.

Hegde, R., & Shanbhag, N. R. (2000). Toward achieving energy efficiency in presence of deep submicron noise. *IEEE Trans. Very Large Scale Integr. Syst., 8*(4), 379-391.

Hemani, A., Jantsch, A., Kumar, S., Postula, A., Öberg, J., Millberg, M., and Lindqvist, D. (2000). Network on a Chip: An architecture for billion transistor era. In *Proceeding of the IEEE NorChip Conference*.

Hemani, J. A., Kumar, S., Postula, A., Oberg, J., Millberg, M., & Lindqvist, D. (2000). Network on Chip: An Architecture for Billion Transistor Era. In *Proceedings of the IEEE NorChip Conference*. Turku, Finland.

Hennessy J. L., & Patterson, D. A. (2006). *Computer architecture: a quantitative approach*, 4th. ed. San Francisco, CA: Morgan Kaufmann Publishers.

Hetch, R., Kubisch, S., Herrholtz, A., & Timmermann, D. (2005). Dynamic Reconfiguration With hardwired Networks-on-Chip on Furure FPGAs. In Tero Rissa et al. (Ed.), *Field Programmable Logic and Application FPL, 15th International Conference* (pp. 527- 530). Washington, DC: IEEE.

Hetch, R., Kubisch, S., Michelsen, H., Zeeb, E., & Timmermann, R. (2006). A distributed object system approach for dynamic reconfiguration. In *Parallel and Distributed Processing IPDPS*, (pp.8). Washington, DC: IEEE.

Hilton, C., & Nelson, B. (2005, Auguest). A flexible circuit-switched NOC for FPGA-based systems. In *Proceedings of the 15th IEEE International Conference on Field Programmable Logic and Applications*, (pp. 191–196).

Hollstein, T., Ludewig, R., Zimmer, H., Mager, C., Hohenstern, S., & Glesner, M. (2006). Hinoc: A hierarchical generic approach for on-chip communication, testing and debugging of socs. *VLSI-SOC: From Systems to Chips, 200*(4), 39–54.

Hommais, D., P'etrot, F., & Aug'e, I. (2001, June). A tool box to map system level communications on Hardware/Software architectures. In *Proceedings of the 12th International Workshop on Rapid System Prototyping*, (pp. 77–83).

Horta, E., Lockwood, J., & Kofuji, S. (2002). Using PARBIT to Implement Partial Run-Time Reconfigurable Systems. In *Proceedings of the 12th Conference on Field Programmable Logic and Application* (pp. 182-191). Montpellier, France.

Hoskote, Y., Vangal, S., Singh, A., Borkar, N., & Borkar, S. (2007, September- October). *A 5-ghz mesh interconnect for a teraflops processor* (volume 27, pp. 51–61).

Hu, J, Ogras, U., & Marculescu, R. (2006). System-Level Buffer Allocation for Application-Specific Networks-on-Chip router Design. *IEEE Transactions on computer-aided-design of integrated circuits and systems, 25*(12), 2919-2933.

Hu, J. & Marculescu, R. (2005, April 4). Energy- and performance-aware mapping for regular NoC architectures. In *Proceedings of IEEE Transactions on Computer-Aided Design of Integrated Circuits and Systems* (pg 24).

Hu, J., & Marculescu, R. (2003). Energy-aware mapping for tile-based NoC architectures under performance constraints. In *Proceedings of the Design Automation Conference, 2003* (ASP-DAC 2003), (pp. 233–239).

Hu, J., & Marculescu, R. (2003). Exploiting the routing flexibility for energy/performance aware mapping of regular NoC architectures. In *Proceedings of the Design,*

Automation and Test in Europe Conference and Exhibition (pp. 688–693).

Hu, J., & Marculescu, R. (2003a). Energy-aware mapping for tile-based NoC architectures under performance constraints. In *Proceedings of Asia and South Pacific Design Automation Conference,* (233-239).

Hu, J., & Marculescu, R. (2003b). Exploiting the routing flexibility for energy/performance aware mapping of regular NoC architectures. In *Proceedings of the Design Automation and Test in Europe* (DATE).

Hu, J., & Marculescu, R. (2004). Application specific buffer space allocation for networks on chip router design. In *Proceedings of the International Conference on Computer Aided Design.*

Hu, J., & Marculescu, R. (2004). DyAD – smart routing for networks-on-chip. In *proceedings of the 41st Annual Conf. Design and Automation,* (pp. 260–263).

Hu, J., & Marculescu, R. (2004, June). DyAD - Smart Routing for Networks-on-Chip. In *Proceedings of DAC 2004.* San Diego, CA.

Hu, J., & Marculescu, R. (2005). Energy- and Peformance-Aware Mapping for Regular NoC Architectures. *IEEE Transactions on Computer-Aided Design f Integrated Circuuits and Systems, 24*(4), 551-562.

Hu, J., Marculescu, R. (2003, January 21-24). Energy-aware mapping for tile-based NoC architectures under performance constraints. In *Proceedings of Design Automation Conference, 2003* (ASP-DAC 2003), (pp. 233-239).

Huang, S. S., Hormati, A., Bacon, D. F., & and Rabbah, R. (2008, July 7-11). Liquid Metal: Object-Oriented Programming Across the Hardware/Software Boundary. In *Proceedings of the 22nd European Conference on Object-Oriented Programming.* (Paphos, Cypress., July 07 - 11, 2008). J. Vitek, Ed. Lecture Notes In Computer Science, vol. 5142. Springer-Verlag, Berlin, Heidelberg, 76-103.

Hübner, M., Braun, L., Göhringer, D., & Becker, J. (2008). Run-time reconfigurable adaptive multilayer network-on-chip for fpga-based systems. In Proceedings of *Parallel and Distributed Processing IPDPS*, (pp.1–6). Washington, DC: IEEE.

Hudak, P., Jones, S., Wadler, P., Boutel, B., Fairbairn, J., Fasel, J., Guzmán, M., Hammond, K., Hughes, J., Johnsson, T., et al. (1992). Report on the programming language Haskell: a non-strict, purely functional language version 1.2. *ACM Sigplan Notices, 27*(5), :1–164.

Huebner, M., Paulsson, K., & Becker, J. (2005). Parallel and flexible multiprocessor system-on-chip for Hwang, K., & Briggs, F. A. (1984) *Computer architecture and parallel processing.* New York: McGraw-Hill.

ITRS (2008). ITRS. In *Proceedings of the Winter Public Conference Presentations.* Retrieved from http://www.itrs.net/Links/2008Winter/2008WinterPresentationsITWG/Presentations.html

Jalabert, A., Murali, S., Benini, L., & De Micheli, G. (2004). XpipesCompiler: A tool for instantiating application specific networks-on-chip. In *Proceedings of Design, Automation and Testing in Europe Conference (DATE).* (884-889).

James-Roxby, P., Cerro-Prada, E., & Charlwood, S. (1999). A Core-Based Design Method for Reconfigurable Computing Applications. In *Proceedings of the IEE Colloquium on Reconfigurable Systems.* Glasgow, Scotland.

Jantsch, A., & Lu, Z. (2009). Resource Allocation for QoS On-Chip Communication. In F. Gebali, H. Elmiligi, and H. Watheq el-Kharashi (eds.), *Networks-on-Chip, Theory and Practice.* Boca Raton, FL: CRC Press.

Jantsch, A., & Tenhunen, H. (2003). *Networks on Chip.* Boston: Kluwer Academic Publishers.

Jantsch, A., Tenhunen, H. (2003). *Will Networks on Chip Close the Productivity Gap?* Boston: Kluwer Academic Publishers

Jerger, N. E., Lipasti, M., & Peh, L. S. (2007). Circuit-switched coherence. *IEEE Computer Architecture Letters, 6*(1), 5-8.

Jovanovic, S., Tanougast, C., Weber, S., & Bobda, C. (2007) Cunoc: A scalable dynamic noc for dynamically re-

configurable fpgas. In *International Conference on Field Programmable Logic and Applications*, (pp. 753–756).

Joven, J., Font-Bach, O., Castells-Rufas, D., Martinez, R., Teres, L., & Carrabina, J. (2008). xenoc - an experimental network-on-chip environment for parallel distributed computing on noc-based mpsoc architectures. In *Proceedings of Parallel and Distributed Processing* (pp. 141–148). Washington, DC: IEEE Computer Society.

Kalte, H., Lee, G., Porrmann, M., & Rückert, U. (2005). REPLICA: A Bitstream Manipulation Filter for Module Relocation in Partial Reconfigurable Systems. In *Proceedings of the 19th International Parallel and Distributed Processing Symposium* (pp. 151b-151b). Denver, CO.

Kangmin, et al. (n.d.). A 51 mW 1.6 GHz on-chip network for low-power heterogeneous SoC platform. In *Proceedings of the IEEE Int. Solid-States Circuits Conference (Digest of Technical papers)*, (pp. 152–518).

Kao, C. (2005). , "Benefits of Partial Reconfiguration. In *Proceedings of ", XCell*, (pp. 65-67)., 2005

Kelsey, R., Clinger, W., et al. and (Editors), J. R. (1998). Revised report on the algorithmic language Scheme. *ACM SIGPLAN Notices, 33*(9), :26–76.

Keutzer, K., Newton, A. R., Rabaey, & J. M., Sangiovanni-Vincentelli, A. (2000). System-level design: Orthogonalization of concerns and platform-based design. *IEEE Transactions on CAD of Integrated Circuits and Systems, 19*(12), 1523-1543.

Khoa Duc Tran (2005). Elitist non-dominated sorting GA-II (NSGA-II) as a parameter-less multi-objective genetic algorithm. *In Proceedings of SoutheastCon*, (pp. 359–367).

Kim, D., Kim, M., & Sobelman, G. E. (2005, October 20-21). Design of a High-Performance Scalable CDMA Router for On-Chip Switched Networks. In *Proceedings of the International SoC Design Conference*, Korea.

Kim, E.-J., Link, G., Yum, K. H., Narayanan, V., Kandemir, M., Irwin, M. J., & Das, C. (2005). A Holistic Approach to Designing Energy-Efficient Cluster Interconnects. In *Proceedings of the IEEE Trans. on Computers* (pp. 660-671).

Kim, J., Dally, W. J., Scott, S., & Abts, D. (2008). Technology-Driven, Highly-Scalable Dragonfly Topology. In *35th International Symposium on Computer Architecture (ISCA)*, 2008.

Kim, J., Nicopoulos, C., Park, D., Narayanan, V., Yousif, M. S., & Das, C. R. (2006). A gracefully degrading and energy-efficient modular router architecture for on-chip networks. In *Proceedings of International Symposium on Computer Architecture* (ISCA), (4–15).

Kim, K., Lee, S. J., Lee, K., & Yoo, H. J. (2005). An arbitration look-ahead scheme for reducing end-to-end latency in networks-on-chip. In *International Symposium on Circuits and Systems* (ISCAS), (2357-2360).

Kim, M. M., Davis, J. D., Oskin, M., & Austin. T. (2008). Polymorphic on-chip networks. In *Proc. of the 35th International Symposium on Computer Architecture,* (ISCA-2008).

Kim, M., Davis, J., Oskin, M., & Austin, T. (2008). Polymorphic on-chip networks. In *Proceedings of ISCA*.

Kim, M., Kim, D., & Sobelman, E. (2006). NoC link analysis under power and performance constraints. In *Proceedings of ISCAS*.

Kim, W., Gupta, M. S., Wei, G.-Y., & Brooks, D. (2008). System level analysis of fast, per-core DVFS using on-chip switching regulators. In *Proceedings of the 14th International Symposium on High-Performance Computer Architecture* (HPCA).

Koch, et al. (2006). An adaptive system-on-chip for network applications. In *Proceedings of IPDPS 2006*.

Koester, M., Luk, W., Hagemeyer, J., & Porrmann, M. (2009). Design Optimizations to Improve Placeability of Partial Reconfiguration Modules. In *Proceedings of Design, Automation and Test in Europe Conference and Exposition* (pp. 976-981). Nice, France.

Kogel, T., Haverinen, A., & and Aldis, J. (2005). OCP TLM for Architectural Modeling. (*White Paper.*).

Koo, J. J., Fernandez, D., Haddad, A., Gross, W. J. , (2007, July 9-11). Evaluation of a High-Level-Language Methodology for High-Performance Reconfigurable Computers, Application -specific Systems. In Proceedings of the , *Architectures and Processors, 2007. ASAP. IEEE International Conf. on Architectures and Processors, 2007,* vol., no., (pp.30-35)., 9-11 July 2007

Koudri, et al. (2008). Using MARTE in the MOPCOM SoC/SoPC Co-Methodology. In *Proceedings of MARTE Workshop at DATE'08*.

Kramer, J., & Magee, J. (1990). The evolving philosophers problem: Dynamic change management. In Proceedings of IEEE Transactions on Software Engineering, (vol. 16).

Kramer, J., & Magee, J.(1985 April). Dynamic configuration for distributed systems. In *Proceedings of IEEE Transactions on Software Engineering,* (vol. 11).

Krasteva, Y., Jimeno, A. B., de la Torre, E., & Riesgo, T. (2005). Straight Method for Reallocation of Complex Cores by Dynamic Reconfiguration in Virtex II FPGAs. In *Proceedings of the IEEE International Workshop on Rapid System Prototyping RSP* (pp. 77-83). Washington, DC: IEEKrasteva, Y., Jimeno, A., Torre, E., & Riesgo, T. (2005). Straight Method for Reallocation of Complex Cores by Dynamic Reconfiguration in Virtex II FPGAs. In *Proceedings of the International Symposium on Rapid System Prototyping* (pp. 77-83). Montreal, Canada.

Kreutz, M. et al. (2005). Energy and Latency Evaluation of NoC Topologies. In *Proceedings of the Int. Symp. on Circuits and Systems,* (pp. 5866–5869).

Kreutz, M., Marcon, C., Carro, L., Wagner, F., & Altamiro A. (2005). Design Space Exploration Comparing Homogeneous and Heterogeneous Network-on-Chip Architectures. In *Proceedings of the 18th Symposium on Integrated Circuits and Systems Design* (pp. 190-195).

Kumar, A., Hansson, A., Huisken, J., & Corporaal, H.. (2007). An fpga design flow for reconfigurable network-based multi-processor systems on chip. In *Proceedings of the Design, Automation and Test in Europe Conference and Exhibition,* (pp. 1–6).

Kumar, A., Kundu, P., Singh, A., Peh, L. S, &. Jha, N. K. (2007). A 4.6Tbits/s 3.6GHz single-cycle NoC router with a novel switch allocator in 65nm CMOS. In *Proceedings of the International Conference on Computer Design*.

Kumar, A., Peh, L. S., Kundu, P., &. Jha, N. K. (2007). Express virtual channels: towards the ideal interconnection fabric. In *Proceedings of the International Symposium on Computer Architecture*, (150-161).

Kumar, R., Zyuban, V., & Tullsen, D. M. (2005). Interconnections in multi-core architectures: Understanding mechanisms, overheads and scaling. *SIGARCH Comput. Archit. News, 33*(2), 408–419.

Kumar, S., Jantsch, A., Millberg, M., Öberg, J., Soininen, J. P., Forsell, M., Tiensyrjä, K., & Hemani, A. (2002). A network on chip architecture and design methodology. In *Proceedings of the IEEE Computer Society Annual Symposium on ISVLSI* (pp. 117–124). Washington, DC: IEEE Computer Society.

Labrosse, J. J. (2002). *MicroC OS II: The Real Time Kernel*. San Francisco: CMP Books.

Lampinen, H., Perala, P., & and Vainio, O. (2006). Design of a scalable asynchronous dataflow processor. In *Proceedings of the IEEE Design and Diagnostics of Electronic Circuits and systems, 2006 IEEE*, vol., no., (pp.85-86).

Lanuzza, M., Perri, S., & and Corsonello, P. (2007). MORA - A New Coarse Grain Reconfigurable Array for High Throughput Multimedia Processing. In *Proceedings of the. International Symposium on Systems, Architecture, Modeling and Simulation, (SAMOS07)*. (LNCS 4599, pp. 159–168)., 2007. Springer-Verlag Berlin Heidelberg 2007

Lattard, et al. (2007). A Telecom Baseband Circuit based on an Asynchronous Network on Chip (Digest of Technical Papers). In *Proceedings of the IEEE Intl. Solid State Circuits Conf.*, (pp. 258–601).

Lee, S.- J., et al. (2003). An 800 MHz star-connected on-chip network for application to systems on a chip. In *Proceedings of the IEEE Int. Solid-States Circuits Conf.*, (Digest of Technical papers), (pp. 468–469).

Lei, T., & Kumar, S. (2003). Algorithms and Tools for Network-on-Chip Based System Design. In *Proceedings of the 16th Symposium on Integrated Circuits and Systems Design* (pp. 163-168).

Lei, T., & Kumar, S. (2003). A Two-step Genetic Algorithm for Mapping Task Graphs to a NoC Architecture. In *Euromicro Symposium on Digital System Design* (pp.180-187).

Leijen, D., & and Meijer, E. (2001). Parsec: Direct style monadic parser combinators for the real world. (Technical Report UU-CS-2001-27). The Netherlands: , Department of Computer Science, Universiteit Utrecht.

Li, B., Peh, L.-S., & Patra, P. (2008). Impact of Process and Temperature Variations on Network-on-Chip Design Exploration. In *Proceedings of NOCS* (pp. 117-126).

Li, L., Vijaykrishnan, N., Kandemir, M., & Irwin, M. J. (2003). Adapative Error Protection for Energy Efficiency. In *Proceedings of the 2003 IEEE/ACM international conference on Computer-aided design* (ICCAD '03), (pp. 2) Washington, DC: IEEE Computer Society.

Liebeherr, J., & Beam, T. K. (1999). HyperCast: A protocol for maintaining multicast group members in a logical hypercube topology. In *Proceedings First International Workshop on Networked Group Communication (NGC '99)*, (Lecture Notes in Computer Science, Volume 1736, pp. 72-89).

Lim, D., & Peattie, M. (2002). Two flows for partial reconfiguration: Module Based or small bit manipulations. *Xilinx Application Note 290 (v1.0)*.

Mack, R. J. (1996). , "VLSI physical design automation: theory and practice. ," *Electronics & Communication Engineering Journal*, , vol.*8*, no. (2), pp.56, Apr 1996

Magnusson, P. S., Christensson, M., Eskilson, J., Forsgren, D., Hallberg, G., Hogberg, J., Larsson, F., Moestedt, A., Werner, B., & Werner, B. (2002). Simics: A full system simulation platform. *Computer, 35*(2), 50-58.

Mahadevan, S., Virk, K., & Madsen, J. (2006). Arts: A systemc-based framework for modelling multiprocessor systems-on-chip. *Design Automation of Embedded Systems, 11*(4), 285-311.

Mak, T. S. T., ; Sedcole, P., ; Cheung, P. Y. K., ; Luk, W. (2006, August 28-30). , "On-FPGA Communication Architectures and Design Factors. In *Proceedings of the ," International Conference on Field Programmable*

Logic and Applications, 2006. (FPL '06),. International Conference on, (pp.1-8)., 28-30 Aug. 2006

Marescaux, T., Bricke, B., Debacker, P., Nollet, V., & Corporaal, H. (2005, September). Dynamic time-slot allocation for qos enabled networks on chip. In *Proceedings of Embedded Systems for Real-Time Multimedia Workshop.*

Marescaux, T., Mignolet, J.-Y., Bartic, A., Moffat, W., Verkest, D., Vernalde, S., & Lauwereins, R. (2003, September). Networks on chip as hardware components of an OS for reconfigurable systems. In *Proceedings of the 13th International Conference on Field Programmable Logic and Applications, Lecture Notes in Computer Science (LNCS),* (volume 2778, pp. 595–605).

Marescaux, T., Nollet, V., Mignolet, J., Bartic, A., Moffat, W., Avasare, P., Coene, P., Verkest, D., Vernalde, S., and Lauwereins, R. (2004). Run-time support for heterogeneous multitasking on reconfigurable SoCs. *Integration, the VLSI journal, 38*(1), :107–130.

Martini, F., Bertozzi, D., & Benini, L. (2007) Assessing the Impact of Flow Control and Switching Techniques on Switch Performance for Low Latency NoC Design. In *the First Workshop on Interconnection Network Architectures On-Chip.*

McCarthy, J. (1960). Recursive functions of symbolic expressions and their computation by machine, part i. *Commun. ACM, 3*(4), :184–195.

McMillan, S., & Guccione, S. (2000). Partial Run-Time Reconfiguration Using JRTR. In *Proceedings of the 10ᵗʰ Conference on Field Programmable Logic and Application* (pp. 352-360). Villach, Austria.

McUmber, et al. (1999). UML-based analysis of embedded systems using a mapping to VHDL. In *Proceedings of the IEEE International Symposium on High Assurance Software Engineering* (HASE'99), (pp.56–63).

Medardoni, S., Bertozzi, D., Benini, L., & Macii, E. (2007). Control and datapath decoupling in the design of a NoC switch: area, power and performance implications. In *Proceedings of the International Symposium on System-on-Chip* (pp. 1-4).

Meloni, P., Murali, S., Carta, S., Camplani, M., Raffo, L., & De Micheli, G. (2006). Routing aware switch hardware customization for networks on chips. In *Proceedings of NanoNet.*

Mens, T., & Van Gorp, P. (2006). A taxonomy of model transformation. In *Proceedings of the International Workshop on Graph and Model Transformation, GraMoT 2005* (Vol. 152., pp. 125–142).

Mignolet, J., Nollet, V., Coene, P., Verkest, D., Vernalde, S., & Lauwereins, R. (2003). Infrastructure for design and management of relocatable tasks in a heterogeneous reconfigurable system-on-chip. In *Proceedings of Design, Automation and Test in Europe Conference and Exposition* (pp. 986-991). Munich, Germany.

Millberg, M., Nilsson, E., Thid, R., & Jantsch, A. (2004). Guaranteed bandwidth using looped containers in temporally disjoint networks within the nostrum network-on-chip. In *Proceedings of Design, Automation and Testing in Europe Conference* (DATE), (890–895).

Millberg, M., Nilsson, E., Thid, R., Kumar, S., & Jantsch, A. (2004). The nostrum backbone - a communication protocol stack for networks on chip. In *Proceedings of the International Conference on VLSI Design* (pp. 693–696). Washington, DC: IEEE Computer Society.

Milner, R., Tofte, M., & and Harper, R. (1990). *The definition of Standard ML.* Cambridge, MA: MIT Press., Cambridge, MA, USA.

Mirsky, E., & and DeHon, A. (1996). MATRIX: a reconfigurable computing architecture with configurable instruction distribution and deployable resources. In *Proceedings of IEEE Symposium on FPGAs for Custom Computing Machines, 1996. Proceedings. IEEE Symposium on,* (pages 157–166).

Mishra, M., Callahan, T., Chelcea, T., Venkataramani, G., Goldstein, S., & and Budiu, M. (2006). Tartan: evaluating spatial computation for whole program execution. In *Proceedings of the 12th international conference on Architectural support for programming languages and operating systems,* (pages 163–174). New York: ACM. New York, NY, USA.

Moadeli, M., Vanderbauwhede, W., & and Shahrabi, A. (2008b). Quarc: A novel network-on-chip architecture. In *Proceedings of ICPADS '08: Proceedings of The 2008 14th IEEE International Conference on Parallel and Distributed Systems*, (pages 705–712). , Washington, DC: , USA. IEEE Computer Society.

Modarressi, M., & Sarbazi-Azad, H. (2007). Power-aware mapping for reconfigurable NoC architectures. In *Proceedings of the International Conference on Computer Design*.

Modarressi, M., Sarbazi-Azad, H., Tavakkol, A. (2009). Low-power and high-performance on-chip communication using virtual point-to-point connections. In *Proceedings of the IEEE/ACM International Symposium on Network-on-Chip* (NoCS'09).

Mohanty, et al. (2002). Rapid design space exploration of heterogeneous embedded systems using symbolic search and multi-granular simulation. In *LCTES/Scopes 2002*.

Möller, L., & Moraes, F. (2005). Sistemas Dinamicamente Reconfiguráveis com Comunicação Via Redes Intra-Chip (Master Thesis). *Catholic University of Rio Grande do Sul* (142 p.). In Portuguese. Porto Alegre, Brazil.

Möller, L., Calazans, N., Moraes, F., Brião, E., Carvalho, E., & Camozzato, D. (2004). FiPRe: An implementation model to enable self-reconfigurable applications. In *Proceedings of the 14th Conference on Field Programmable Logic and Application* (pp. 1042-1046). Leuven, Belgium.

Möller, L., Grehs, I., Calazans, N., & Moraes, F. (2006). Reconfigurable Systems Enabled by a Network-on-Chip. In *Proceedings of the International Conference on Field Programmable Logic and Application FPL, 16th International Conference* (pp.1-4). Washington, DC: IEEE

Möller, L., Grehs, I., Carvalho E., Soares R., Calazans, N., & Moraes, F. (2007). A NoC-based Infrastructure to Enable Dynamic Self Reconfigurable System. In *Proceedings of ReCoSoC*. Montpellier, France.

Möller, L., Grehs, I., Carvalho, E., Soares, R., Calazans, N., & Moraes, F. (2007). A noc-based infrastructure to enable dynamic self reconfigurable systems. In *Pro-ceedings of the Reconfigurable Communication Centric System on Chip* ReCoSoC (pp. 23–30).

Moore, G. E. (1975). Progress in Digital Integrated Electronics. In *Proceedings of Digest of the 1975 International Electron Device Meeting* (p. 1113). New York

Moraes, F., Calazans, N., Mello, A., Möller, L., & Ost, L. (2004). HERMES: An infrastructure for low area overhead packet-switching networks on chip. In *Proceedings of the 17th Symposium on Integrated Circuits and Systems Design* (pp. 69-93). Porto de Galinhas, Brazil.

Moraes, F., Mesquita, D., Palma, J., Möller, L., & Calazans, N. (2003). Development of a Tool-Set for Remote and Partial Reconfiguration of FPGAs. In *Proceedings of Design, Automation and Test in Europe Conference and Exposition* (pp. 1122-1123). Munich, Germany.

Morandi, M., Novati, M., Santambrogio, M. D., & Sciuto, D. (2008, April). Core allocation and relocation management for a self dynamically reconfigurable architecture. *In Proceedings of the IEEE Computer Society Annual Symposium on VLSI* (ISVLSI 08), (pp. 286 – 291).

Moussa, H., Baghdadi, A., & Jezequel, M. (2008, June). Binary de bruijn on-chip network for a flexible multiprocessor ldpc decoder. In *Proceedings of the 45th ACM/IEEE Conference on Design Automation*.

Mrabet, H., Marrakchi, T., Mehrez, H., & Tissot, A. (2006). Implementation of Scalable Embedded FPGA for SoC. In *Proceedings of the International Conference on Design and Test of Integrated Systems in Nanoscale Technology* (pp. 74-77). Gammarth, Tunisia.

Mukherjee, S., Bannon, P., Lang, S., Spink, A., & Webb, D. (2002, January-February). The Alpha 21364 network architecture. In *Proceedings of IEEE Micro* (pp. 26–35).

Mullins, R. et al. (2006). The Design and Implementation of Low-Latency on-Chip Network. In *Proceedings of the Asia and South Pacific Design Automation Conference* (pp. 164-169).

Mullins, R., & Moore, W. S. (2004). Low-latency virtual-channel routers for on-chip networks. In *Proceedings of the International Symposium on Computer Architecture*. (188–197).

Murali S., Coenen, M., Radulescu, A., Goossens, K., De Micheli, G. (2006, March 6-10). A Methodology for Mapping Multiple Use-Cases onto Networks on Chips. In *Proceedings of Design, Automation and Test in Europe, 2006* (DATE '06), (vol.1, pp.1-6).

Murali, S., & De Micheli, G. (2004). Bandwidth-Constrained Mapping of Cores onto NoC Architectures. In *Proceedings of Design, Automation and Test in Europe* (pp.896-901).

Murali, S., & De Micheli, G. (2004b). SUNMAP: A tool for automatic topology selection and generation for NoCs. In *Proceedings of Design Automation Conference (DAC)*.

Murali, S., Benini, L., dc Micheli, G. (2005, January 18-21). Mapping and physical planning of networks-on-chip architectures with quality-of-service guarantees. In *Proceedings of the Design Automation Conference, 2005* (ASP-DAC 2005).(vol.1, pp. 27-32).

Murali, S., Coenen, M., Radulescu, R., Goossens, K., & De Micheli, G. (2006b). A methodology for mapping multiple use-cases onto networks on chips. In *Proceedings of Design Automation and Test in Europe* (DATE), (118-123).

Murali, S., De Micheli, G. (2004, February 16-20). Bandwidth-constrained mapping of cores onto NoC architectures. In *Proceedings of Design, Automation and Test in Europe Conference and Exhibition, 2004* (vol.2, pp. 896-901).

Murali, S., et al. (2004). SUNMAP: A Tool for Automatic Topology Selection and Generation for NOCs. In *Proceedings of the Design Automation Conf., 2004*, (pp. 914–919).

Murali, S., Theocharides, T., Vijaykrishnan, N., Irwin, M. J., Benini, L., & Ovanni De Micheli, G. (2005). Analysis of Error Recovery Schemes for Networks on Chips. In *Proceedings of the IEEE Des. Test*.

Muralimanohar, N., & Balasubramonian. R. (2007). Interconnect design considerations for large nuca caches. In *Proceedings of the 34th Annual International Symposium on Computer Architecture*, (ISCA '07), (pp. 369–380).

Najjar, W. A., Böhm, W., Draper, B. A., Hammes, J., Rinker, R., Beveridge, J. R., Chawathe, M., & and Ross, C. (2003). High-Level Language Abstraction for Reconfigurable Computing. *Computer, 36*, (8), (Aug. 2003), 63-69.

Nayfeh, B. A. (1999). *The case for a single-chip multi-processor* (PhD Thesis). Stanford, CA.

Nicopoulos, C. A., Park. D., Kim, J., Vijaykrishnan, N., Yousif, M. S., & Das, C. R. (2006 December). Vichar: A dynamic virtual channel regulator for network-on-chip routers. In *Proceedings of MICRO-39, the 39th Annual IEEE/ACM International Symposium*, (pp. 333–346).

Nicopoulos, C., Yanamandra, A., Srinivasan, S., Narayanan, V., & Irwin, M. J. (2007). Variation-Aware Low-Power Buffer Design. In Proceedings *of The Asilomar Conference on Signals, Systems, and Computers*.

Noguera, J., ; Badia, R. M. (2006, July), System-level power-performance tradeoffs for reconfigurable computing., *IEEE Transactions on Very Large Scale Integration (VLSI) Systems, , IEEE Transactions on* , (vol.14, no.7, pp.730-739)., July 2006

Nollet, V., Coene, P., Verkest, D., Vernalde, S., & and Lauwereins, R. (2003). Designing an operating system for a heterogeneous reconfigurable soc. In *Proceedings of IPDPS '03: Proceedings of The 17th International Symposium on Parallel and Distributed Processing*, (page 174.1), Washington, DC,: USA. IEEE Computer Society.

Nollet, V., Coene, P., Verkest, D., Vernalde, S., & Lauwereins, R. (2003). Designing an operating system for a heterogeneous reconfigurable SoC. In *Proceedings of the Parallel and Distributed Processing Symposium IPDPS* (pp. 7).

Nollet, V., Marescaux, T., Avasare, P., & and Mignolet, J. (2005). Centralized run-time resource management in a network-on-chip containing reconfigurable hardware tiles. In *Proceedings of the conference on Design, Automation and Test in Europe* -(volume 1, pages 234–239). Washington, DC: IEEE Computer Society. Washington, DC, USA.

Nollet, V., Marescaux, T., Verkest, D., Mignolet, J., & and Vernalde, S. (2004). Operating-system controlled network on chip. In *Proceedings of the 41st annual conference on Design automation*, (pages 256–259). ACM New York: ACM., NY, USA.

OCPIP. (2003). *Open Core Protocol (OCP) Specification, Release 2.0*. Retrieved from http://www.ocpip.org

Ogras, U., & Marculescu, R. (2005a). Application-specific network-on-chip architecture customization via long-range link insertion. In *Proceedings of the Design Automation Conference (DAC)*.

Ogras, U., & Marculescu, R. (2005b). Energy and performance-driven customized architecture synthesis using a decomposition approach. In *Proceedings of the Design Automation and Test in Europe Conference*, (352–357).

Ogras, U¨. Y., Marculescu, R., Lee, H. G., Choudhary, P., Marculescu, D., Kaufman, M., & Nelson, P. (2007, September). Challenges and promising results in NoC prototyping using FPGAs. *IEEE Micro, 27*(5), 86–95.

OMG. (2005). *MOF Query /Views/Transformations*. Retrieved from http://www.omg.org/cgi-bin/doc?ptc/2005-11-01

OMG. (2007a). *OMG MARTE Standard*. Retrieved from http://www.omgmarte.org

OMG. (2007b). *OMG Unified Modeling Language* (OMG UML), Superstructure, V2.1.2. Retrieved from http://www.omg.org/spec/UML/2.1.2/Superstructure/PDF/

OSI. (2007). *SystemC*. Retrieved from http://www.systemc.org/

Ost, L., Mello, A., Palma, J., Moraes, F., & Calazans, N. (2005). Maia: a framework for networks on chip generation and verification. In Ting-Ao Tang (Ed), *ASP-DAC* (pp49–52). New York: ACM Press.

Osterloh, B., Michalik, H., Fiethe, B., & Kotarowski, K. (2008). SoCWire: A Network-on-Chip Approach for Reconfigurable System-on-Chip Designs in Space Applications. In *Proceedings of the NASA/ESA Conference on Adaptive Hardware and Systems* (AHS-2008), (pp. 51-56). Noordwijk, The Netherlands.

Owens, J., Dally, W. J., Ho, R., Jayasimha, D. N., Keckler, S. W., & Peh, L. S. (2007). Research challenges for on-chip interconnection networks. *IEEE Micro, 27*(5), 96-108.

Palesi, M., Holsmark, R., Kumar, S., & Catania, V. (2009). Application Specific Routing Algorithms for Networks on Chip. *IEEE Transactions on Parallel and Distributed Systems, 20*(3), 316-330.

Palma, J., Mello, A., Möller, L., Moraes, F., & Calazans, N. (2002). Core communication interface for FPGAs. In *Proceedings of the 15th Symposium on Integrated Circuits and Systems Design* (pp. 183-188). Porto Alegre, Brazil.

Panades, I., & Greiner, A. (2007). Bi-synchronous FIFO for synchronous circuit communication well suited for Network-on-Chip in GALS architectures. In *Proceedings of the 1st IEEE/ACM Int. Symp. On Networks-on-Chip*, (pp. 83–92).

Pande, P., Grecu, C., Jones, M., Ivanov, A., & Saleh, R. (2005). Performance evaluation and design trade-offs for network-on-chip interconnect architectures. *IEEE Trans. Computers, 54*(8), 1025–1040.

Park, D., Nicopoulos, C., Kim, J., Narayanan, V., & Das, C. R. (2006). Exploring Fault-Tolerant Network-on-Chip Architectures. In *Proceedings of DSN 2006*.

Peh, L. S., & Dally, W. J. (2001). A delay model for router microarchitectures. *IEEE Micro 2*(1), 26-34.

Peh, L.-S., & Dally, W. J. (2001). A Delay Model and Speculative Architecture for Pipelined Routers. In *Proceedings of the 7th International Symposium on High-Performance Computer Architectur*PetaLogix Developer Portal (2009). *UserGuide - PetaLogix Developer Portal*. Retrieved from http://developer.petalogix.com/wiki/UserGuide

Pham, D. C., & and Aipperspach, T. (2006). Overview of the architecture, circuit design, and physical implementation of a first-generation cell processor. *IEEE Journal of Solid-State Circuits, 41*(1), :179–196.

Phi-Hung, P., Kumar, Y., & Chulwoo, K. (2006). High Performance and Area-Efficient Circuit-Switched Network on Chip Design. In *Proceedings of The sixth IEEE International Conference on Computer and Information Technology, 2006* (pp. 243 -243)

Pinto, A., Carloni, L., & Sangiovanni-Vincentelli, A. (2003). Efficient Synthesis of Networks on Chip. In *Proceedings of the 21st International Conference on Computer Design*, (pp. 146-150).

Pionteck, T., ; Albrecht, C.,.; Koch, R., ; Maehle, E., ; Hubner, M., & ; Becker, J. (2007, March 26-30). , "Communication Architectures for Dynamically Reconfigurable FPGA Designs. In *Proceedings of the IEEE International ," Parallel and Distributed Processing Symposium, 2007.* (IPDPS 2007),. IEEE International, (pp.1-8)., 26-30 March 2007

Pionteck, T., ; Koch, R., & ; Albrecht, C. (2006, August 28-30). , Applying partial reconfiguration to networks-on-chips. In *Proceedings of the International Conference on Field Programmable Logic and Applications, 2006.* (FPL '06),. International Conference on, (pages 1–6)., 28-30 Aug. 2006

Pionteck, T., Albrecht, C., & Koch, R. (2006). A Dynamically Reconfigurable Packet-Switched Network-on-Chip. In *Proceedings of the conference on Design, automation and test in Europe*, (pp. 136-137).

Pionteck, T., Koch, R., & Albrecht, C. (2006). Applying Partial Reconfiguration to Networks-on-Chips. In *Proceedings of the 16th International Conference on Field Programmable Logic and Applications* (155-160). Madrid, Spain.

Planet MDE. (2007). *Portal of the Model Driven Engineering Community.* Retrieved from http://www.planetmde.org

Platform studio and the EDK (2008). *Platform studio and the EDK.* Retrieved from http://www.xilinx.com/ise/embedded design prod/platform studio.htm

Purohit, S., Chalamalasetti, S. R., Margala, M., & and Corsonello, P. (2008). Power-Efficient High Throughput Reconfigurable Datapath Design for Portable Multimedia Devices. In *International Conference on Reconfigurable Computing and FPGAs* (Reconfig08), (pages 217–222).

Quadri, et al. (2009). A Model Driven design flow for FPGAs supporting Partial Reconfiguration. *International Journal of Reconfigurable Computing.* New York: Hindawi Publishing. In Press.

Radulescu, A., Dielissen, J., Goossens, K., Rijpkema, E., & Wielage, P. (2004). An efficient on-chip NI offering guaranteed services, shared-memory abstraction, and flexible network configuration. In *Proceedings of IEEE TCAD.*

Raghavan, A., & Sutton, P. (2002). JPG – A Partial Bitstream Generation Tool to Support Partial Reconfiguration in Virtex FPGAs. *In Proceedings of the 16th International Parallel and Distributed Processing Symposium* (pp. 155-160). Florida, USA.

Raghunathan, V., Srivastava, B., & Gupta, R. K. (2003). A survey of techniques for energy efficient on-chip communication. In *Proceedings of the Design Automation Conference* (DAC).

Rana V., Atienza, D., Santambrogio, M. D., Sciuto, D., & De Micheli, G. (2008, October). A Reconfigurable Network-on-Chip Architecture for Optimal Multi-Processor SoC Communication. In *Proceedings of the 16th IFIP/IEEE International Conference on Very Large Scale Integration* (pp. 321-326).

Rana, V., Atienza, D., Santambrogio, M. D., Sciuto, D., & De Micheli, G. (2008, October). A reconfigurable network-on-chip architecture for optimal multi-processor SoC communication. In *Proceedings of the 16th IFIP/IEEE International Conference on Very Large Scale Integration*, (pp. 321–326). New York: Springer Press.

Randal, A., Sugalski, D., & and Toetsch, L. (2004). *Perl 6 and Parrot Essentials*, Second Edition. Sebastopol, CA: O'Reilly Media, Inc.

Reed, I. S. (1960). *Error-Control Coding for Data Networks.* Boston: Kluwer Academic Publishers.

Resano, J., Mozos, D., Verkest, D., Catthoor, F., & Vernalde, S. (2004). Specific scheduling support to minimize the reconfiguration overhead of dynamically reconfigurable hardware. In *Proceedings of Design Automation Conference* (pp. 119-121). San Diego, CA.

Rezgui, S., George, J., Swift, G., Somervill, K., Carmichael, C., & Allen, G. (2005). *SEU Mitigation of a Soft Embedded Processor in the Virtex-II FPGAs.* San Joese, CA: North American Xilinx Test Consortium, Xilinx Inc.

Rijpkema, E. (2003). Trade offs in the design a router with both guaranteed and best-effort services for networks on chip. In Proceedings of Design, Automation and Test in Europe Conference and Exhibition, (pp. 350–355).

Sabbaghi, R., Modarressi, M., & Sarbazi-Azad, H. (2008-1). The 2d DBM: an attractive alternative to the simple 2d mesh topology for on-chip networks. In *Proceedings of the International Conference on Computer Design.*

Sabbaghi, R., Modarressi, M., & Sarbazi-Azad, H. (2008-2). A novel high-performance and low-power mesh-based NoC. In *Proceedings of the 7th. IPDPS Workshop on Performance Modeling, Evaluation, and Optimization of Ubiquitous Computing and Networked Systems* (PMEO).

Salminen, E., Kulmala, A., & Hämäläinen, T. (2008). *Survey of Network-on-Chip Proposals.* (WHITE PAPER, OCP-IP). Tampere University of Technology, Finland.

Samira, S., Ahmad, K., Mehran & Armin. (2007). Smap: An intelligent mapping tool for network on chip. In Proceedings of *Signals, Circuits and Systems* (ISSCS), (pp. 1–4, 13-14_.

Santambrogio, M. D., Rana, V., & Sciuto, D. (2008, September). Operating system support for online partial dynamic reconfiguration management. In *Proceedings of the 18th International Conference on Field Programmable Logic and Applications,* (pp. 455–458).

Sassatelli, G., Torres, L., Benoit, P., Gil, T., Diou, C., Cambon, G., & Galy, J. (2002). Highly scalable dynamically reconfigurable systolic ring-architecture for DSP applications. *In Proceedings of Design, Automation and Test in Europe Conference and Exposition* (pp. 553-558). Paris, France.

Scholz, S. (2003, November). Single Assignment C: efficient support for high-level array operations in a functional setting. *J. Funct. Program. 13,* (6) (Nov. 2003), 1005-1059.

Semeraro, G., Magklis, G., Balasubramonian, R. D. H., Albonesi, S. D., & Scott, M. L. (2002). Energy-efficient processor design using multiple clock domains with dynamic voltage and frequency scaling. In *Proceedings of the International Symposium on High-Performance Computer Architecture.*

Sendall, S., & Kozaczynski, W. (2003). Model transformation: The heart and soul of model driven software development. *IEEE Software, 20*(5), 42–45.

Shang, L., Dick, R. P., & Jha, N. K. (2007, March). SLOPES: Hardware–Software Cosynthesis of Low-Power Real-Time Distributed Embedded Systems With Dynamically Reconfigurable FPGAs. In *Proceedings of the IEEE Transactions on Computer-Aided Design of Integrated Circuits and Systems* (vol.26, no.3, pp.508-526).

Shang, L., Peh, L.-S., & Jha, N. K. (2003). Dynamic Voltage Scaling with Links for Power Optimization of Interconnection Networks. In *Proceedings of HPCA.*

Shang, L., Peh, L.-S., & Jha, N. K. (2003). PowerHerd: Dynamic Satisfaction of Peak Power Constraints in Interconnection Networks. In *Proceedings of the 17th Annual International Conference on Supercomputing* (ICS '03), (pp. 98-108).

Shashi, K., et al. (2002). A Network on Chip Architecture and Design Methodology. In *Proceedings of the IEEE Computer Society Annual Symposium on VLSI.* Pittsburgh, PA.

Shin, K. G., & Daniel, S. W. (1996). Analysis and implementation of hybrid switching. In *Proceedings of the IEEE Tran. on Computers.*

Singh, H., Lee, M., Lu, G., Kurdahi, F., Bagherzadeh, N., & and Chaves Filho, E. (2000). MorphoSys: An Integrated Reconfigurable System for Data-Parallel and Computation-Intensive Applications. *IEEE Transactions On Computers,* (pages 465–481).

Singh, H., Lee, M., Lu, G., Kurdahi, F., Bagherzadeh, N., & Filho, E. (1998). MorphoSys: A reconfigurable architecture for multimedia applications. In *Proceedings of the 11th Symposium on Integrated Circuits and Systems Design* (pp. 134-139). Rio de Janeiro, Brazil.

Singhal L., & Bozorgzadeh, E. (2006, August 28-30). Multi-layer Floor planning on a Sequence of Reconfigurable Designs. In Proceedings of the International Conference on Field Programmable Logic and Applications, 2006 (FPL '06), (pp.1-8).

So, H. K.-H., & Brodersen, R. W. (2008, February). A unified hardware/software runtime environment for FPGA-based reconfigurable computers using BORPH. *ACM Transactions on Embedded Computing Systems, 7*(2), 14.

Soares, R., ; Silva, I. S., & ; Azevedo, A. (2004, September 7-11). , "When reconfigurable architecture meets network-on-chip. In *Proceedings of the 17th Symposium on ," Integrated Circuits and Systems Design, 2004.* (SBCCI 2004). 17th Symposium on, (pp. 216-221)., 7-11 Sept. 2004

Srinivasan, K. & Chatha., K. (2005). ISIS: A genetic algorithm based technique for synthesis of on-chip interconnection networks. In *Proceedings of VLSI Design Conference.*

Srinivasan, K., & Chatha, K. (2006). A low complexity heuristic for design of custom network-on-chip architectures. In *Proceedings of Design Automation and Test in Europe* (DATE).

Srinivasan, K., Chatha, K. S. (2006, March 27-29). A methodology for layout aware design and optimization of custom network-on-chip architectures. In *Proceedings of the 7th International Symposium on Quality Electronic Design* (ISQED '06), (pp.6-357).

Srinivasan, K., Chatha, K. S. (2006, October 22-25). Layout aware design of mesh based NoC architectures. In Proceedings of the 4th international conference on Hardware/software codesign and system synthesis (CODES+ISSS '06), (pp.136-141).

Srinivasan, K., Chatha, K.S. & Konjevod, G. (2007). Application Specific Network-on-Chip Design with Guaranteed Quality Approximation Algorithms. In

Proceedings of the *Asia and South Pacific Design Automation Conference ASP-DAC* (pp.184-190).

Srinvasan, K., Chatha, K., & Konjevod, G. (2006). Linear programming-based techniques for synthesis of network-on-chip architectures. In *IEEE Transaction on VLSI, 14*(4), 407-420.

Stensgaard, M. B. & Sparso,J. (2008, April). *ReNoC: A network-on-chip architecture with reconfigurable topology*. In *Proceedings of NoCs 2008, the Second ACM/IEEE International Symposium*, (pp. 55–64).

Stevens, P. (2007). A landscape of bidirectional model transformations. In *Generative and Trans- formational Techniques in Software Engineering 2007*, (GTTSE'07).

Su, C. C., & Shin, K. G. (1996). Adaptive Fault tolerant dead lock free routing in meshes and hypercubes. *IEEE Trans. Computers, 45*(6), 666-683.

Sullivan, C., Wilson, A., & Chappell, S. (2005). Deterministic Hardware Synthesis for Compiling High-Level Descriptions to Heterogeneous Reconfigurable Architectures. In *Proceedings of the Hawaii International Conference on System Sciences* (1-9), Big Island, HI.

Sun, Y., Kumar, S., & Jantsch, A. (2002). Simulation and Evaluation for a Network on Chip Architecture Using Ns-2. In *Proceeding of NorChip*. Copenhagen, Denmark.

Sussman, G. J., & and Steele, G. L. (1975). *An interpreter for extended lambda calculus,*. (Technical report). , Cambridge, MA: MIT Press., USA.

Taylor, M. B., Kim, J., Miller, J., Wentzlaff, D., Ghodrat, F., Greenwald, B., Hoffmann, H., Johnson, P., Lee, J.-W., Lee, W., Ma, A., Saraf, A., Seneski, M., Shnidman, N., Strumpen, V., Frank, M., Amarasinghe, S., & Agarwal, A. (2002). The Raw Microprocessor: A Computational Fabric for Software Circuits and General Purpose Programs. *IEEE Micro, 22*(2), 25-35.

Taylor, M., Psota, J., Saraf, A., Shnidman, N., Strumpen, V., Frank, M., Amarasinghe, S., Agarwal, A., Lee, W., Miller, J., et al. (2004). Evaluation of the Raw microprocessor: an exposed-wire-delay architecture for ILP and streams. In *Proceedings of the 31st Annual Inter-*

national Symposium on Computer Architecture, 2004. Proceedings. 31st Annual International Symposium on, (pages 2–13).

Treleaven, P. C., Brownbridge, D. R., & and Hopkins, R. P. (1982). Data-driven and demand-driven computer architecture. *ACM Comput. Surv.*, *14*(1), :93–143.

ul Abdin, Z., & and Svensson, B. (2008). Evolution in architectures and programming methodologies of coarse-grained reconfigurable computing. *Microprocessors and Microsystems*, In Press., Uncorrected Proof:–.

Ullmann, M., Huebner, M., Grimm, B., & Becker, J. (2004). An FPGA run-time system for dynamical on-demand reconfiguration. In *Proceedings of the 18th International Parallel and Distributed Processing Symposium* (pp. 135-142). Santa Fe, NM.

Vaidyanathan, R., Trahan, J. (2004). *Dynamic Reconfiguration: Architectures and Algorithms*. New York: Springer.

Van den Branden, G., Touhafi, A., & Dirkx, E. (2005). A design methodology to generate dynamically self-reconfigurable SoCs for Virtex-II Pro FPGAs. In *Proceedings of the IEEE International Conference on Field-Programmable Technology* (pp. 325-326). Singapore.

Vanderbauwhede, W. (2006a). Gannet: a functional task description language for a service-based SoC architecture. In *Proceedings of the. 7th Symposium on Trends in Functional Programming (TFP06)*.

Vanderbauwhede, W. (2006b). The Gannet Service-based SoC: A Service-level Reconfigurable Architecture. In *Proceedings of 1st NASA/ESA Conference on Adaptive Hardware and Systems* (AHS-2006), (pages 255–261), Istanbul, Turkey.

Vanderbauwhede, W. (2007, September 30). Gannet: a Scheme for Task-level Reconfiguration of Service-based Systems-on-Chip. In *Proceedings of the 2007 Workshop on Scheme and Functional Programming* (pages 129–137). , September 30th, 2007, Freiburg, Germany., pages 129–137. N/A.

Vanderbauwhede, W. (2008). A Formal Semantics for Control and Data flow in the Gannet Service-based System-on-Chip Architecture. In Proceedings of *The International Conference on Engineering of Reconfigurable Systems and Algorithms, ERSA 2008*. N/A.

Vanderbauwhede, W., Mckechnie, P., & and Thirunavukkarasu, C. (2008). The Gannet Service Manager: A Distributed Dataflow Controller for Heterogeneous Multi-core SoCs. In Proceedings of the *NASA/ESA Conference on Adaptive Hardware and Systems, 2008*. (AHS'08),. *NASA/ESA Conference on*, (pages 301–308).

Vangal, et al. (2007). On An 80-Tile 1.28TFLOPS Network-on-Chip in 65 nm CMOS (Digest of Technical Papers) *In Proceedings of IEEE Intl. Solid State Circuits Conference*, (pp.98–589).

Varatkar, G., & Marculescu, R. (2004). On-chip traffic modeling and synthesis for mpeg-2 video applications. *IEEE Trans. VLSI Syst.* *12*(1), 108–119.

Vassiliadis, S., & Sourdis, I. (2006). Flux networks: Interconnects on demand. In *Proceedings of International Conference on Embedded Computer Systems: Architectures, Modeling and Simulation* (IC-SAMOS), (160-167).

Veen, A. (1986). Dataflow machine architecture. *ACM Computing Surveys (CSUR)*, *18*(4), :365–396.

Vegdahl, S. (1984). A survey of proposed architectures for the execution of functional languages. *IEEE transactions on computers*, *100*(33), :1050–1071.

Véstias, M., & Neto, H. (2006). Area and Performance Optimization of a Generic Network-on-Chip Architecture, In *Proceedings of the 19th Symposium on Integrated Circuits and Systems Design* (pp.68-73).

Véstias, M., & Neto, H. (2007). Router design for application specific networks-on-chip on reconfigurable systems. In *Proceedings of the International Conference on Field Programmable Logic and Applications* (pp.389.394).

Vieira, A., Ost, L., Moraes, F., & Calazans, N. (2004). *Evaluation of routing algorithms on Mesh Based NoCs. (Technical Report). Porto Alegre, Brasil.*

Waingold, E., Taylor, M., Srikrishna, D., Sarkar, V., Lee, W., Lee, V., Kim, J., Frank, M., Finch, P., Barua, R., Babb, J., Amarasinghe, S., & Agarwal, A. (1997). Baring it all to software: Raw machines. *IEEE Computer, 30*(9), 86-93.

Walder, H., & Platzner, M. (2004). A runtime environment for reconfigurable hardware operating systems. In *Proceedings of the 14th Conference on Field Programmable Logic and Application* (pp. 831-835). Leuven, Belgium.

Walker, P. (1999). *IEEE 1355 Why yet another high-speed serial interface standard?* IEE Electronics and Communications Open Forum. Washington, DC: IEEE.

Wang, H., et al. (2005). A Technology-aware and Energy-oriented Topology Exploration for On-chip Networks,. In Proceedings of the Conf. on Design Automation and Test in Europe, (pp. 1238–1243).

Wang, H., Peh, L.-S., & Malik, S. (2003). Power-driven Design of Router Microarchitectures in On-chip Networks. In *Proceedings of the 36th annual IEEE/ACM International Symposium on Microarchitecture* (MICRO 36), (pp. 105). Washington, DC: IEEE Computer Society.

Wang, H., Zhu, X., Peh, L. S., & Malik, S. (2002). Orion: A power-performance simulator for interconnection networks. In *Proceedings of the 35th International Symposium on Microarchitecture* (MICRO).

Wang, K., Gu, H., & Wang, C. (2007). Study on Hybrid Switching Mechanism in Network on Chip. In *Proceedings of the 7th International Conference on ASIC.* (pp. 914-917).

Wenbiao, Z., Zhang, Y., & Mao, Z. (2007). Link-load balance aware mapping and routing for NoC. *WSEAS Trans. Cir. and Sys., 6*(11), 583–591.

Wilkinson, B. (1996). Computer architecture: design and performance., chapter 10, pages 434–437. Upper Saddle River, NJ: Prentice-Hall, 2nd edition.

William, J. D., & Brian, T. (2001). Route packets, not wires: On-chip interconnection networks. In *Proceedings of the Design Automation Conference DAC* (pp. 684–689). ACM.

Williams, J. A., & Bergmann, N. W. (2004, June). Embedded Linux as a platform for dynamically self-reconfiguring Systems-on-Chip. In *Proceedings of International Conference on Engineering of Reconfigurable Systems and Algorithms*, (pp. 163–169). CSREA Press.

Wirthlin, M., & Hutchings, B. (1995). A dynamic instruction set computer. In *Proceedings of the 3rd IEEE Symposium on Field-Programmable Custom Computing Machines* (pp. 99-107). Napa, CA.

Wirthlin, M., & Hutchings, B. (1997). Improving functional density through run-time constant propagation. In *Proceedings of the 5th International Symposium on Field Programmable Gate Array* (pp. 86-92). Monterey, CA.

Wirthlin, M., Hutchings, B., & Gilson, K. (1994). The Nano Processor: a Low Resource Reconfigurable Processor. In *Proceedings of the IEEE Symposium on Field-Programmable Custom Computing Machines* (pp. 23-30). Napa, CA.

Wittig, R., & Chow, P. (1995). OneChip: An FPGA processor with reconfigurable logic. In *Proceedings of the 3rd IEEE Symposium on Field-Programmable Custom Computing Machines* (pp. 126-135). Napa, CA.

Wolkotte, P. T., Holzenspies, P. K. F., & Smit, G. J. M. (2007). Fast, accurate and detailed NoC simulations. In *Proceedings of Network on Chips.*

Woo, S. C., Ohara, M., Torrie, E., Singh, J. P., & Gupta, A. (1995). The SPLASH-2 programs: Characterization and methodological considerations. In *Proceedings of the 22nd Annual International Symposium on Computer Architecture* (pages 24-37). New York: ACM Press.

Wu, J. (2002). A deterministic fault-tolerant and deadlock-free routing protocol in 2-D meshes based on odd-even turn model. In *Proceedings of the 16th international conference on Supercomputing.*

Xie, F., Martonosi, M., & Malik, S. (2003). Compile-time Dynamic Voltage Scaling Settings: Opportunities and Limits. In Proceedings of PLDI: The Conference on Programming Language Design and Implementation.

Xilinx Inc. (2006, March). , *Early Access Partial Reconfiguration User Guide*. San Jose, CA: Xilinx Inc., March 2006.

Xilinx (2008). *Xilinx: PlanAhead*. Retrieved from http://www.xilinx.com/ise/optional prod/planahead.htm

Xilinx (2009). *Xilinx Corporation, Inc.* Retrieved on March 30, 2009, from http://www.xilinx.com

Xilinx Inc. (2003). *Virtex Series Configuration Architecture User Guide*. (Technical Report XAPP151), (MXilinx Inc. (2004). XAPP290 - Two flows for partial reconfiguration: Module based or Difference based, September 2004.

Xilinx Inc. (2006), Early Access Partial Reconfiguration User Guide, March 2006.

Xilinx Inc. (2007). *Virtex-4 user guide*. Technical Report UG70, Xilinx Inc., March 2007.

Xilinx Inc. (2007). Virtex-5 configuration user guide. Technical Report UG191, Xilinx Inc., February 2007.

Xilinx Inc. (2007). *Virtex-II Pro and Virtex-II ProX Virtex-II Pro and Virtex-II Pro X FPGA User Guide*. Xilinx Inc., March 2007.

Xilinx. (2006). Early access partial reconfiguration user guide. *ISE 8.1.01i*.

Xilinx. (2007). *Virtex-II Platform FPGAs:Complete Data Sheet* (3.5). San Jose, CA: Xilinx Inc.

Xilinx. (2008). *Virtex-4 FPGA Configuration User Guide* (1.1). San Jose, CA: Xilinx Inc.

Xmulator NoC Simulator (2008). *Xmulator NoC Simulator.* Retrieved December 20, 2008, from http://www.xmulator.org

Ye, T. T., & De Micheli, G. (2003, June 24-26). Physical planning for on-chip multiprocessor networks and switch fabrics. In *Proceedings of IEEE International Conference on Application-Specific Systems, Architectures, and Processors, 2003* (pp. 97-107).

Yeo, S., Lyuh, C., Roh, T., & Kim, J. (2008). High Energy Efficient Reconfigurable Processor for Mobile Multimedia. In *Proceedings of the IEEE International Conference on Circuits and Systems for Communications* (pp. 618-622). Shanghai, China.

Yi, L., Marconi, T., Gaydadjiev, G., Bertels, K., & Meeuws, R. (2008, April 14-18). A self-adaptive on-line task placement algorithm for partially reconfigurable systems. In *Proceedings of the IEEE International Symposium on Parallel and Distributed Processing* (IPDPS 2008), (pp.1-8).

Zhong, M. (2005, June). Evaluation of deflection-routed on-chip networks (Master's thesis), In *Proceedings of KTH*. Stockholm.

Zhou, B., Qiu, W., & Peng, C. (2005, September). An operating system framework for reconfigurable systems. In *Proceedings of the 15th International Conference on Computer and Information Technology*, (pp. 781–787). IEEE Computer Society Press.

Zhu, H., Pande, P. & Grecu, C. (2007). Peformance Evaluation of Adaptive Routing Algorithms for Achieving Fault Tolerant in NoC Fabrics. In *IEEE International Conference on Application-Specific Systems, Architectures and Processors*, (pp. 42-47).

Ziegler, J. F. (1996). IBM Experiments in Soft Fails in Computer Electronics. *IBM J.Res.Develop., 40*(3).

Zimmermann, H. (1980). OSI Reference Model—the ISO Model of Architecture for Open Systems Interconnection. *IEEE Transactions on Communications, 28*(4), 425–432 (arch 2003).*e* (HPCA '01).E.

About the Contributors

Jih-Sheng Shen received his B.S. and his M.S. in Computer Science and Information Engineering from the I-Shou University and the National Chung Cheng University, Taiwan, ROC, in 2003 and 2004, respectively. His M.S. thesis was on the design and implementation of on-chip crossroad communication architectures for low power embedded systems. He is currently pursuing his Ph.D. in the Department of Computer Science and Information Engineering at the National Chung Cheng University, Taiwan, ROC. His research interests include the theories and the architectures of reconfigurable systems, machine learning strategies, Network-on-Chip (NoC) designs, encoding methods for minimizing crosstalk interferences and dynamic power consumption.

Pao-Ann Hsiung, Ph.D., received his B.S. in Mathematics and his Ph.D. in Electrical Engineering from the National Taiwan University, Taipei, Taiwan, ROC, in 1991 and 1996, respectively. From 1996 to 2000, he was a post-doctoral researcher at the Institute of Information Science, Academia Sinica, Taipei, Taiwan, ROC. From February 2001 to July 2002, he was an assistant professor and from August 2002 to July 2007 he was an associate professor in the Department of Computer Science and Information Engineering, National Chung Cheng University, Chiayi, Taiwan, ROC. Since August 2007, he has been a full professor. Dr. Hsiung was the recipient of the 2001 ACM Taipei Chapter Kuo-Ting Li Young Researcher for his significant contributions to design automation of electronic systems. Dr. Hsiung was also a recipient of the 2004 Young Scholar Research Award given by National Chung Cheng University to five young faculty members per year. Dr. Hsiung is a senior member of the IEEE, a senior member of the ACM, and a life member of the IICM. He has been included in several professional listings such as Marquis' Who's Who in the World, Marquis' Who's Who in Asia, Outstanding People of the 20th Century by International Biographical Centre, Cambridge, England, Rifacimento International's Admirable Asian Achievers (2006), Afro/Asian Who's Who, and Asia/Pacific Who's Who. Dr. Hsiung is an editorial board member of the International Journal of Embedded Systems (IJES), Inderscience Publishers, USA; the International Journal of Multimedia and Ubiquitous Engineering (IJMUE), Science and Engineering Research Center (SERSC), USA; an associate editor of the Journal of Software Engineering (JSE), Academic Journals, Inc., USA; an editorial board member of the Open Software Engineering Journal (OSE), Bentham Science Publishers, Ltd., USA; an international editorial board member of the International Journal of Patterns (IJOP). Dr. Hsiung has been on the program committee of more than 50 international conferences. He served as session organizer and chair for PDPTA'99, and as workshop organizer and chair for RTC'99, DSVV'2000, and PDES'2005. He has published more than 170 papers in international journals and conferences. He has taken an active part in paper refereeing for international journals and conferences. His main research interests include reconfigurable computing and system design, multi-core programming, cognitive radio architecture, System-on-Chip (SoC) design and verification, embedded software synthesis and verification, real-time system design and verification, hardware-software codesign and coverification, and component-based object-oriented application frameworks for real-time embedded systems.

* * *

Balal Ahmad has received his BEng (Hons) in Computer and Electronic System Engineering from the University of Strathclyde, Glasgow, UK, and MSc in System Level Integration from the Institute of System Level Integration, Livingston, UK. Balal became a member of System Level Integration Group in University of Edinburgh, UK in 2004, where he also did his dissertation in "Communication Centric Platforms for Future High Data Intensive Applications. At present he is working as an associate professor in Government College University, Lahore, Pakistan in the Faculty of Engineering. His interests include Network-on-chip, on-chip communication, SoC design methodologies, reconfigurable SoC architectures, Real-time Operating Systems and data communication.

Ali Ahmadinia received the B.Sc. degree in Computer Engineering from Tehran Polytechnics University, Tehran, Iran, in 2000, and the M.Sc. degree from Sharif University of Technology, Tehran, Iran, in 2002. In 2003, he joined the chair of Hardware/Software codesign, University of Erlangen-Nuremberg, Erlangen, Germany, as a research assistant; in 2004, he became a member of the electronic imaging group of the Fraunhofer Institute for Integrated Circuits (IIS), Erlangen, where he also finished his Ph.D. dissertation on "Optimization algorithms for dynamically reconfigurable embedded systems". In 2006-2008, he was working as a research fellow with the School of Engineering and Electronics, University of Edinburgh, Edinburgh, U.K. Since 2008, he is a lecturer in embedded systems at the School of Engineering and Computing, Glasgow Caledonian University, Glasgow, U.K. His main research interests include system-on-chip architectures, reconfigurable computing, and DSP applications on embedded systems.

Tughrul Arslan holds the Chair of Integrated Electronic Systems in the School of Engineering and Electronics in the University of Edinburgh and is also a co-founder and the Chief Technical Officer of Spiral Gateway Ltd. He is a member of the Integrated Micro and Nano Systems (IMNS) Institute and Leads the System Level Integration group (SLIg) in the University. His research interests include: low power system design, integrated micro and nano systems, secure and long life wireless sensor networks, autonomous systems, System-On-Chip (SoC) Architectures, Evolvable intelligent systems, Multi-objective optimisation, and studying autonomous self-adaptive behaviour in general. He has published more than 300 articles and is the inventor of a number of patents in these areas

Ney Laert Vilar Calazans received the bachelor's degree from the Federal University of Rio Grande do Sul (UFRGS), Brazil, in Electrical Engineering in 1985, the M.Sc. degree in Computer Science in 1988, also from UFRGS, and the Ph.D. degree in Microelectronics in 1993, from the Université Catholique de Louvain (UCL), Belgium. He is currently a Professor at the Catholic University of Rio Grande do Sul (PUCRS). His research interests include intrachip communication networks, non-synchronous circuit design and implementation, and computer-aided design techniques and tools. Professor Calazans is a member of the IEEE and of the Brazilian Computer Society (SBC).

Ewerson Luiz de Souza Carvalho was born in Rio Grande, Brazil, on September, 7, 1978. He received the bachelor's degree in computer engineering from the Federal University of Rio Grande (FURG), Brazil, in 2002, received the M.Sc. degree in computer science from Catholic University of Rio Grande do Sul (PUCRS), Brazil, in 2004, and received the Ph.D. degree in computer science in 2009, also from the PUCRS. His research interests include Multi-processor System-on-Chip (MPSoC) design, System-on-Chip (SoC) design, and also Fast Prototyping of Digital Systems with FPGAs. He is currently a hardware designer at the DATACOM, Brazil, a company specialized in Metro Ethernet and SDH solutions for telecom operators.

Simone Corbetta received the B.Sc. and M.Sc. degrees in Computer Engineering from Politecnico di Milano in September 2006 and April 2009 respectively, with a thesis on the design and implementation of dynamically reconfigurable architectures and communication infrastructures based on FPGA devices; he received his M.Sc. degree in Computer Science from the University of Illinois at Chicago in 2008, working on Network-on-Chip architectures. He is now a research assistant at Politecnico di Milano, working on multicore embedded systems optimization.

Rachid Dafali received the computer-science engineer degree in 2005 from EMSI (Morocoo), and the M.S degree from European University of Brittany-UBS (France) in 2007. His research interests include Reconfigurable Network-on-Chip concept and design methodology. He is currently finishing a Ph.D. degree in electronics from the European University of Brittany-UBS (France).

Jean-Luc Dekeyser received his PhD degree in computer science from the University of Lille 1 in 1986; afterwards, he was a fellowship at CERN Geneva. After a few years at the Supercomputing Computation Research Institute in Florida State University, where he worked on high performance computing for Monté-Carlo methods in High Energy Physics, he joined the University of Lille 1 in France as an assistant professor, in 1988. There he worked on data parallel paradigm and vector processing. He created a research group working on High Performance Computing in the CNRS lab in Lille. He is currently Professor in computer science at University of Lille 1 and is also heading the DaRT INRIA project at the INRIA Lille Nord Europe research center. His research interests include embedded systems, System on Chip co-design, synthesis and simulation, performance evaluation, High Performance Computing and Model Driven Engineering.

Jean-Philippe Diguet received the computer-science engineer degree in 1992 from ESEO (France), the M.S degree and the Ph.D degree from Rennes University (France) in 1993 and 1996 respectively. In 1997, he has been a visitor researcher at IMEC (Leuven, Belgium). He has been an associated professor at UBS University (Lorient, France) from 1998 until 2002. In 2003, he initiated a technology transfer and co-funded the dixip company in the domain of wireless embedded systems. Since 2004 he is a CNRS researcher at Lab-STICC (Lorient, Brest, France). His current work focuses on various topics in the domain of embedded system design: high-level synthesis, design space exploration, power / energy modeling, NOC architectures and CAD tools, reconfigurable and self-adaptive HW/SW architectures, secured architectures and pervasive computing components.

Soumya Eachempati completed her undergraduate studies majoring in Computer Science and Engineering from Indian Institute of Technology, Madras in the year 2005. She is pursuing Doctoral studies in Computer Science and Engineering since August 2005 under the guidance of Dr Vijaykrishnan Narayanan and Dr Yuan Xie. her research interests include exploring the architectural impact of emerging interconnect such as Carbon Nanotube bundles, upcoming post Silicon devices such as quantum devices, reconfigurable gate arrays, and on-chip interconnection networks for CMPs.

Majdi Elhaji was born in Kasserine, Tunisia in 1982. He received the Masters degree in Physics (Electronics option) from Faculty of Sciences of Monastir, Tunisia in 2008. Currently he is engaged in research for his Ph.D. degree in the EµE laboratory within the same Faculty in collaboration with the LIFL laboratory at the INRIA Institute in France. His researches interests include hardware/software co-design of video encoder on GALS NoC.

Björn Fiethe is senior research assistant at the Institute of Computer and Network Engineering at the Technical University of Braunschweig. He obtained his Dipl.-Ing. from the Technical University of Braunschweig (1993) in the field of electrical engineering. He joined the institute in 1993. Since then, he worked on design, development and implementation of hardware and software for high reliable spaceborne computers and test equipment for space applications, e.g. for ESA Rosetta ROSINA and OSIRIS instruments. In recent years he performs system engineering of various projects, e.g. ESA/JAXA BepiColombo MMO MSA, NASA Dawn Framing Camera and ESA Venus Express VMC. His research focuses on systems for future spaceborne data processing units and Wavelet-based data compression for space instruments.

Ismael Augusto Grehs was born in Santa Cruz do Sul, Brazil, on October, 28, 1985. He received the bachelor's degree in computer engineering from the Universidade Federal do Rio Grande do Sul (UFRGS), Porto Alegre, Brazil, in 2009. His research interests include dynamic reconfigurable systems, multi-processor system-on-chip (MPSoC) design and intrachip communication networks (NoCs).

Mary Jane Irwin received the MS and PhD degrees in computer science from the University of Illinois, Urbana-Champaign, in 1975 and 1977, respectively, and an honorary doctoral degree from the Chalmers University of Technology, Goteborg, Sweden, in 1997. Since 1977, she has been with the faculty of the Pennsylvania State University, University Park, where she is currently the Evan Pugh Professor and the A. Robert Noll Chair in Engineering with the Computer Science and Engineering Department. Her research and teaching interests include computer architecture, embedded and mobile computing systems design, and power-aware and reliable systems design. Her research is supported by grants from the MARCO Gigascale Systems Research Center, the US National Science Foundation, and the Semiconductor Research Corp. She is currently a co-editor in chief of the ACM Journal of Emerging Technologies in Computing Systems. She is currently a co-chair of the publications board of the ACM and the Steering Committee of the CRA Committee on the Status of Women in Computing Research. She is a fellow of the IEEE, the ACM, and member of the National Academy of Engineering.

Yana Esteves Krasteva was born in Lima, Perú, in 1979, she receive the M.Sc degree from the Technical University of Sofia, Bulgaria, in 2002 and the Ph.D. degree in Electronic engineering from the Universidad Politecnica de Madrid (UPM), Spain, in 2009. Since 2006 she is a researcher in the Center of Industrial Electronics of the Universidad Politecnica de Madrid (UPM). Her research interests are focused on reconfigurable and control systems, networks on chip and wireless-sensor network.

She has worked in several research projects with European and national funding, as well in projects for the industry. She is also involved in the center academic activity co-tutoring final degree projects and master thesis.

Wei-Wen Lin received his M.S. degree in Computer Science and Information Engineering from the National Chung Cheng University, Taiwan, ROC, in 2009. His M.S. thesis was on the analysis and implementation of reconfigurable Network-on-Chip architectures. His research interests include dynamically partially reconfigurable systems and Network-on-Chip designs.

Samy Meftali obtained his PhD in computer sciences from University of Grenoble 1 in 2001. Since then, he is associate professor at University of Lille 1 in France. His research interests are mainly modeling, simulation and FPGA implementation of intensive signal processing embedded systems. He is member of PCs of several conferences, and has published many articles on embedded systems design methodologies.

Alessandro Meroni received his Laurea Specialistica (M.Sc. equivalent) degree in Computer Engineering from the Politecnico di Milano in 2009, and his second M.Sc. degree in Computer Science from the University of Illinois at Chicago (UIC) in 2009. He is a member of the DRESD research group, with particular interests in the study and development of communication infrastructures tailored to embedded systems.

Harald Michalik received a degree in Electrical Engineering in 1982 and a PhD in 1991 from Technical University of Braunschweig, Germany. Since 2001, after employments in Industry and as University Professor, he holds a Professorship on Spaceborne Computers Design at the Institute for Computer and Network Engineering (IDA) at Technical University of Braunschweig. He is leading a group of 18 scientists and engineers at IDA. Current fields of research are System-on-Chip architectures, communication processors, mass memory systems and software systems. All these topics are mainly related to Spaceborne applications but also to other fields like robotics.

Mehdi Modarressi received the B.Sc. degree from Amirkabir University of Technology, Tehran, Iran, in 2003, and the M.Sc. degree from Sharif University of Technology, Tehran, Iran, in 2005, both in computer engineering. He is currently working toward the Ph.D. degree in the Department of Computer Engineering, Sharif University of Technology. His research focuses on different aspects of high-performance computing and digital design automation with a particular emphasis on network-on-hip architectures. He is a student member of ACM and IEEE.

Leandro Möller is currently a PhD student at the Technical University of Darmstadt, Germany. He is a member of the Microelectronic Systems group in this same University. He received his master and bachelor degrees in Computer Science, in 2005 and 2003, respectively, both from the Catholic University of Rio Grande do Sul, Brazil. Since 2001 he is also a member of the Hardware Design Support Group (GAPH) at Catholic University of Rio Grande do Sul. Further he received two awards, the first place in the Xilinx Student Contest given during the SBCCI conference in 2005 and the Europractice Design Contest Award in the conceptual design category in 2006 given during the DATE conference. In the summer of 2006, he was an intern at Xilinx Research Labs, San Jose, US. His research interests are partial and dynamic reconfiguration of FPGAs, Networks-on-Chip and High Level Modeling of Systems-on-Chip.

Fernando Gehm Moraes received the Electrical Engineering and M.Sc. degrees from the Universidade Federal do Rio Grande do Sul (UFRGS), Porto Alegre, Brazil, in 1987 and 1990, respectively. In 1994 he received the Ph.D. degree from the Laboratoire d´Informatique, Robotique et Microélectronique de Montpellier (LIRMM), France. He is currently a Professor at the Catholic University of Rio Grande do Sul (PUCRS). . He has authored and co-authored 12 peer refereed journal articles in the field of VLSI design, comprising the development of networks on chip and telecommunication circuits. He has also authored and co-authored more than 140 conference papers on these topics. His research interests include intrachip communication networks (NoCs), and MPSoC design. Professor Moraes is a member of the IEEE and Brazilian Computer Society, SBC.

Vijaykrishnan Narayanan is a professor in the Department of Computer Science and Engineering, Pennsylvania State University, University Park. His research interests include energy aware reliable systems, embedded Java, nano VLSI systems, and computer architecture. He is a member of the IEEE.

Horácio Neto received the electrical engineering degree in 1980, and the PhD in electrical and computer engineering in 1992, both from the Technical University of Lisbon, Portugal. He is an Associated Professor at the Technical University of Lisbon, School of Engineering (IST), Department of Electrical and Computer Engineering (DEEC), where he is responsible for undergraduate and graduate courses on computer architecture and digital systems design. He is also a senior researcher at the ESDA (Electronic Systems Design and Automation) group at INESC-ID, a research institute associated with the Engineering University. His current research interests are in the area of reconfigurable systems, namely on the use of reconfigurable hardware for computing purposes and on the design of on-chip system-level architectures with reconfigurable hardware/software components.

Björn Osterloh is research assistant at the institute of Computer and Network Engineering at the Technical University of Braunschweig. He obtained his BEng. from the Manchester Metropolitan University (2001) and his Dipl.-Ing.(FH) from the University of Applied Sciences Wilhelmshaven (2002) and Dipl.-Ing. from the Technical University of Braunschweig (2005) all in the field of electrical and electronic engineering. He joined the Spaceborne Data Processing Units workgroup of Prof. Michalik 2001. Since then, he works on high reliable systems for space applications and was involved in the development of Data Processing Units e.g. for the Venus Express mission with the Venus Monitoring Camera (VMC). His research focuses on fault-tolerant dynamic reconfigurable System-on-Chip and Network-on-Chip approaches for future space applications.

Imran Rafiq Quadri is currently doing his PhD degree in computer science from the University of Lille 1. He completed his B.E (Bachelors of Engineering) from NED University in Karachi, Pakistan in the field of Computers and Information Systems Engineering in February 2004. Afterwards, he achieved his Master's degree from University of Lille 1 in the field of Embedded Systems. He is currently involved in the DaRT INRIA project at the INRIA Lille Nord Europe research centee. His research interests include embedded systems, Systems on Chip co-design, high abstraction levels, Model Driven Engineering, synthesis and Partial Dynamic Reconfiguration in state of the art FPGAs.

Vincenzo Rana is a Ph. D. student at the Politecnico di Milano. He received his Laurea in Computer Engineering, in 2004, and his Laurea Specialistica in Computer Engineering, in 2006, from the Politecnico di Milano. His research interests include embedded systems design methodologies and architectures, with particular attention to dynamic reconfiguration aspects and communication infrastructures design.

Teresa Riesgo was born in Madrid, Spain, in 1965. She received the M.Sc. and Ph.D. degrees in Electrical Engineering from Universidad Politecnica de Madrid (UPM), in 1989 and 1996, respectively. Since 2003, she is Full Professor of Electronics at UPM. Her research interests are focused on embedded-system design, wireless-sensor networks, configurable systems, and power estimation in digital systems. She has published a large number of papers in these fields and has participated and acted as main researcher in several European Union-funded projects. She is currently the Director of the Center of Industrial Electronics of Universidad Politecnica de Madrid (CEI-UPM)

Marco Domenico Santambrogio received his laurea (M.Sc. equivalent) degree in Computer Engineering from the Politecnico di Milano in 2004, his second M. Sc. degree in Computer Science from the University of Illinois at Chicago (UIC) in 2005 and his PhD degree in Computer Engineering from the Politecnico di Milano in 2008. Dr Santambrogio is now at the Computer Science and Artificial Intelligence Laboratory (CSAIL) at Massachusetts Institute of Technology (MIT) as postdoc fellow. He has also held visiting positions at the EECS Department of the Northwestern University (2006 and 2007) and Heinz Nixdorf Institut (2006 and 2009). He has been with the Micro Architectures Laboratory at the Politecnico di Milano, where he founded the Dynamic Reconfigurability in Embedded System Design (DRESD) project in 2004. He conducts research and teaches in the areas of reconfigurable computing, Self-aware and autonomic systems, hardware/software codesign, embedded systems, and high performance processors and systems. Dr. Santambrogio has published more than 70 papers in peer-reviewed international journals and conferences, one books and several book chapters.

Hamid Sarbazi-Azad received his B.Sc. in electrical and computer engineering from Shahid-Beheshti University, Tehran, Iran, in 1992, his MS.c. in computer engineering from Sharif University of Technology, Tehran, Iran, in 1994, and his Ph.D. in computing science from the University of Glasgow, Glasgow, UK, in 2002. He is currently associate professor of computer engineering at Sharif University of Technology, and heads the IPM School of Computer Science, Tehran, Iran. His research interests include high-performance computer architectures, NoCs and SoCs, parallel and distributed systems, performance modelling/evaluation, graph theory and combinatorics, wireless/mobile networks, on which he has published more than 200 refereed conference and journal papers. He received Khwarizmi International Award in 2006, and TWAS Young Scientist Award in engineering sciences in 2007. He is a member of managing board of Computer Society of Iran (CSI), and has served as the editor-in-chief for the CSI Journal on Computer Science and Engineering since 2005. He is editorial board member, and guest editor for several special issues on high-performance computing architectures and networks in related journals. Dr Sarbazi-Azad is a member of ACM and CSI.

Rafael Iankowski Soares was born in Pelotas, Brazil, on May, 18, 1977. He received the bachelor's degree in computer engineering from the Federal University of Rio Grande (FURG), Brazil in 2004, received the M.Sc. degree in computer science from Catholic University of Rio Grande do Sul (PUCRS),

Brazil, in 2006. He is currently working toward Ph.D degree in computer science from PUCRS under the supervision of Prof. Ney Calazans. His research interests include System-on-Chip design, Reconfigurable Systems, Non-Synchronous circuits design and Side-Channels Attacks (SCAs).

Eduardo de la Torre got his MsC in Industrial Engineering (1989) and PhD in Industrial Electronics (2000) by Universidad Politécnica de Madrid, where he is presently working as an Associate Professor, and doing his research at the Industrial Electronics Centre. He has worked in integrated circuit design for ASICs and FPGAs, mostly for control applications, as well as embedded systems design. He has also worked in tools for test and debug of these systems. Recently, his main research interests are reconfigurable systems and Networks on Chip.

Wim A. Vanderbauwhede is Lecturer at the Dept of Computing Science of the University of Glasgow and an EPSRC Advanced Research Fellow. His research focuses on novel architectures for Networks-on-Chip and dynamically reconfigurable Systems-on-Chip; his other research interests include processor architectures and programming languages. He has published widely and is reviewer for several international journals and program committee member for several international conferences. Dr. Vanderbauwhede has a degree in Electrotechnical Engineering (Physics) and a PhD in Integrated Optoelectronics from the University of Gent, Belgium. Before returning to academic research, he worked as a Mixed-mode Design Engineer and Senior Technology R&D Engineer for a telecommunications ASIC (integrated circuits) design and manufacturing company.

Mário Véstias received the electrical engineering degree in 1993, and the PhD in electrical and computer engineering in 2002, both from the Technical University of Lisbon, Portugal. He is an Coordinate Professor at the Polytechnic Institute of Lisbon, School of Engineering (ISEL), Department of Electronic, Telecommunications and Computer Engineering (DEETC), where he is responsible for undergraduate and graduate courses on computer architecture and digital systems design. He is also a senior researcher at the ESDA (Electronic Systems Design and Automation) group at the research institute INESC-ID. His current research interests are in the area of reconfigurable systems, namely on the use of reconfigurable hardware for computing purposes and on the design of on-chip system-level architectures with reconfigurable hardware/software components.

Aditya Yanamandra received his Bachelors of Technology degree in Computer Science and Engineering from the Indian Institute of Technology, Madras, in 2005. Currently, he is working towards his Ph.D. in the Department of Computer Science and Engineering at the Pennsylvania State University, USA, where he is a research assistant. His research interests include reliability of systems, Networks-on-Chip and scratchpad memories.

Index

Symbols

2D 313

A

abstraction level 135, 136, 137, 145, 146, 147, 148, 149, 150, 155
adaptive algorithms 32, 40, 43
ad-hoc 166, 169, 180, 184
allocation policy 90
application characterization graph (APCG) 124
application layer 168, 171
application modeling 144
application-specific 309, 310, 311, 312, 313, 315, 324
a priori 158, 161, 170
arbitration algorithms 11
architecture characterization graph (ARCG) 124
arithmetic logic unit (ALU) 223
assembly language 208
asynchronous clocking 136

B

bandwidth 28, 29, 31, 32, 34, 35, 36, 37, 39, 40, 41, 42, 44, 50, 52, 53, 54, 61, 63, 64, 123, 125, 126, 160, 161, 162
BCH codes 291, 292
BEE2 hardware platform 91, 107
behavioural model 208
Berkeley operating system for reprogrammable hardware (BORPH) 90, 91, 108
best effort (BE) 75, 140
bitstreams 14, 15, 16, 17, 20, 27, 94
bottleneck 29, 44, 49, 188, 189, 202, 203, 204, 256

branch-and-bound (BnB) 123, 125, 130
buffer size 31, 32, 46, 52
bus macro 6, 8, 10, 112, 113, 158, 159, 160, 177, 178, 179, 223, 230, 231, 232, 233, 234, 237, 243
bytecode 194, 196, 197, 208
bytewords 197

C

cache coherency 68
cache policy 90
chip multi-processors (CMP) 277, 278, 281, 284, 285, 287, 288, 289, 292, 309, 313, 317, 323
circuit blocking 11
circuit switching 143
CLB macro 6
clock managers 3
clock synchronization 140
CMOS 186
coarse-grained reconfigurable architectures (CGRA) 187, 218
coarse grain reconfigurable devices 3
code packet 193, 197, 199
coherency 68, 70, 71
communication architecture 85, 86, 87, 88, 89, 92, 93, 99, 100
communication dependency graph (CDG) 76
communication flow 167, 175
communication graph 122, 123, 124, 125, 126, 128
communication infrastructure 2, 4, 5, 6, 8, 9, 10, 12, 110, 111, 112, 119, 120, 121, 122, 125, 126, 127, 129, 130, 131, 133, 158, 159, 161, 163, 164, 165, 167, 174, 182